ARCHAEOLOGY AND LANGUAGE IV

ONE WORLD ARCHAEOLOGY
Series Editor: P. J. Ucko

Animals into Art
H. Morphy (ed.), vol. 7

Archaeological Approaches to Cultural Identity
S. J. Shennan (ed.), vol. 10

Archaeological Heritage Management in the Modern World
H. F. Cleere (ed.), vol. 9

The Archaeology and Anthropology of Landscape: shaping your landscape
P. J. Ucko & R. Layton (eds), vol. 30

Archaeology and Language I: theoretical and methodological orientations
R. Blench & M. Spriggs (eds), vol. 27

Archaeology and Language II: archaeological data and linguistic hypotheses
R. Blench & M. Spriggs (eds), vol. 29

Archaeology and Language III: artefacts, languages and texts
R. Blench & M. Spriggs (eds), vol. 34

Archaeology and the Information Age: a global perspective
P. Reilly & S. Rahtz (eds), vol. 21

The Archaeology of Africa: food, metals and towns
T. Shaw, P. Sinclair, B. Andah & A. Okpoko (eds), vol 20

Centre and Periphery: comparative studies in archaeology
T. C. Champion (ed.), vol. 11

Conflict in the Archaeology of Living Traditions
R. Layton (ed.), vol. 8

The Constructed Past: experimental archaeology, education and the public
P. G. Stone & P. Planel (eds), vol. 36

Cultural Resource Management in Contemporary Society: perspectives on managing and presenting the past
F. P. McManamon & A. Hatton (eds), vol. 33

Domination and Resistance
D. Miller, M. J. Rowlands & C. Tilley (eds), vol. 3

Early Human Behaviour in Global Context: the rise and diversity of the Lower Palaeolithic Record
M. D. Petraglia & R. Korisettar (eds), vol. 28

The Excluded Past: archaeology in education
P. Stone & R. MacKenzie (eds), vol. 17

Foraging and Farming: the evolution of plant exploitation
D. R. Hams & G. C. Hillman (eds), vol. 13

From the Baltic to the Black Sea: studies in medieval archaeology
D. Austin & L. Alcock (eds), vol. 18

Historical Archaeology: back from the edge
P. P. A. Funari, M. Hall & S. Jones (eds), vol. 31

Hunters of the Recent Past
L. B. Davis & B. O. K. Reeves (eds), vol. 15

The Meanings of Things: material culture and symbolic expression
I. Hodder (ed.), vol. 6

The Origins of Human Behaviour
R. A. Foley (ed.), vol. 19

The Politics of the Past
P. Gathercole & D. Lowenthal (eds), vol. 12

The Prehistory of Food: appetites for change
C. Gosden & J. G. Hather (eds), vol. 32

The Presented Past: heritage, museums and education
P. G. Stone & B. L. Molyneaux (eds), vol. 25

Sacred Sites, Sacred Places
D. L. Carmichael, J. Hubert, B. Reeves & A. Schanche (eds), vol. 23

Signifying Animals: human meaning in the natural world
R. G. Willis (ed.), vol. 16

Social Construction of the Past: representation as power
G. C. Bond & A. Gilliam (eds), vol. 24

State and Society: the emergence and development of social hierarchy and political centralization
J. Gledhill, B. Bender & M. T. Larsen (eds), vol. 4

Time, Process and Structured Transformation in Archaeology
S. E. van der Leeuw & J. McGlade (eds), vol. 26

Tropical Archaeobotany: applications and development
J. G. Hather (ed.), vol. 22

The Walking Larder: patterns of domestication, pastoralism and predation
J. Clutton-Brook (ed.), vol. 2

What is an Animal?
T. Ingold (ed.), vol. 1

What's New? A closer look at the process of innovation
S. E. van der Leeuw & R. Torrence (eds), vol. 14

Who Needs the Past? Indigenous values and archaeology
R. Layton (ed.), vol. 5

ARCHAEOLOGY AND LANGUAGE IV

Language change and cultural transformation

Edited by

Roger Blench and Matthew Spriggs

Routledge
Taylor & Francis Group
NEW YORK AND LONDON

First published 1999
by Routledge
2 Park Square, Milton Park, Abingdon, Oxfordshire OX14 4RN

Simultaneously published in the USA and Canada
by Routledge
711 Third Avenue, New York, NY 10017

First issued in paperback 2014
Routledge is an imprint of the Taylor and Francis Group, an informa company

Typeset in Bembo by The Florence Group, Stoodleigh, Devon

British Library Cataloguing in Publication Data
A catalogue record for this book is available from the British Library.

Library of Congress Cataloging in Publication Data
A catalog record for this book has been requested.

ISBN 13: 978-0-415-11786-9 (hbk)
ISBN 13: 978-1-138-00682-9 (pbk)

Cum remotae gentium origines historiam transcendant, linguae nobis praestant veterum monumentorum vicem.

Gottfried Wilhelm Leibniz, *De originibus gentium*

There is no tracing the connection of ancient nations but by language; and therefore I am always sorry when any language is lost, because languages are the pedigree of nations. If you find the same language in distant countries, you may be sure that the inhabitants of each have been the same people; that is to say, if you find the languages are a good deal the same; for a word here and there the same will not do.

Samuel Johnson, quoted in Boswell
1785

If we possessed a perfect pedigree of mankind, a genealogical arrangement of the races of man would afford the best classification of the various languages now spoken throughout the world; and if all the extinct languages, and all intermediate and slowly changing dialects had to be included, such an arrangement would, I think, be the only possible one . . . this would be strictly natural, as it would connect together all languages extinct and modern, by the closest affinities, and would give the filiation and origin of each tongue.

Charles Darwin, *On the Origin of Species*

To seek, by the multiple routes of anatomy, physiology, history, archaeology, linguistics and even palaeontology, what have been in historic times and in the ages which preceded the most ancient remains of humanity, the origins, the affiliations, the migrations, the mixtures of the numerous and diverse groups which make up the human species.

Paul Broca, 'La linguistique et l'anthropologie'

Für mich est jedes Wort ein sprechendes Lebewesen, das seine Geschichte erzählt, sobald ich es kennen gelernt habe. Ich sehe die Zeit kommen, wo man von einer etymologischen Biologie sprechen wird.

Gottlieb Adolf Krause
'Die Stellung des Temne innerhalb der Bantu-Sprachen', 1895

Contents

Figures

Tables

Contributors

Václav Blažek, Ryneček 148, 26101 Pribram III, Czech Republic.

Roger Blench, Cambridge Institute for the Study of Pacific and African Languages, 8 Guest Road, Cambridge CB1 2AL, United Kingdom.

Jane Conroy, Department of Archaeology, University College, Galway, Ireland.

John Lynch, University of the South Pacific, Pacific Languages Unit, P.O. Box 12, Vila, Vanuatu.

Waruno Mahdi, Faradayweg 4–6, 14195 Berlin, Germany.

Jeff Marck, Department of Linguistics, Research School of Pacific and Asian Studies, Australian National University, Canberra, ACT 0200, Australia.

Chinyere Ohiri-Aniche, Faculty of Education, University of Lagos, Akoka, Lagos, Nigeria.

Matthew Spriggs, Department of Archaeology and Anthropology, Faculty of Arts A.D. Hope Building, Australian National University, Canberra, ACT 0200, Australia.

Bertil Tikkanen, Department of Asian and African Studies, University of Helsinki, P.O. Box 13, FIN-00014, Finland.

John Waddell, Department of Archaeology, University College, Galway, Ireland.

Preface

The relation between the present volumes and the Third World Archaeological Congress held in New Delhi in December, 1994, is complex. Events at the Congress have been described in some detail (e.g. Bernbeck and Pollock 1995; Colley 1995; Golson 1995; Hassan 1995) and need not be further touched upon. Some chapters were presented as papers in the Congress, as part of a five-day session containing some eighty papers on Language and Archaeology, whilst others were commissioned for the present volumes. In some cases, scholars who presented papers at the Conference have substantially revised their work or even divided it into several chapters. The object has been to develop as comprehensive a coverage as is practical of the issues raised in this area, both geographically and methodologically. These books should therefore be regarded not as proceedings, but as ideas stimulated following that meeting.

Issues of nomenclature, style of data presentation and editorial principles are dealt with below. The introduction is divided into two parts: a generic introduction, dealing with the broad issues raised by the interface of archaeology and language, and an introduction specific to the volume in hand.

TERMINOLOGY AND METHOD: SOME EDITORIAL PRINCIPLES

Terminology

An issue thrown into sharp relief by pulling together chapters that in principle undertake the same enterprise in very different intellectual traditions is the wide variety of terminology used to describe the same phenomena. This is nowhere more apparent than in the case of language subgrouping. The terms 'phylum', 'stock', 'family', 'branch', 'section', 'group', 'subgroup', 'language', 'lect', 'communalect' and 'dialect' are thrown freely around without any clear definition that could assist someone in another region to apply them consistently. This is not to say that the literature is not well endowed with

Table 1 Definitions of language groupings

Term	*Percentage range*
Phylum	5–12
Stock	13–28
Family	29–45
Subfamily	46–69
Language	70–81
Dialect	above 81

Source: adapted from Wurm and McElhanon (1975: 152)

attempts to define these categories. The most common of these are in terms of lexicostatistics. Lexicostatistics provides mathematical definitions of the relations between one language and another, and therefore would seem very suitable for concrete definitions. For example, one well-known use of this system was applied to the languages of Papua New Guinea (Table 1).

The use of such a table depends heavily on the faith of individual linguists in lexicostatistics. If it is possible for languages to be 'mixed', i.e. to draw a significant proportion of basic vocabulary from two or more unrelated languages, then lexicostatistics will give contradictory results. It used to be denied that mixed languages existed; then, when this view became untenable, it was said that they were very rare. Mbugu (Ma'a) in Tanzania appears frequently in the literature exemplifying this sort of rarity (Mous 1994). However, Oceania has supplied some of the most striking examples of 'mixed' languages, such as Maisin or Magori (Dutton 1976; Ross 1984), which create problems in applying the lexicostatistic method. Since the work of Thomason and Kaufman (1988), it is increasingly accepted that this type of language mixing may in fact be quite common. The effect of a synchronic perspective on language description is that extraneous elements in the lexicon have been assimilated and are no longer evident. If we identify a mixed language in the present, it is because we can still identify its components. Assuming that these types of language mixture occurred in the past (and probably did with greater frequency), it may well be that many languages today are 'mixed' but that their elements are no longer so easily discerned.

As more syntheses of world languages appear (notably Ruhlen 1991), a consensus on terminology is slowly emerging. The most important of these is the use of 'phylum', now applied to the large, well-known and reasonably established families of languages such as Indo-European or Uralic, but more controversially extended to any language grouping whose external affiliations are not well established or remain highly controversial. This can mean that an individual language may represent a phylum; thus Japanese/Ryukyuan is generally considered an isolate and is usually referred to as 'Japonic'. Indeed, Northeast Asia represents an intriguing cluster of either very small language groups or individual isolates; these are generally considered to be phyla (see Janhunen, Volume II).

The term 'stock' has remained in discussions of Pacific, especially Papuan, languages but has not been widely adopted outside; most linguists probably use 'family' as the next level of relationship below phylum. Indeed, Indo-European scholars, the most conservative subgroup of historical linguists, remain unused to referring to Indo-European as a phylum. Between stock and language something of a free-for-all obtains; branch, section, group, subgroup are used quite freely, and no fiat from individual scholars is likely to change this situation. 'Language' is generally considered to be a group of speech-forms whose speakers can all understand one another without considerable effort. Below 'language' in the hierarchy of classification either dialect or communalect are commonly used. However, recently, the term 'lect' has been adopted to capture the ambiguity between language and dialect and in part also to avoid the pejorative overtones of dialect.

Reconstructions and conventions

Reconstructions form a particular focus of historical linguistics, usually denoted by an asterisk * and often referred to as 'starred forms'. These are abstract forms, derived from attested languages, supposedly part of a hypothetical proto-language. Thus an author citing * plus a formula for a word is implying that it was part of the proto-language spoken by the particular reconstructed group. Terms such as 'proto-Indo-European' are common enough to be standard terminology. However, not all authors use the same standards of evidence to derive these proto-forms. Problems arise

1 when the data set is defective, i.e. lexical attestations are known only from some languages in the proposed subgroup;
2 when a reconstruction is built indirectly, i.e. on the back of other reconstructed forms whose status is doubtful.

Proto-forms can be cited for defective data sets; this is an inevitable part of hypothesis building. Problems arise when speculative reconstructions of this type are quoted as solid results by specialists from another area.

In some domains of African language research a distinction has been adopted between a 'quasi-reconstruction' or 'pseudo-reconstruction' and a 'regular reconstruction' (e.g. in Bendor-Samuel 1989). Quasi-reconstructions are essentially well-informed guesses based on partial data sets, as opposed to regular reconstructions which are based on a thorough analysis of historical sound correspondences. Quasi-reconstructions are marked '#' in contrast to regular reconstructions which retain the asterisk *. This distinction is difficult to enforce as authors are inevitably touchy about the reality of their reconstructions. This is particularly true of deep-level macrophylum reconstructions such as the hypothetical Nostratic; the claim by Hegedüs (Ch. 4, Volume I) that it is based on regular sound correspondences would be disputed by many historical linguists. However, as variations arise in the reconstruction and subgrouping of the language phyla of the world, historical linguists will gradually be compelled to become more critical of proposed reconstructions.

Phonetic characters and orthographic conventions

These books make no apology for making use of the technical conventions of linguistics; unless authors can back up their results in a way credible to linguists, their assertions will remain speculative. As far as possible, authors have been encouraged to shift their data tables to an appendix and to establish a clear flow of argument independent of these. The tables have been left in place, however, where argument and data are inextricably intertwined.

In an ideal world, all linguists would switch to a standard set of conventions for representing phonetic characters and these would be internationally agreed upon and developed or expanded as research continues. The conventions of the IPA (International Phonetic Association) largely serve this function in the case of basic phonetic research and often in the description of undescribed languages. However, where an old-established research tradition exists, as in Indo-European, Kartvelian or Sino-Tibetan, phylum-specific conventions have been established and writers are often loath to break away from these and shift their whole data set to IPA. In addition, orthographies that have been developed in this century for mission or other literacy purposes often reflect the technology of the period. Where authors were expecting to produce primers or Bible translations, they developed conventions that were effective on typewriters. In some cases, these have become well established, and now that printed materials are produced by computer, word-processors have to mimic these conventions.

In the chapters that follow, most authors use IPA phonetic symbols, but in the case of well-established traditions, they follow disciplinary orthographic conventions. Where these might be obscure they are explained in endnotes.

Editorial policy

Approximately half the contributions in these volumes were written by scholars whose first language is not English. These books are not intended to present a façade of ideological homogeneity; indeed, as an overview of the field, they include many contradictory points of view. A particular effort has been made to include research by Russian and East European scholars, the importance of whose work is only gradually being recognized. This has involved the editors in very extensive rewriting in places and it is not always easy to ensure that the full meaning of the original has been retained. An endnote following relevant chapters indicates the extent of the changes that have been made. Some of the flavour of Russians writing in English has been maintained, partly because it is also important to understand the parameters of their strikingly different style of argumentation.

REFERENCES

Bendor-Samuel, J. (ed.) 1989. *The Niger-Congo Languages*. Lanham: University Press of America.

Bernbeck, R. and S. Pollock 1995. Ayodhya, Archaeology and Identity. *Current Anthropology* 37 Supplement: 138–142.

Colley, S. 1995. What really happened at WAC-3? *Antiquity* 69(26), 15–18.

Dutton, T. 1976. Magori and similar languages of Southeast Papua. In *New Guinea Area Languages and Language Study, Volume II. Austronesian Languages*, 581–636. Pacific Linguistics C-39. (ed.) S. Wurm. Canberra: Australian National University.

Golson, J. 1995. What went wrong with WAC 3 and an attempt to understand why. *Australian Archaeology* 41, 48–54.

Hassan, F. 1995. The World Archaeological Congress in India: politicizing the past. *Antiquity* 69(266), 874-877.

Mous, M. 1994. Ma'a or Mbugu. In *Mixed Languages*, P. Bakker and M. Mous (eds), 175–200. Amsterdam: IFOTT [Institute for Functional Research into Language and Language Use].

Ross, M. 1984. Maisin: a preliminary sketch. In *Papers in New Guinea Linguistics No. 23*. 1–82. Pacific Linguistics A-69. Canberra: Australian National University.

Ruhlen, M. 1991. *A Guide to the World's Languages. Volume I.* Stanford: Stanford University Press.

Thomason, S.G. and T. Kaufman 1988. *Language Contact, Creolization and Genetic Linguistics.* Berkeley: University of California Press.

Wurm, S.A. and K. McElhanon 1975. Papuan language classification problems. In *Papuan Languages and the New Guinea Linguistic Scene* S.A. Wurm (ed.), 145–164. Pacific Linguistics C-38. Canberra: Australian National University.

General acknowledgements

To avoid repetition these have not been included in previous volumes and so should be taken to refer here to all four volumes of *Archaeology and Language*. First of all we would like to thank the authors of the papers included for their contributions and assistance in the production of the four volumes. Almost without exception we have found them professional, tolerant, fair-minded in response to the editorial process and ready to reply in a timely manner to deadlines, requests for information and correction of page proofs. The volumes resulted from the World Archaeological Congress held in New Dehli, India, in December 1994, although, as noted in the Preface, the relation between the Congress and the publication is a complex one. In the somewhat difficult conditions of the Congress, Professor Colin Renfrew, a co-organiser in the Language, Anthropology and Archaeology Theme, made a critical contribution by mobilising the resources of the British High Commission to provide much-needed material assistance at a critical stage of the programme.

Chairs and secretaries of our sessions at the Congress were John Hines, Jean-Marie Hombert, Lis Hudson, Jim Mallory (another co-organiser), S.S. Misra, Chinyere Ohiri-Aniche, Colin Renfrew, Malcolm Ross, Margaret Sharpe, Victor Shnirelman, Darrell Tryon and Kay Williamson. The Australia–India Council of the Department of Foreign Affairs and Trade in Canberra through its Executive Officer, Ian Black, provided financial assitance for Australian National University participants including one of the editors, and the World Archaeological Congress itself provided financial assistance for several of the participants in the Language, Anthropology and Archaeology Theme of the Congress. Dr Makkham Lal did his best in stressful circumstances to assist at the New Delhi end in facilitating the attendance of participants and making them welcome. WAC President Jack Golson was also most helpful both before and during the Congress. Jerry Taki's attendance was facilitated by Prof. John Lynch of the Emalus Campus of the University of the South Pacific in Vanuatu and Dr Lissant Bolton of ANU. Professor Peter Ucko introduced the editors to each other prior to the Congress.

On occasion we sought editorial opinion from specialists in particular fields beyond our own expertise and would like to acknowledge here especially the assistance of Peter Bellwood, Geoff Hope, Mark Hudson, Jim Mallory, Jeff Marck, Andy Pawley and Malcolm Ross. Several of the figures in these volumes were re-drawn for publication by Ian Faulkner of the Cartography Unit, RSPAS at the Australian National University. Other material assistance was given by Terry Crowley, Jim Fox and Glenn Summerhayes. Finally we would like to acknowledge the institutional support provided to the editors by the Overseas Development Institute, London, where Roger Blench has been since 1996, and the Australian National University, specifically the Department of Prehistory of the Research School of Pacific and Asian Studies, which subsequently became the Division of Archaeology and Natural History, where Matthew Spriggs was based from 1987 until the end of 1996, and its sister Department of Archaeology and Anthropology in the Faculty of Arts, where he has been since the beginning of 1997.

General introduction

ROGER BLENCH AND MATTHEW SPRIGGS

PRINCIPAL THEMES IN ARCHAEOLOGY AND LANGUAGE

The relationship between linguistics and archaeology has been affected by both the internal dynamic of the disciplines in question and external political and social trends. Many archaeologists still feel that archaeology and linguistics do not share much common ground; some of the reasons for that are internal to archaeology, whilst others can be traced to the sometimes startling misuse of these linkages by earlier scholars.

The idea of a relationship between a linguistic prehistory and an archaeological prehistory is a seductive one, but in the past it has often led to dangerous liaisons. The data from both disciplines are open to constant reinterpretation as new evidence comes in and new models are adopted. Linguists or archaeologists who interpret their data by tying it to a particular statement of 'fact' for another discipline in one year may well find that 'fact' discredited the next and the interpretation of their evidence undermined. If circularity of argument is to be avoided, these two databases for constructing prehistory must be assembled quite separately, and compared only at a subsequent stage of synthesis.

For many areas of the world, such as the Pacific and Africa, it is common for an overview of linguistic prehistory to be available before an equivalent archaeological picture has been produced. The newly arrived archaeologist should not completely ignore hypotheses of culture history derived from linguistic data, but should treat them as just that, hypotheses that may or may not provide a realistic model for a region's prehistory. An explanation derived solely from archaeological data may turn out to have greater explanatory power, or the original linguistic model may provide a plausible narrative that adequately encompasses the evidence of both disciplines. In this latter case, the archaeological data is not so much explained by the linguistic as consonant with it, as both are linked to the same broad social processes. They may, of course, not be in any particular case.

The comparison of archaeological and linguistic evidence has not proved very popular in the post-1945 era, partly because of the stigma derived from the misuse of both disciplines by the Nazis to construct their 'master race' ideology, but also because of flaws in the method of comparison. Theories of language affiliation were often developed without the use of a critical or orthodox methodology to reconstruct human history. Isolated archaeological observations were being explained by equally isolated linguistic ones.

Another reason that archaeology and linguistics have been kept apart has been because of internal developments in archaeological theory, particularly the trend of the discipline towards a sort of 'archaeology is archaeology is archaeology' position. This has acted to exclude data from multiple sources:

> Yet there is little general awareness of the value of combining the study of archaeological data with that of historical linguistics, oral traditions, historical ethnography and historical records although it is clear that many archaeological problems can be resolved in this way . . . the resistance seems to come from the view, widely held by processual archaeologists, that their discipline must be based as exclusively as possible on the study of material culture.
>
> (Trigger 1989: 356)

Partly in response to earlier theoretical excesses, the 'sceptical' generation of post-war western archaeologists was extremely aware of the limitations of their discipline for reconstructing a rounded prehistory. In 1956–7, Glyn Daniel could write:

> We must alas, for the most part, keep the builders and bearers of our prehistoric cultures speechless and physically neutral. This may seem to you an unsatisfying conclusion. And so it is but then much of our prehistory is unsatisfying and difficult, tantalisingly meagre and sketchy. We can appreciate this and accept the limitations of prehistory along with its excitements.
>
> (Daniel 1962: 114–115)

Hawke's 1954 'ladder of inference' was climbed by archaeologists with increasing fear of heights. Details of prehistoric technology could be learned, economy could be investigated with some success, but the higher rungs of prehistoric socio-political organisation would always remain shaky, and an understanding of prehistoric ideology remained forever beyond the reach of a sensible archaeologist (Hawkes 1954). Trigger (1989: 327, 392) notes that despite the optimistic assertions of the 'new archaeologists' of the 1960s such as Binford (1962), the processualist agenda, as it developed in subsequent decades, has remained firmly on the lower rungs.

From the early 1980s onwards, increasing concern was expressed by archaeologists over the seemingly limited goals of processual archaeology. A variety of approaches, often lumped together as 'contextual archaeology', have returned again to the optimistic aim of earlier generations to construct a more rounded prehistory. Attempting to identify past social and linguistic groupings is part of this project. As is perhaps the case with all such developments in social and historical disciplines, this is reflective of broader changes in contemporary society rather than being internal to archaeology.

We are in a period of growing interest in 'roots'. When personal identities are under a bewildering array of pressures, the certainties of the past are combed for answers to the question 'Who am I?' In justifying his interest in the old question of the origins of the Indo-Europeans, Colin Renfrew (1987) did not claim purely disinterested motives for wishing to know 'What songs the sirens sang':

> You may ask, who cares? What on earth does it matter what language was spoken by long-dead people? . . . But language and identity are closely linked and there are few things more personal than the language one speaks. Indeed language and national identity are today very widely equated. One's 'ethnic' affinity is often determined much more by language than by any identifiable physical characteristics, and elections are won or lost by Flemish or Walloons, bombs detonated by Welsh nationalists and Basque separatists, and massacres perpetrated in many parts of the world – most recently in Sri Lanka – on the basis of distinctions which are linguistic and cultural more than anything else.
>
> (Renfrew 1987: 2)

And so he feels it must have been in the past: 'if we are interested in the origins of the modern world, we must understand the nature of past societies; this includes the social organisation of these ancient peoples and their sense of self-identity, which brings us to the questions of ethnicity and language' (ibid.: 3).

Trigger (1989: 376) sees this interest in the past of specific groups of people as part of a growing humanist trend in archaeology, in opposition to the goals of neo-evolutionist processual archaeology which saw case studies of particular regions as merely testing grounds for general theories of human behaviour and cultural change. When carried out in the developing world and/or with native peoples, such archaeology can be seen as both neocolonialist and insulting. As archaeologists have become more sensitized to the needs and aspirations of the peoples among whom they work, and whose ancestors they may be studying, they have responded by providing histories that are relevant to the lives of local populations and that seek to answer the 'where do we come from?' questions that help to anchor identity in a world in flux.

STREAMS IN LINGUISTIC PREHISTORY

Tlon, Uqbar, Orbis Tertius: fringe theories of linguistic affiliation

As the epigraphs on p. v indicate, the view that historical linguistics has something to contribute to the history of peoples has existed for more than two centuries. Indeed, Johnson appears to be already reacting to an aspect of historical linguistics that has often caused it to be regarded with the gravest suspicion by other disciplines: the tendency for some of its practitioners to develop unusual models of world prehistory based on apparent links between geographically remote languages.

One of the earliest theories to develop along these lines was the version of Amerindian history that claimed that the inhabitants of the New World were the Lost Tribes of Israel. This interpretation was advanced as early as 1650, when Menasseh ben Israel published his account of the traveller Aaron Levi who reported that he had encountered Hebrew-speaking Amerindians in the mountains near Quito. This type of linguistics is often broadly referred to as Voltairean linguistics, from his famous characterization 'Etymology is a science in which the vowels count for nothing and the consonants for very little.'[1]

This type of theorizing, usually the province of amateurs, is often linked with bolder cultural hypotheses that usually involve long-distance migration, and often have a religious or political agenda. It is easily caricatured and may often provide a well-founded excuse for archaeologists and prehistorians to avoid this type of excursus. Such theories are, of course, not exclusively based on linguistic evidence, but lexical connections are generally claimed to support the comparison of material culture. Two key themes of this body of scholarship relate to specific regions of the world: Ancient Egypt and the Pacific.

The notion that civilization was somehow invented in Ancient Egypt and spread out through the remarkable navigations of its inhabitants has a pedigree as far back as Classical Greece (Bernal 1987), and the ascription of Egyptian origins to African peoples was well under way by the beginning of the twentieth century. Johnson (1921 but manuscript prepared in 1897) wrote an influential history of the Yoruba, arguing against an Arabian origin for the Yoruba and promoting their migration from Egypt. Such theorizing continues today in the works of the followers of Cheikh Anta Diop and is often promulgated in luxuriously produced handbooks of hieroglyphics. However, claims for such land migrations were relatively restrained compared with the deepwater navigation proposed in classics such as Perry's (1923) 'Children of the Sun'. Elliot Smith and later Thor Heyerdahl were eloquent proponents of long-distance migrations, and much curious scholarship was adduced in support of such hypotheses.

The substantial literature on pre-Portuguese Trans-Pacific contacts originated as early as the seventeenth century (Wauchope 1962: 83 ff.). Although recent DNA research may be taken to suggest that such contacts did indeed occur at least sporadically, this is far from accepting that some of Kublai Khan's ships, still carrying elephants, were driven eastwards to the New World after a failed invasion of Japan (Ranking 1827), or that fragments of the fleet of Alexander the Great reached the Americas in 323 BC (Gladwin 1947).

Exponents of such ideas are typically aggrieved when the predictably cautious academic establishment fails to take on board their ideas. One of the advocates of trans-Pacific contact took a robust view of their caution:

> All the lights in the House of the High Priests of American Anthropology are out, all the doors and windows are shut and securely fastened (they do not sleep with their windows open for fear that a new idea might fly in); we have rung the bell of Reason, we have banged on the door with Logic, we have thrown the gravel of evidence against their windows; but the only sign of life in the house is an occasional snore of dogma.
>
> (Gladwin 1947)

There is probably a useful distinction to be drawn between fringe ideas that draw the attention of more cautious scholars to possible, previously unsuspected, connections and similarities (Heyerdahl, for example) and those that are nothing more than an encumbrance to scholarship (Atlantis, Von Daniken, Velikovsky). The moral is that we should keep Gladwin's windows open but look out through them rather than simply sleeping by them.

Links with nationalist ideologies

One of the more troubling aspects of the history of this discipline has been its links with nationalist ideologies. Linguistic nationalism still engenders a rich emotional harvest at present, often for good reason, since the suppression of minority languages is commonly a prominent feature of totalitarian governments. Democracies sometimes encourage voluntary euthanasia among minorities through neglect. Nonetheless, when a national language is linked to a national culture, it is a short step to linking that to archaeological entities and thence to broader historical claims on territory and political authority (see Kohl and Fawcett 1995).

Throughout the nineteenth century, these ideas would have been considered acceptable by many researchers, and links between nationalist ideologies and scientific research were unproblematic. However, somewhere in the early twentieth century, a split developed between the rationalist, academic tradition and the promotion of certain types of archaeology in support of nationalist goals. This has been well documented in Germany and the former Soviet Union, where linguistic ideologues developed theories of the relation between particular language groups and specific types of material culture and were ruthless with those tempted to disagree (Trigger 1989). Nonetheless, evidence is mounting that there is a European-wide tradition of rewriting the past in pursuit of nationalist goals (Díaz-Andreu and Champion 1996).

Nikolay Marr (1865–1934), who has been called the 'Lysenko of anthropology' in Russia, had a comparable influence on all types of linguistic, ethnographic and archaeological research in his tenure as Director of the Russian Academy of Material Culture. His career and influence are described in Slezkine's (1994) account of Russian imperial relations with the minority peoples of Siberia. Central to Marr's ideas were evolutionary or 'Japhetic' theories of language, whereby languages developed in stages from 'primitive' to advanced. Primitive societies had 'mollusc-like' speech forms that had to develop 'upwards', until at the conclusion of history all language would merge into a single Communist speech. This eventually led him to the conclusion that both ethnography and archaeology were anti-Marxist, and these were formally condemned at the All-Russian Conference on Archaeology and Ethnography in 1932. The practical consequence of Marr's tenure of authority was the destruction of much of the academic infrastructure around these subjects: museums, journals and learned societies were disbanded and non-Marxist teachers persecuted. Marr's work was explicitly rejected by no less a figure than Stalin, who wrote an essay in 1950 examining the relation of Marxism to linguistics (Stalin 1950; Slezkine 1994: 314). Shnirelman (Chapter 10, Volume I), describing Russian 'linguo-archaeology', warns that links with nationalist ideologies are still alive today although their structure is less formalised than in an era of centralized state control.

German linguists played an important role in the development of Indo-European scholarship, and as early as the mid-nineteenth century, Jacob Grimm was to explain the distribution of various sound changes by referring to the ethnic character of speakers. Gustaf Kossinna (1858–1931), whose principal work, *Die Herkunft der Germanen*, published in 1911, became a key text in Nazi Germany, provided an important ideological plank for German territorial expansion. Kossinna argued that specifically Germanic material culture could be identified in archaeological sites and that where such material was found, this was evidence of the original extent of Germany.

The positivist tradition

It is tempting to dismiss both marginal historical linguistics and nationalist ideology as forgotten errors of a past epoch. Historically, however, they have had an important influence on archaeologists, making them wary of all types of correlation with linguistic theories, no matter how carefully couched.

Another, more sceptical, tradition of historical linguistics has existed for several centuries and indeed persisted through a long period of neglect. For example, precursors to historical linguistics exist both among the Sanskrit grammarians and in the works of the rabbinical scholars. Most striking is the work of Yehuda Ibn Quraysh, who lived in Fez, Morocco, in the tenth century, and was the first to compare the phonology and morphology of Hebrew, Aramaic and Arabic in his book *'Risāla* (Téné 1980). Such precursors seem to have had little influence on their successors, and an intellectual tradition developed only after historical linguistics was put on a more scientific footing. This event is conventionally attributed to Sir William Jones' famous lecture in 1786 demonstrating the links between Sanskrit and the classical languages of Europe, but it has become clear in recent years that Jones' perception was far from original (Muller 1986). Bonfante (1953) quotes a reference to an unpublished manuscript by Marcus Boxhorn (1612–1653) hypothesizing a 'Scythian' origin for all the major languages of Europe, whilst in Saumasius' *De Hellenistica*, published in 1643, reconstructed proto-forms for European numerals are proposed. The concept of reconstruction of an

Indo-European proto-language appears as early as 1713 in the works of the English divine William Wotton:

> My argument does not depend on the difference of Words, but upon the Difference of Grammar between any two languages; from whence it proceeds, that when any Words are derived from one Language into another, the derived Words are then turned and changed according to the particular Genius of the Language into which they are transplanted. [. . .] I can easily suppose that they might both be derived from one common Mother, which is, and perhaps has for many Ages been entirely lost.
>
> (Wotton 1730 [1713]: 57)

Wotton had related Icelandic ('Teutonic'), the Romance languages and Greek, which are certainly as convincing a demonstration of Indo-European affinities as Jones' demonstration of the links of classical languages with Sanskrit. Moreover, Wotton developed some estimates of the speed of language change and was concerned about the apparent contradiction with the widely accepted 'Biblical' age of the earth. Jones, in contrast, erroneously believed that Egyptian, Japanese and Chinese were part of Indo-European while Hindi was not, which suggests that his method had serious flaws.

Outside Indo-European, Uralic classification had been virtually completed prior to Jones. As Ruhlen observes: 'The basic structure of the Uralic family had thus been roughly worked out at least six years before William Jones's celebrated address, which opened the era of I-E [Indo-European] studies' (Ruhlen 1991: 66).

The nineteenth century was a major period for the development of historical linguistics, and indeed most of the debates that still characterize the discipline today have their origin in the work of scholars of the previous century. Throughout the nineteenth century, there was a strong conviction that language could be analysed to establish historical results. Donaldson commented in the 1830s:

> There is in fact no sure way of tracing the history and migrations of the early inhabitants of the world except by means of their languages; any other mode of enquiry must rest on the merest conjecture and hypothesis. It may seem strange that anything so vague and arbitrary as language should survive all other testimonies, and speak with more definiteness, even in its changed and modern state, than all other monuments however grand and durable.
>
> (Donaldson 1839: 12)

and Craik in the 1860s: 'Each language has a life of its own, and it may be made to tell us its own life, so to speak, if we set the right way to work about it' (Craik 1861: 1).

Just as Finno-Ugric (i.e. Uralic) and Indo-European were earliest on the scene in terms of historical reconstruction, so their scholars began the tradition of reconstructing history through lexical reconstruction. Early attempts to do this, such as those by Pictet[2] (1859–63), evolved convoluted theories of the migrations of the Aryan race that we should now consider highly suspect; however, this should not distract attention from the significance of the enterprise.

These efforts continued throughout the late nineteenth century and they served to establish the conventions that were to be adopted and developed elsewhere in the world. Historical linguistics of this type requires a certain density of research to be credible; without adequate lexical materials for language classification and reconstruction, no amount of methodological sophistication will fill the lacuna.

The pattern of research

Research concentrations are often reflections of political accessibility and funding. Research on the Andamanese and Nicobarese languages has remained largely static due to the refusal of the Indian government to issue research permits. Although they coexist in the same part of the world, Papuan has lagged far behind Austronesian due to the inaccessibility of many Papuan languages. Comparative Australian has taken off following the efforts of relatively few highly motivated individuals. Bantu is far better known than Niger-Congo due to early interest in the topic, accessibility of many of the languages and relatively unproblematic transcription.

Despite these problems, a global picture of the disposition and relations of language phyla is slowly beginning to emerge. The established phyla assigned to the world's languages now appear to be relatively stable (although the analysis of macrophyla is highly controversial; see next section). Data are beginning to be less of a problem than collating them. Few regions of the world are entirely without archaeology, although the density of excavated sites is highly variable. In consequence, crackpot theorizing and the promotion of nationalist ideologies are at a lower level, and the volume of papers and books exploring the links between language and archaeology is on the increase. The major threat to this area of scholarship is probably now its old-fashioned empiricist allegiance and a positivist commitment to data; to avoid strangulation at the hands of the post-modern devotees of Kali, it will have to develop more sophisticated public relations. Lenin is reputed to have said that the express train of history cannot be stopped; all that revolutionaries can do is grease the wheels.

THEMES IN THE INTERACTION OF LINGUISTICS AND ARCHAEOLOGY

Historical linguistics, lexicostatistics and glottochronology

The single most important theme of these books is the interaction of historical linguistics with archaeology. Historical linguistics may be defined as the analysis of the relationship between languages that are assumed to be genetically related, that is to 'have sprung from some common source', such as English and German. Historical linguists attempt to establish the rules that have allowed each language to evolve from the common source and from this information to reconstruct hypothetical proto-forms. Usually this is based on the comparison of two or more languages, but the 'internal reconstruction' of a single language is also possible, using indications within a language, such as dialect variation or fossil morphology, to build up a picture of an earlier stage of that language. In the case of isolates such as Basque or Burushaski, this is the only procedure possible. Historical linguists are also increasingly concerned with the sociological aspects of the construction of a modern speech form: to establish the patterning of loanwords, the extent of former dialect variation and possible social distinctions in former stages of reconstructed languages.

Linguists are concerned to develop testable rules by which specific languages can be related to one another, relating to phonology, morphology and lexicon. These rules generate a tree-like genetic structure that allows the modelling of the relative antiquity of splits between different languages or other more complex aspects of their inter-relations (see Ross, Chapter 13, Volume I). Proto-forms predicted by the rules that relate two or more languages and a sequence of proto-languages can be reconstructed for nodal points in the genetic tree.

Lexicostatistics – the counting of cognate words in a standardized list, and assigning a numerical degree of relationship – seems to have been first used in the early nineteenth century. Dumont d'Urville (1834) compared a number of Oceanic languages (which would today be called Austronesian) and proposed a method for calculating a coefficient of relationship. He extended his comparison to some Amerindian languages and concluded that there was no evident relationship with the Oceanic languages in his sample. Hymes (1983) provides a detailed history of the further development of lexicostatistics in the nineteenth and twentieth centuries.

Another aspect of historical linguistics is glottochronology. Writers such as Wotton (1730) had the idea of calculating how rapidly languages change by comparing ancient texts of known date with the modern form of those languages. Robert Latham (1850) was probably the first author to sketch the possibility of assigning a precise date to the split of two languages through applying a mathematical algorithm. Hymes (1983: 73 ff.) cites other tentative experiments in the nineteenth century but these seem not to have been developed until Swadesh (1952).

Lexicostatistics and glottochronology have the attractive aspect of quantification: they seem to represent a scientific approach to the dating and genetic classification of languages. However, very few historical linguists now accept the premises of such approaches. In part this may reflect a wave of criticism of the mathematics underlying these methods (see discussion in Hymes 1983: 75). More important, however, has been the realization that languages undergo a variety of changes in interacting with one another. Lexicostatistics must assume a standard of lexical purity that allows languages to change at a regular rate, especially in their core vocabulary. Using the generally accepted methods of historical linguistics, only relative dating is possible; for absolute dating linguists now turn to archaeology.

Historical linguistics as a discipline

Archaeology is taught as a method that can be applied to any situation, rather like economics, and although archaeologists divide into theoretical schools and schools develop their own terminologies, this is usually not location-specific. Indeed, within a single institution different methods may well be promulgated by individual scholars. In other words, the archaeology of, for example, Japan or Australia does not appear to have a large technical vocabulary that would not be immediately comprehensible to a regional outsider.

Although theoretical linguistics has comparable intellectual subdivisions, there is only a limited interface between historical linguists and the larger linguistic establishment. This is partly because historical linguistics remains a minority interest in a world dominated by syntax, phonology and, to a lesser extent, sociolinguistics. Historical linguists are often partly self-taught or take their cue from individual teachers. The consequence is that there can be striking disagreements over method and standards of evidence; this debate is most apparent in the case of the sometimes bitter disputes that have ranged over macrophyla.

Scholars of the older-established phyla often take a patronizing attitude to results from those phyla more recently recognized. This is particularly striking in the case of Indo-European, where the conviction that the phylum is well founded and that its reconstructions are accurate and convincing appears to be widespread among its adherents. A darkly humorous version of this can be seen in the comments of Hopper (1989), reviewing Thomason and Kaufman (1988), who contrasted the 'factually established genetic categories' such as Indo-European with 'broad-based guesses' such as Niger-Congo, Afroasiatic and Nilo-Saharan. The view taken in these volumes is that the major language phyla of the world that are accepted by the scholarly community are all equally well founded.

The Indo-Europeanist habit of ignoring what are strangely called 'minor languages' has resulted in a virtual lacuna in research on Indo-European languages of India with only small numbers of speakers. One of the more evident tendencies in Indo-European linguistics is to give primacy to written languages, such as Sanskrit. Thus, reconstruction of the Indo-Aryan languages is in terms of relating the present-day forms to attested Sanskrit (cf. Turner 1966), rather than subjecting the body of Indo-Aryan languages to the usual procedures of historical linguistics. The consequence has been a striking inadequacy of fieldwork to describe the more than 300 unwritten Indo-European languages spoken in the India–Pakistan region today (see the assessment of research needs in Grimes 1996). A rather similar procedure for Dravidian has allowed the assemblages of cognates compiled in Burrow and Emeneau (1984) to be cited as 'proto-Dravidian', even though their work is very Tamilcentred. The conventional practice of historical linguistics in the region is thus in a rather backward state. Applying the standards of proof common, say among Austronesianists, would of course reduce Indo-European to a 'broad-based guess'.

Geographical coverage

All types of research have a patchy coverage when considered globally, but linguistics and archaeology have proven especially sensitive to political and economic constraints (see above). Different disciplinary traditions also lead to uneven emphases with particular regions. For example, although East Asian archaeology is well represented in terms of excavated sites, specific digs seeking the origins of food production are a relatively new phenomenon. The incidence of monuments can be in inverse relationship to an emphasis on economic prehistory. Countries with a dominant culture often discourage work on regional languages for fear of encouraging local aspirations. Until recently, the languages of China were poorly known, and research on minority languages unaccountably spoken by peoples not part of an officially recognized 'nationality' was strongly discouraged (Ramsey 1992: 162 ff.).

In addition, intellectual traditions and the organization of scholarship affect interdisciplinary work. Countries with national research centres that unite scholars from different intellectual areas, such as France, the former Soviet Union and Australia, are far more likely to produce interdisciplinary scholarship than England and America, where experts are ghettoized in university departments. Generally speaking, where careers depend upon publications, and only publications in a specific discipline are highly valued, there is every incentive to concentrate in one intellectual area to the exclusion of others. Indeed, in both linguistics and archaeology, intellectual justifications for excluding other approaches have been explicitly developed, as witness the example of generativism (Chomsky 1988).

The consequence has been that both historical linguistics and its combination with archaeology are developed to very different degrees in different parts of the world. The areas where the focus has been most significant are Eurasia and Oceania: Eurasia because of the Indo-Europeanist tradition and its remarkable survivals in the former Soviet Union, and Oceania because of the fortunate support for this type of approach in a few key institutions. India represents a curious lacuna in Eurasia, since, despite its importance in the early decades of the twentieth century and the production of the massive 'Linguistic Survey of India' during the 1920s, restrictions on research permits have led to an almost complete cessation of research by outside scholars on its some 500 unwritten languages. The New World and Africa have been marked by relatively small amounts of research. In Africa this may be due to nothing more than time-depth (convincing amounts of data have only recently become available) and lack of dedicated institutions. In the case of the Americas, however, despite the all-embracing tradition of anthropology, which conjoined archaeology, cultural anthropology, and linguistics, the absence of a major tradition of synthesis suggests that the reality has been academic isolationism.

Texts and pretexts

One of the earliest interfaces between archaeology and language has remained distinct from the type of historical linguistics discussed here: the interpretation of ancient written documents and the decipherment of scripts. This story has been rehearsed too many times (e.g. Simpson 1985) to need further recounting, beginning with the decipherment of hieroglyphics and cuneiform, through to Hittite and other epigraphic languages of the Ancient Near East. In this century, decipherment has been extended to India, China and Central America, and continues today with recent proposals for the decipherment of the Olmec script of the Yucatan (Wichmann, Volume II). Epigraphy is also equipped with its own eccentric fringe: a Harvard Professor of Zoology tells us that inscribed rocks in Texas record the journey of migrant Zoroastrians from Iberia some 2,000 years ago (Fell 1980: 164).

Interpreting epigraphy and relating it both to known historical events and to excavated sites has been a major theme of archaeology, especially in the Near East. Indeed, the prominence accorded to written texts has obscured other types of interpretation of linguistic data. Thus, although a considerable amount of work is done translating, transcribing and interpreting ancient texts in a variety of Semitic languages, overall models of the evolution and dispersal of this language family barely exist. An example of this is the attempt by Zohar (1992) to interpret the spread of Semitic in the Near East. African Semitic languages (which are considerably more numerous and diverse than those of the Near East) are referred to as 'minor languages' in the text and excluded entirely from the family tree of Semitic (Zohar 1992: Figure 1).

There is a strong argument for supposing that much of the most innovative work in using historical linguistics has been brought about by the *absence* of ancient texts. Just as North American archaeology developed innovative analytic techniques to analyse the sites of hunter-gatherer communities, modelling in historical linguistics has been stimulated in regions of the world where there are no early texts.

Testable hypotheses

One of the attractive aspects of linking historical linguistics with archaeology is that it is possible to generate testable hypotheses. Linguists are usually way ahead of archaeologists in their speculations. Finding an informant for a language is easier and far less costly than mounting an archaeological expedition to search, for example, for the origins of food production. An experienced linguist can often elicit a range of basic and key cultural vocabulary in a few hours, whereas excavations often take many months and sometimes years. Historical linguists are often tempted to throw off hypotheses on the origins of food production far more quickly and perhaps more casually than would be permissible within other academic frameworks.

When a prediction is made, however, it can at least be tested. So, for example, if a historical linguist claims that certain species of domestic animal can be reconstructed back to the proto-language of a particular phylum, and at the same time makes a proposal for the homeland of the speakers of the proto-language, then excavations should ideally be able to confirm the presence of those species. An example of such a correlation is presented in the chapter by Green and Pawley (Volume III) where linguistics is used both to pinpoint a proposed homeland of Oceanic languages and to suggest the structural features of house-forms that should be present. Archaeology suggests that structures of the predicted type are indeed present. Such correlations are rare in practice, especially when only a small number of sites have been identified, but as the density of well-investigated sites increases, hypotheses can be subjected to a reasonable test.

Phyla and macrophyla

There are some language phyla whose existence is generally accepted, such as Indo-European or Austronesian, as a result of the weight of scholarly opinion. In a few cases, such as Nilo-Saharan, despite its introduction in the 1950s and a series of conferences since then, a body of scholarly comment exists questioning either its unity as a phylum or the families that compose it. In addition, there are regions of the world where a large number of languages exist that show common features but that have not been shown to be related to the satisfaction of most researchers. These 'geographical' names are often shown as phyla in works of synthesis. The most important of these are Papuan, Australian and Amerind: zones of languages with common features and coherent subgroups where overall genetic relations have proved resistant to the methods of historical linguistics. Similarities of phonology or other features do suggest a common origin, but it is possible that they have diversified so far from a common proto-language that proof will remain a chimera. Finally, in one case, Andamanese, inadequate data makes any final judgement impossible at present. Table 2 sets out the language phyla of the world and their status in this hierarchy.

It is not possible to order the class of 'accepted' phyla by degree of acceptance. In recent years, numerous publications have advanced the case for macrophyla, that is, the uniting of several accepted phyla into one genetic group. The best known example is Nostratic, a macrophylum that brings together most of the phyla of the Eurasian landmass, whose membership varies according to different authors. The journal *Mother Tongue* has published the speculations of 'long-rangers' who wish to promote continent-spanning comparisons. With increasing awareness of the traditions of such scholarship in the former Soviet Union, and the publication of some major texts (e.g. Bomhard 1994), this type of large-scale comparison has reappeared. Other more controversial proposals include Indo-Pacific and Amerind (Greenberg 1987) and Sino-Caucasian from the Soviet School, especially Starostin (e.g. Shevoroshkin 1992). These proposals have excited considerable scepticism, although most linguists do not

Table 2 Language phyla of the world and their status

Phylum	*Usual acronym*	*Where spoken*	*Status/comment*
Niger-Congo	NC	Western, central and southern Africa	Accepted
Afroasiatic	AA[a]	NE Africa and the Middle East	Accepted
Indo-European	IE	Eurasia	Accepted
Uralic	U	Eurasia	Accepted
Kartvelian	K	Caucasus	Accepted
North Caucasion	NC	Caucasus	Accepted
Chukchi-Kamchatkan	CK	Siberia	Accepted
Yeniseic	Y	Siberia	Accepted
Eskimo-Aleut	EA[b]	Bering Strait	Accepted
Dravidian	DR	India	Accepted
Sino-Tibetan	ST	Central Asia	Accepted
Miao-Yao	MY	China	Accepted
Daic (=Tai-Kadai)	D	SE Asia	Accepted
Austroasiatic	AS[a]	SE Asia	Accepted
Austronesian	AN	Pacific	Accepted
Trans-New-Guinea	TNG[b]	Papua New Guinea	Accepted
Pama-Nyungan	PNY	Australia	Accepted
Na-Dene	ND[b]	North America	Accepted though affiliation of Haida is debated
Khoisan	KH	Eastern and southern Africa	Usually accepted
Nilo-Saharan	NS	Eastern and central Africa	Usually accepted
Altaic	AT	Eurasia	Usually accepted although the affiliation of Korean is debated
Papuan	PP[b]	New Guinea	Consists of a large number of accepted groups but their unity is not considered proven
Australian	AU[b]	Australia	Consists of a large number of accepted groups but their unity is not considered proven
Amerind	AM[b]	Americas	Consists of a large number of accepted groups but their unity is not accepted
Andamanese	AD[b]	Andaman islands	Inadequate data make effective historical linguistics impractical

This table excludes a number of well-known isolates such as Basque, Burushaski, Ghilyak, Ainu and Japanese, as well as African isolates (see Blench this volume) and problematic languages of Asia such as Nahali and Kusunda.

[a]AA is unfortunately used for both Afroasiatic and Austroasiatic. AS is adopted here for Austroasiatic to eliminate confusion. PN is applied to Polynesian, hence the use of PP for Papuan here.

[b]Proposed acronym

command the vast range of data that would be necessary to give them a full evaluation (see Blench, this volume, for discussion of African examples). Ruhlen (1991: 270 ff.) gives a lengthy bibliography of 'alleged connections between families usually assumed to be unrelated', which suggests that almost any two or more of the world's language phyla have been related by some researcher.

Behind such enterprises is an intriguing and controversial agenda: the reconstruction of proto-World, or 'Proto-Sapiens' as Ruhlen (1994: 192) has it. The hypothesis that all human language has a common origin is certainly emotionally persuasive; the myth of the Tower of Babel still exerts a powerful pull. However, conviction is not proof and enthusiasm not demonstration. Although one of the most eloquent advocates of proto-World, Vitaly Shevoroshkin, has recited poems in this remarkable language on radio and television, this cannot yet conjure it into reality.

The exploration of long-range comparison has aroused considerable opposition; historical linguists working on a smaller scale are frequently outraged at the misuse of language data by non-specialists. Trask (1995), for example, has recently analysed in considerable detail the evidence for a traditional hypothesis linking Basque to Caucasian languages, and concludes that it depends in almost every case on a misuse or defective analysis of the Basque language materials. Thurgood (1994) has shown that the hypotheses, such as Benedict's Austro-Tai, that link together the major language phyla of SE Asia are based on ancient loanwords.

Between near-global hypotheses and accepted phyla stand more modest proposals that link together two phyla that already have a history of observed similarities. Two recent examples are Austric (linking the Austronesian and Austroasiatic phyla; Reid 1994) and Niger-Saharan (Niger-Congo with Nilo-Saharan; Blench 1995). The linking of Japanese (or Japonic) to the Altaic phylum has a venerable pedigree but still continues to generate controversy and cannot be regarded as accepted.

Intriguing as these planet-spanning proposals are, they remain to be critically evaluated by the body of historical linguists and thus cannot easily be used by archaeologists. Indeed, there are still few wholly convincing models to explain the origin and diversification of accepted phyla; to interpret the more doubtful macrophyla would be over-egging an already rich pudding.

Linguistics and genetics: 'The New Synthesis'

An aspect of the reconstruction of prehistory that has come to the fore since the mid-1980s is the use of evidence from genetics, especially from analysis of mitochondrial DNA. However, the reputations of traditional biological anthropologists have stood recently at an all-time low following analyses such as that of Gould (1982), who accurately skewered the underlying racial preoccupations of the supposedly scientific physical anthropologists of the nineteenth and early twentieth centuries. It should be noted that osteometrics remain acceptable in many European traditions, especially in France, as witness a standard text on human remains in the Sahara (Dutour 1989).

A major break with traditional biological anthropology occurred, however, with the development of modern techniques of DNA analysis, both because DNA could potentially be recovered from archaeological material and because DNA analysis seemed to offer a way of relating present human populations to one another and to past materials. Linguistic classifications of human populations seemed to offer a way beyond simple racial models; more abstract, they seemed to provide an ideal analogue to the classificatory trees from DNA. If DNA trees and language trees were to correspond, then this would provide striking mutual confirmation for models of human prehistory. Indeed, the links between them were enthusiastically promoted at the end of the 1980s and into the early 1990s as 'The New Synthesis' (see, for example, Cavalli-Sforza *et al.* 1988; Renfrew 1992). The culmination of this trend was the appearance of *The History and Geography of Human Genes* (Cavalli-Sforza *et al.* 1994), which promotes a major revision of the methodology for exploring human history.

Some archaeologists are among those disturbed by the implications of 'The New Synthesis' for encouraging narrow nationalistic readings of history, and restoring the discredited view of race, language and culture as generally coterminous (Pluciennik 1996). Linguistic and archaeological naiveté aside, the new data of genetics are not being inserted into a political vacuum as geneticists sometimes seem to assume. A more self-critical awareness is clearly required when dealing with the implications of broad genetic generalizations linked most uncertainly, as Pluciennik points out, to archaeological entities.

Such entities themselves are sometimes subject to divisive claims by putative descendant groups. For instance, the continuing dispute over who are the 'real' Macedonians with a claim to the heritage of Alexander's symbols of power nearly brought Greece and the former Yugoslav Republic of Macedonia to war recently (Brown 1994). In such circumstances, genetic data are more than likely to be seized upon and misused to stir up feelings of enmity between the rival claimants. Language and archaeology have already been misused in this way.

More recently, there has been a distinct withdrawal from some of the claims of this type of work. The 'fit' between language trees and DNA results has been seen not to be quite as close as suggested in earlier publications. Chen *et al.* (1995: 610) compare genetic and language trees on a global basis and conclude that: 'The consensus between language trees and genetic trees is low . . . so low as to make the trees incomparable.' This will probably remain the case on the scale of phylic and macrophylic relations that they analyse. With very large landmasses such as Eurasia, language shift is an extremely common process, as the disappearance of Basque-related languages suggests. To find a people speaking their 'original' language may prove to be the exception. In contrast, much of the Pacific has seen expansion of populations into otherwise uninhabited territory. Almost certainly, Oceania will again prove an important testing-ground for the methods of DNA analysis as it has with linguistics and archaeology, because the parameters of population movement and contact can be simplified.

CONCLUSIONS: AN AGENDA PAST 2000

With the publication of these volumes, we hope that the process of synthesizing historical linguistics and archaeology will have largely shaken off its previously negative image. Many archaeologists still hold the view, either explicitly or implicitly, that linguistic and human biological evidence are either inadmissible or irrelevant in the discussion of archaeologically defined entities such as 'cultures'. At one level they are right: much confusion has occurred in the past by mixing the investigation of concepts and terms between the disciplines involved in researching the history of particular regions at too early a stage. If, however, it is history one is after, rather than simply a narrow archaeology, then archaeologists cannot ignore important sister disciplines such as historical linguistics, genetics and human biology when attempting to synthesize the evidence.

An encouraging trend of the last few years, represented by the interest shown in the language and archaeology sessions held at the New Delhi WAC Congress, is the increasing number of linguists and archaeologists who *are* interested in what multi-disciplinary research has to offer.

We must remain aware, however, of the abuses of the earlier part of this century, when biological, linguistic and archaeological data were combined wilfully to create extreme nationalist fantasies that race, culture and language are always coterminous. There are enough examples of this from recent and indeed contemporary history to necessitate critical self-awareness of how interpretations can come to be used in ways never intended, by people to whose views we may not wish to subscribe. The alarm bells sounded in some quarters over 'The New Synthesis' of archaeology, genetics and language need to be heeded.

As with all types of scientific change, paradigm shifts occur over time, though with a less revolutionary time-scale than that advocated by Kuhn (1962). Universities and academic institutions have been able to keep dominant schools of method coherent through control of publishing and because a relatively small circle of individuals were in power. As these networks of power increasingly fragment, as publishing becomes cheaper and more accessible (particularly with advances in information technology) and as more research takes place outside the academy, then more diverse approaches to interdisciplinary studies will be able to flourish.

With this added diversity of approach, the current unfortunate distinction between prehistory and history should lessen or even disappear. Both the study of the archaeological evidence of the past and the modelling of social change through historical linguistics should be considered valid approaches to the past. The result should be the study of the broad outlines of a human history that allows for a complexity in the past that is so evident in the present.

ACKNOWLEDGEMENTS

We would like to thank Peter Christian, Kevin MacDonald, Malcolm Ross, Mark Thomas and Peter Ucko for commenting on this introduction.

NOTES

1 Although quoted in Leonard Bloomfield's 'Language' (1935: 6), the direct source in Voltaire's writings has yet to be uncovered, and there is more than a suspicion that this is a piece of convenient linguistic folklore.
2 Pictet also first used the expression 'linguistic palaeontology', often attributed to more recent authors.

REFERENCES

Bernal, M. 1987. *Black Athena: the Afroasiatic roots of Classical civilization. Volume 1. The Fabrication of Ancient Greece 1785–1985.* London: Free Association Books.
Binford, L.R. 1962. Archaeology as anthropology. *American Antiquity* 28, 217–225.
Blench, R.M. 1995. Is Niger-Congo simply a branch of Nilo-Saharan? In *Proceedings: Fifth Nilo-Saharan Linguistics Colloquium, Nice, 1992*, R. Nicolai and F. Rottland (eds), 83–130. Köln: Rudiger Köppe.
Bloomfield, L. 1935. *Language*. London: Allen & Unwin.
Bomhard, A. 1994. *The Nostratic Macrofamily*. Berlin and New York: Mouton de Gruyter.
Bonfante, G. 1953. Ideas on the kinship of the European Languages from 1200 to 1800. *Cahiers d'Histoire Mondiale* 1, 679–699.
Broca, P. 1862. La linguistique et l'anthropologie. *Bulletin de la Société d'Anthropologie de Paris*, 264–319.
Brown, K.S. 1994. Seeing stars: character and identity in the landscapes of modern Macedonia. *Antiquity* 68, 784–796.
Burrow, T. and M.B. Emeneau 1984. *A Dravidian Etymological Dictionary*. 2nd edition. Oxford: Clarendon.
Cavalli-Sforza, L.L., A. Piazza and P. Menozzi *et al.* 1988. Reconstruction of human evolution: bringing together genetic, archaeological and linguistic data. *Proceedings of the National Academy of Science, USA* 85, 6002–6006.
Cavalli-Sforza, L.L., A. Piazza and P. Menozzi 1994. *The History and Geography of Human Genes*. Princeton: Princeton University Press.
Chen, J., R.R. Sokal and M. Ruhlen 1995. Worldwide analysis of genetic and linguistic relations of human populations. *Human Biology* 67(4), 595–612.
Chomsky, N. 1988. *Language and Problems of Knowledge. The Managua Lectures*. Cambridge, MA: MIT Press.
Craik, G. L. 1861. *A Compendious History of English Literature and of the English Language*. London: Griffin, Bohn & Co.
Daniel, G. 1962 *The Idea of Prehistory*, reissued 1971. Harmondsworth: Penguin.
Darwin, C. 1859. *On the Origin of Species*. London: John Murray.
Díaz-Andreu, M. and T.C. Champion (eds) 1996. *Nationalism and Archaeology in Europe*. London: UCL Press.
Donaldson, J.W. 1839. *The New Cratylus*. Cambridge: Deighton.
Droixhe, D. 1990. Le voyage de *Schreiten*: Leibniz et les débuts du comparatisme finno-ougrien. In *Leibniz, Humboldt and the Origins of Comparativism*, Tulli de Mauro and Lia Formigari (eds), 3–30. Amsterdam: Benjamins.

Dumont d'Urville, J.S.C. 1834. *Philologie, par M. D'Urville. Seconde Partie. Les autres vocabulaires de langues ou Dialectes océaniens recueillies durant la voyage, et le Vocabulaire comparatif des languages françaises, madekass, malaio, mawi, tonga, taiti et hawaii, suivis de quelques considérations générales sur ces langues.* Paris: Ministère de la Marine.

Dutour, O. 1989. *Les hommes fossiles du Sahara.* Marseille: CNRS.

Fell, B. 1980. *Saga America.* New York: Times Books.

Gladwin, H.S. 1947. *Men out of Asia.* New York: McGraw Hill.

Gould, S.J. 1982. *The Mismeasure of Man.* New York: Norton.

Greenberg, J. 1987. *Language in the Americas.* Stanford: Stanford University Press.

Grimes, B.F. 1996. *Ethnologue.* (13th edn). Dallas: Summer Institute of Linguistics.

Hawkes, C.F. 1954. Archaeological theory and method: some suggestions from the Old World. *American Anthropologist* 56, 155–68.

Hopper, P.J. 1989. Review of *Language Contact, Creolization and Genetic Linguistics,* by S.G. Thomason and T. Kaufman. *American Anthropologist* 91, 817–818.

Hymes, D.H. 1983. Lexicostatistics and glottochronology in the nineteenth century (with notes towards a general history). In *Essays in the History of Linguistic Anthropology* D.H. Hymes (ed.), 59–113. Amsterdam and Philadelphia: John Benjamins.

Johnson, Revd. S. 1921. *The History of the Yorubas.* Lagos: CMS (Nigeria) Bookshops.

Kohl, P.L. and C. Fawcett 1995. *Nationalism, Politics and the Practice of Archaeology.* Cambridge: Cambridge University Press.

Kuhn, T.S. 1962. *The Structure of Scientific Revolutions.* Chicago: University of Chicago Press.

Latham, R.G. 1850. *The Natural History of the Varieties of Man.* London: John van Voorst.

Muller, J.-C. 1986. Early stages of language comparison from Sassetti to Sir William Jones (1786). *Kratylos* 31(1), 1–31.

Perry, W.J. 1923. *Children of the Sun.* London: Methuen.

Pictet, A. 1859–63. *Les origines indo-européennes, ou les Aryas primitifs: essai de paléontologie linguistique.* Paris: Chetbuliez.

Pluciennik, M. 1996. Genetics, archaeology and the wider world. *Antiquity* 70, 13–14.

Ramsey, S.R. 1992. *The Languages of China.* Princeton: Princeton University Press.

Ranking, J. 1827. *Historical Researches on the conquest of Peru, Mexico, Bogota, Natchez and Talomeco, in the thirteenth Century, by the Mongols, accompanied with Elephants.* London.

Reid, L. 1994. Morphological evidence for Austric. *Oceanic Linguistics* 33(2), 323–344.

Renfrew, C. 1987. *Archaeology and Language: the puzzle of Indo-European origins.* London: Jonathan Cape.

Renfrew, C. 1992. Archaeology, genetic and linguistic diversity. *Man* 27(3), 445–478.

Ruhlen, M. 1991. *A Guide to the World's Languages. Volume I.* Stanford: Stanford University Press.

Ruhlen, M. 1994. *The Origin of Language.* New York: John Wiley and Sons.

Shevoroshkin, V. (ed.) 1992. *Nostratic, Dene-Caucasian, Austric and Amerind.* Bochum: Brockmeyer.

Simpson, G. 1985. *Writing Systems: a linguistic introduction.* Stanford: Stanford University Press.

Slezkine, Y. 1994. *Arctic Mirrors.* Ithaca and London: Cornell University Press.

Stalin, J. 1950. *Problems of Marxism and Linguistics.* Moscow: Progress Press.

Swadesh, M. 1952. Lexicostatistic dating of prehistoric ethnic contacts. *Proceedings of the American Philosophical Society* 96, 453–462.

Téné, D. 1980. The earliest comparisons of Hebrew with Aramaic and Arabic. In *Progress in Linguistic Historiography* Konrad Koerner (ed.), 355–377. Amsterdam: John Benjamins.

Thomason, S.G. and T. Kaufman 1988. *Language Contact, Creolization and Genetic Linguistics.* Berkeley: University of California Press.

Thurgood, G. 1994. Tai-Kadai and Austronesian: the nature of the relationship. *Oceanic Linguistics* 33(2), 345–368.

Trask, R.L. 1995. Basque and Dene-Caucasian: a critique from the Basque side. *Mother Tongue* 1, 3–82.

Trigger, B.G. 1989. *A History of Archaeological Thought.* Cambridge: Cambridge University Press.

Turner, R. 1966. *An Indo-Aryan Comparative Dictionary.* London: Oxford University Press.

Wauchope, R. 1962. *Lost Tribes and Sunken Continents.* Chicago and London: University of Chicago Press.

Wotton, W. 1730. *A Discourse Concerning the Confusion of Languages at Babel.* London: Austen & Bowyer.

Zohar, M. 1992. Pastoralism and the spread of the Semitic languages. In *Pastoralism in the Levant: archaeological materials in anthropological perspectives* Ofer Bar-Yosef and A. Khazanov (eds), 165–180. Madison: Prehistory Press.

Introduction

ROGER BLENCH AND MATTHEW SPRIGGS

Volumes I and II of this sequence have dealt with the theory of relating archaeology and language in terms of 'broad-brush' correlation, exploring the connections over time and space between regional archaeology and language distribution across large areas of Oceania, Africa or central Asia. Volume III dealt more directly with the concrete linkages between material culture, texts and linguistic and archaeological sequencing. This final volume is intended to explore the more elusive areas of language change and language classification. As was the case with Volume III, it is intended to provide a feeling for the texture of current debate in terms of topics covered and methods used.

RETHINKING LANGUAGE CLASSIFICATION

The General Introduction attempts to emphasize the fluidity of language classification, a feature perhaps more obvious to editors than authors, as individual chapters tend to underline certainties. The first theme of the book deals with some contentious linguistic issues concerning the classification of languages. There is little doubt that much of the classificatory agenda in this century has been set by Joseph Greenberg (1963, 1987), whose hypotheses, even where wrong, have often become a major stimulus to other scholars to amend and expand upon them. Greenberg's first major field of enquiry was African languages and his classification was widely considered to be a success of the method of 'mass comparison' (Greenberg 1966). His results have been widely quoted in other disciplines, including all types of writing about the early history of the continent. Nonetheless, Greenberg's classification has been superseded in many areas and the newer versions are gradually beginning to replace the Greenberg model. Various summaries of the present classificatory situation in Africa have been published (e.g. Blench 1993, 1997), whilst Ruhlen (rev. ed. 1991) remains the best 'global' synthesis.

Apart from refining and adding detail to the internal classification of the major phyla, recent speculation has turned on the possible external links of

Africa's phyla to those outside the continent. This is particularly the case with the Afroasiatic phylum, whose links with Eurasian languages have been proposed by a variety of authors, especially as part of the various versions of the Nostratic hypothesis (see Hegedüs, Volume I, Blažek, this volume). The chapter by Blench synthesizes recent classificatory models and the growing literature on external links, with specific emphasis on the relevance for archaeology and genetics.

Macrophyla proposals are intellectually stimulating but remain outside the mainstream (see General Introduction). The contradictions between them suggest that they are more useful to stimulate discussion and further research than as serious historical proposals at this stage of development. In other words, it would be extremely premature for archaeologists or geneticists to attempt to correlate them with cultural sequences or DNA results. This is less the case with more detailed proposals, such as those reordering languages within phyla, or reassigning languages to specific phyla. Three case studies of this type are presented in this section – Iran, Nigeria and Polynesia.

Blažek sets out a case for a major reorientation of one of the relatively well-documented languages of the Ancient Near East, Elamitic. In a pictographic script, Elamitic appears to predate 3000 BC and is written in decipherable inscriptions for more than two millennia from 2300 BC onwards. Despite this body of data, Elamitic, like Sumerian, has never been classified with any certainty. This partly reflects the difficulties in securely translating many of the texts due to the lack of affiliation between Elamitic and other known languages. The view that has gained widespread acceptance is that of McAlpin (1981), who claimed that Elamitic was Dravidian. The 'Elamo-Dravidian' phylum has appeared in many synoptic language surveys, despite the somewhat shaky evidence to support it.

Blažek proposes that Elamitic would be better seen as a 'seventh branch' of Afroasiatic, although he also assumes that these languages are related at a higher level to others in Eurasia. His theory not only proposes a classification of Elamitic but also assigns Sumerian a role in the patchwork of relationships. The evidence for this is principally lexical, although the cognate word sets do yield sound correspondences. If Blažek is correct, then this will create a minor revolution in Afroasiatic studies and have considerable impact on interpretations of the peopling of the Ancient Near East. At the very least, the evidence he presents suggests substantial contact between Elamitic and early Afroasiatic groupings.

Niger-Congo is the largest language phylum in the world, with some 1,600 languages. Although the broad outlines of its classification are widely accepted, recent research has identified many 'minority' languages whose position is less than clear. A particular region of interest is that of the confluence of the Niger-Benue in central Nigeria, which is a major centre of language diversification, or alternatively a locus where the original diversity of Benue-Congo languages is conserved. Ohiri-Aniche has collected fresh data on these little-known languages, and her chapter makes a preliminary approach to their classification. It is interesting that the affiliations of these languages are unclear and seem to relate to language groups some distance from the confluence. Although archaeology provides some indications of the time-depth of settlement in the region, the lack of radiocarbon dates so far makes any correlation highly speculative at present.

The Polynesian languages may be compared to the Bantu languages of central and southern Africa in that they are relatively closely related to one another, and the simple CV (Consonant-Vowel) syllable structure of most of them makes transcription and cognacy judgements relatively easy. As with Bantu, many Polynesian languages are extremely well documented, beginning with a rich heritage of missionary dictionaries in the nineteenth century. An invaluable tool in the analysis of Polynesian subclassification is *POLLEX*.[1] Marck presents a reanalysis of the position of Eastern Polynesian and its relation to Proto-Polynesian. If this new analysis is correct, then existing models of the proposed settlement pattern of Polynesia will need to be rethought (as argued by Spriggs, Volume II, on the basis of data presented here).

INTERPRETING LANGUAGE CHANGE

The second theme is the interpretation of language change. Fundamental to the practice of historical linguistics is the existence of a diversity of languages and dialects, allowing comparison and illuminating linguistic and social processes. Although language change itself is a truism, the potential range of such processes remains to be fully documented. In Oceania, the importance of maritime trade routes and colonization is taken for granted, and understanding sea travel, its constraints and advantages, is crucial to modelling linguistic processes of diversification. However, as Waddell and Conroy point out, maritime contact may well have played a major role in the 'Celticization' of Ireland. It is generally conceded that Celts represent an incoming group in Ireland, and yet it has proven difficult to advance any obvious archaeological correlation of this presumably dramatic linguistic change, which caused all the pre-existing languages to disappear (compare the situation on Mailu described by Dutton, Volume III). The gradual spread of a prestige culture and the assimilation of presumably diverse substrate languages would explain the divergent features of Irish, which have been the source of much speculation over time. There seems to be a problem with this model in relation to England, since it would be likely that English would be more heavily affected by the Celtic languages than appears to be the case (see also Hines, Volume II).

A concept from historical linguistics that had considerable influence in the earlier phases of historical linguistics was the *Sprachbund*, a region where mutual influence between languages has allowed many features of phonology and syntax to become common to languages of quite different genetic affiliation. The classic region for this is the Balkans, which encompasses languages of quite distinct branches of Indo-European. The usual model for the evolution of such cases is to postulate intense bilingualism over a long period, but in a social context such that one or other of the languages is not simply eliminated. Two other regions of the world are frequently cited as areas of a *Sprachbund*: Ethiopia and the Indian subcontinent (see Tikkanen, Chapter 6 this volume). In many ways these are more convincing examples than the Balkans, since in both cases the languages in contact belong to quite distinct phyla. In Ethiopia, Nilo-Saharan and Afroasiatic languages are in contact, whilst in south Asia, Indo-European, Dravidian, Sino-Tibetan, Austroasiatic and the isolate Burushaski meet.

Tikkanen considers the south Asian case in some detail. He takes a primary feature of south Asian languages – retroflexion (a distinctive type of consonant articulation) – and plots its distribution across language phyla and then correlates it with other features, both syntactic and toponymic. This enables him to infer the genetic affiliation of possible substrate languages in the regions of the original Indo-Aryan expansion. At present, archaeological correlations with these are tenuous, to say the least, but studies of this type make possible new avenues for the interpretation of cultural change in sites in this region. It is, moreover, valuable to begin to speculate on the type of historical process that generates such regions of intense interaction and thus causes the crossover of striking features from one phylum to another.

A study of another, quite different type of historical change raises important questions about interpretive structures and incomplete data. Some languages of southern Oceania have been characterized in the literature as 'aberrant', with its attendant historical implications. If languages are deemed to be extremely distinct from the main body of languages to which they are related, then this can be taken to imply that they form a single early branching and that they have long been isolated or that there is an important substrate influence. Lynch takes on the concept that aberrancy is somehow related to genetic distance and shows that even extremely diverse surface forms in the languages of Vanuatu and New Caledonia can be derived from Proto-Oceanic through the application of established rules.

The implication is that with partial information, it is often possible to assume that isolated data sets represent an aberrant language because the surface appearance of words is so distinctive. With more complete information, the rules deriving the 'aberrant' forms from the 'exemplary' forms can be established. As a cautionary narrative, this has relevance to, for example, fragmentary epigraphic languages. Languages such as Sumerian and Elamitic are now seen as isolated because no matrix of other clearly related language exists. Yet they must certainly have been part of larger groupings, and, were those sister-languages also to be recorded, their genetic affiliation might well be easily discernible.

Mahdi's chapter on 'Linguistic and philological data towards a chronology of Austronesian activity in India and Sri Lanka' is the greater part of an original submission split into its component elements for these volumes. It should be read in conjunction with the chapter in Volume III on Austronesian boat morphology and that in Volume II on the spread of food plants. Mahdi's chapter probably represents the most comprehensive attempt in these volumes to tie together linguistic, archaeological, ethnographic and textual sources. It also incorporates personal communications from those present at the World Archaeological Congress in New Delhi, making use of the interplay between disciplines and regions that occurred.

The topic is of considerable importance, since Mahdi identifies a major migratory and trade movement with important implications for the spread of population, material culture and subsistence techniques that has been overlooked because of the absence of permanent resident populations. The presence of Austronesians in the Indian Ocean has been accepted for a long time because of the unmistakable linguistic and cultural affiliations of the Malagasy. Although it is evident that the Austronesians must have been trading with India, the impact of that trade and its interpretation in Indian historical sources has remained relatively unexplored.

Mahdi's chapter and those by Van Driem and Janhunen in Volume II are reminders of the rich seam of European polyglot scholarship that is in danger of being extinguished by monoglot Anglo-American traditions.[2]

CONCLUSIONS

During the course of editing these books, more detailed surveys of specific regions have appeared, notably the impressive overview of these issues in Australia (McConvell and Evans 1997) and Spriggs' (1997) synthesis of Island Melanesian culture history. From these, as well as other recently published works, and the survey presented here, the issues that will dominate the field in the coming decades emerge quite clearly. These are:

1 building a synthesis and classificatory consensus for the world's language phyla;
2 a clearer understanding of the potential for cross-disciplinary analyses with results from both ancient and synchronic studies of DNA;
3 archaeological research that is explicitly linked with language distribution as opposed to the haphazard connections that must be made at present.

As these issues develop, they should contribute to a changing and enlarged notion of linguistic and archaeological histories, willing to incorporate the results of many disciplines into a broader narrative of the past.

NOTES

1 A continually updated electronic file of the *Comparative Polynesian Lexicon* (*POLLEX*) is kept and maintained by Bruce Biggs of the Department of Maori Studies, Auckland University.

2 It is particularly ironic that the publishing conventions of these volumes, aimed at least partly at linguists, decrees that all chapters and quotations shall be translated into English.

REFERENCES

Blench, R.M. 1993. Recent developments in African language classification and their implications for prehistory. In *The Archaeology of Africa: food, metals and towns*, T. Shaw, P. Sinclair, B. Andah and A. Okpoko (eds), 126–138. London: Routledge.

Blench, R.M. 1997. Language studies in Africa. In *Encyclopaedia of Precolonial Africa*, J.O. Vogel (ed.), 90–100. Walnut Creek, London and New Delhi: Altamira.

Greenberg, J.H. 1966. *The Languages of Africa*. The Hague: Mouton for Indiana University.

Greenberg, J.H. 1987. *Languages in the Americas*. Stanford: Stanford University Press.

McAlpin, D.W. 1981. *Proto-Elamo-Dravidian: the evidence and its implications*. Philadelphia: Transactions of American Philosophical Society 71/3.

McConvell, P. and N. Evans 1997. *Understanding Ancient Australia: perspectives from archaeology and linguistics*. Melbourne: Oxford University Press.

Ruhlen, M. 1991. *A Guide to the World's Languages. Volume I*. Stanford: Stanford University Press.

Spriggs, M. 1997. *The Island Melanesians*. Oxford: Basil Blackwell.

Part I

RETHINKING LANGUAGE CLASSIFICATION

1 The languages of Africa: macrophyla proposals and implications for archaeological interpretation

ROGER BLENCH

INTRODUCTION

The question of the genetic classification of the language phyla of the world, from being a marginal study outside mainstream linguistics, has again begun to command considerable attention from both professional linguists and researchers in related disciplines. There appear to be two major forces behind this change in attitudes: the potential for correlation with genetics, notably mitochondrial DNA, and the opening of the former Soviet Union to the world.

Hypotheses generated by DNA studies need to be confirmed by other types of evidence, and language groupings offer broader, older and more coherent structures than archaeology. Moreover, in sampling terms, more is known about languages and their affiliations than archaeology in almost all parts of the world. Geneticists have therefore looked to macrophyla classifications as evidence for their continent-spanning hypotheses (e.g. Excoffier *et al.* 1987; Cavalli-Sforza *et al.* 1994).

At the same time, the tradition of genetic linguistics in the former Soviet Union, with its often idiosyncratic methods, had remained almost unknown to western scholarship until the frontiers began to open. A concern with large-scale language groupings and long-range comparison has been the thrust of much of this tradition. Although many of the actual results of these researches are properly treated with scepticism by scholars in the western tradition, access to publishing has meant that classification has been placed firmly back on the agenda of linguists. Indeed, the now notorious conflicts over Greenberg's 'Language in the Americas' underlines the newly developed importance of classification.

One consequence has been to reopen many existing questions relating to both the internal classification and the external affiliations of African languages. Older proposals claiming that the major African language phyla are to be united with other, non-African phyla, have again been given prominence. At the same time, what can only be described as disarray rules in relation to the internal arrangement of the principal phyla.

The purpose of this chapter is to summarize the present situation, and to present to archaeologists and prehistorians what can be salvaged from this academic chaos in terms of cross-disciplinary interpretation. It sets out some recent views of the traditionally recognized phyla and explores some recent more wide-ranging and speculative theories. Although DNA studies have been important in returning language classification to the agenda, they are dealt with here only in passing (see summaries in Ruhlen 1992; Renfrew 1991, 1994).

CLAIMS ABOUT RECONSTRUCTION

The recent revival of macrophylum theory has led to some striking claims for the reconstructibility of lexical items of cultural and historical significance in African language phyla. Most notable in the context of this chapter are the claims made for Afroasiatic, for example in Militarev (1990) or Orel and Stolbova (1995). According to these authors, almost a full set of terms connected with both agriculture and livestock production can be reconstructed for Afroasiatic. However, a detailed investigation of terms for domestic animals in Afroasiatic (Blench in press, b) could not substantiate these claims. An investigation of Proto-Omotic by Bender (1988) suggested that the only terms for crops that could be reconstructed in Omotic were those that were already part of the native flora of Ethiopia.

A similar problem has arisen in the case of Nilo-Saharan. Not only is the internal structure of the phylum much disputed, but opposing claims have been made about the reconstruction of food crops. Ehret (1989, 1993, in press) has claimed that cultivated plants are reconstructible to a high (i.e. ancient) level. Bender (1991b, 1996a) has been unable to substantiate such reconstructions.

This poses an important methodological question: if different linguists have opposing views about reconstructions, can others make use of their results? The only conclusion that can be drawn from this is that extreme caution is necessary when using reconstructions of lexical items carrying such a heavy interpretive load. Obviously, serious researchers claim to exercise such caution; the existence, however, of major disagreements must suggest that very different notions of the comparative method coexist. The strategy for dealing with this is:

1. no reconstructions can be accepted without data tables;
2. reconstructions based on isolated occurrences of words must be regarded with extreme scepticism.

These probably seem quite restrictive demands that would exclude the imaginative approach sometimes necessary in historical linguistics. However, where a cultural revolution so major as, for example, the inception of agriculture is being implied, a necessary scepticism is essential.

THE LANGUAGE PHYLA OF AFRICA

In contrast to the New World and Papua, the composition of the major language phyla of Africa is generally agreed within the scholarly community (Blench 1993a, 1997). Their internal classification remains disputed, as does the position of various isolates. However, given that Africa has the highest absolute number of languages of any continent, their classification remains a considerable achievement. Figure 1.1 shows a sketch of the approximate distribution of the major language phyla.

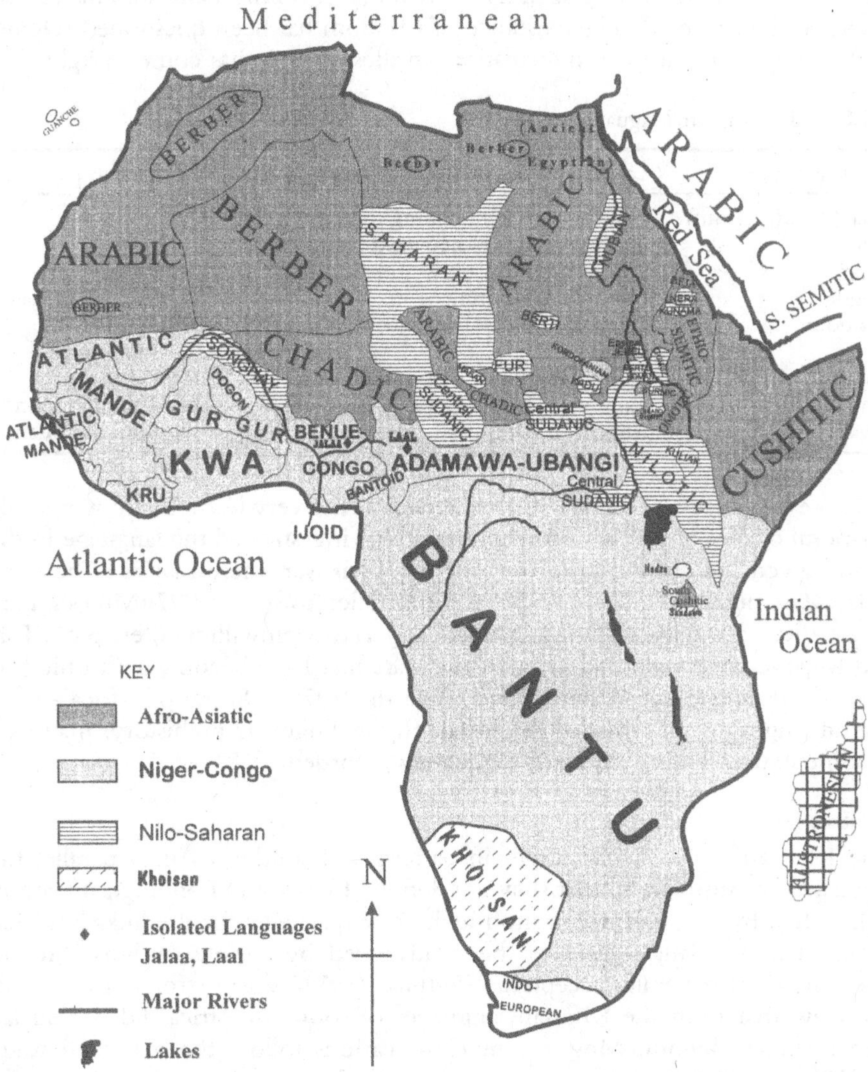

Figure 1.1 Approximate distribution of the major African language phyla

Source: Blench

Isolated languages

The existence and classification of language isolates in Africa remains controversial. Table 1.1 shows the languages often considered to be isolates. The inclusion of Hadza and Sandawe on this list is controversial, because in many quarters these are still considered to be related to Khoisan. Jalaa, like Laal in Chad, has a significant proportion of loanwords from a scatter of neighbouring languages, but a core of apparently unidentifiable lexemes. Little is known about Kujarge, except that Doornbos and Bender report a 29 per cent cognacy with neighbouring Chadic languages. New data has recently become available on Ongota (Fleming *et al.* 1992), and Blažek has argued that it is Nilo-Saharan, but his case has yet to be accepted. The existence of Oropom has been questioned (Heine, p.c.) and certainly no new information on these people has come to light.

Table 1.1 African language isolates

Language name	*Location*	*Source*	*Comments*
Jalaa (=Cuŋ Tuum)	Nigeria	Kleinewillinghöfer (in press)	
Laal	Chad	Boyeldieu (n.d.)	
Kujarge	Sudan	Doornbos and Bender (1983)	Perhaps Chadic
Ongota	Ethiopia	Fleming *et al.* (1992)	Perhaps Afroasiatic
Oropom	Uganda	Wilson (1970)	Existence unconfirmed
Hadza	Tanzania	Sands (1995 and p.c.)	
Sandawe	Tanzania	Sands (1995)	Perhaps Khoisan
Kwadi	Angola	Westphal (1963)	Perhaps Khoisan

These languages are almost all threatened, at the very least. There were only a handful of speakers of Kwadi when Westphal investigated the language in the 1950s; given the disruption of the Angolan civil war, there may well be none today. The number of Ongota speakers had fallen to six in 1997 (Mikesh, p.c.) whilst there are only a few of speakers of Jalaa (Kleinwillinghöfer, p.c.). Laal and Kujarge were recorded prior to the wars that have become a chronic feature of the Sudan/Chad borderland since the 1970s.[1] Some of Africa's most crucial languages, in terms of reconstructing its linguistic prehistory, may well become extinct before they are adequately recorded.

Khoisan

The Khoisan or 'click' languages in eastern and southern Africa parallel the languages of Australia, in that they are defined by shared phonological features rather than by an evident common lexicon. Arguments for the links between all the Khoisan languages have been advanced by various authors, but no one schema is generally accepted. Westphal (1971) was a strong advocate of the view that even the Khoisan languages of southern Africa did not all fall into a single phylum. Most recent classifications follow the extended study of Köhler (1981), who proposed a series of isoglosses linking the major Khoisan families. Traill (1986) has put forward further isoglosses linking Khoi and San, whilst warning that until our understanding of the process of lexical diffusion improves, guaranteeing that these are proof of genetic relationship remains difficult. Central Khoisan is the most well-substantiated family with a significant number of reconstructions (Voßen 1988, 1997).

Bleek, Greenberg and, most recently, Ehret (1986) argue that Khoisan is in turn linked to Sandawe and Hadza, click languages spoken in east Africa. This has been questioned by other Khoisanists (e.g. Elderkin, 1983 for Sandawe). Sands' (1998) study of Khoisan relationships has shown that especially in the case of Hadza, most of the lexical arguments that were advanced to support this case rest on very doubtful correspondences or erroneous lexical citations.

The classification and indeed the inventory of Khoisan speech forms remains in doubt. One recently published 'tree' following Köhler is Grimes and Grimes (1993). This has been combined and corrected in consultation with Rainer Voßen (and compared with Voßen, 1990) to produce the tree given in Figure 1.2. However, the present diagram represents an uneasy compromise, since it does not eliminate the lects mentioned in Grimes and Grimes that could not be reconciled with other known speech-forms.[2]

Nilo-Saharan and a proposed Niger-Saharan macrophylum

Nilo-Saharan was first characterized by Greenberg (1966, 1971), and extended by Bender (1983a, b, 1989, 1991a, b, 1996a, b) and Ehret (1989, 1993, in press) Later studies have confirmed Greenberg's basic hypothesis as to the overall unity of the family. The most elaborate proposals for the subclassification of Nilo-Saharan have been developed by Bender (1996a). Bender bases his classification on grammatical isomorphs and shared innovations, but it is considerably at variance with that of Ehret. Bender (1996a, b) considers at some length the reasons for this dissonance and relates it to rather general philosophies of classification. Figure 1.3 shows the classification proposed by Bender.

The details of the classification remain to be fully worked out, but the essential feature is a split between the outliers (Songhay, Saharan, Kunama-Ilit and Kuliak) and the remaining languages including the Sudanic languages, For, Berta, Koman and Kadu (=Kadugli-Krongo). The membership of Kadu (=Kadugli-Krongo) and Kuliak is not accepted by all researchers (e.g. Ehret 1995a). It has recently been argued that Shabo[3] is part of Nilo-Saharan, although its position is disputed (Blench 1995a). The case for Ongota is even more precarious: Fleming *et al.* (1992) express the opinion that it is Afroasiatic, whilst Blažek (1991) has argued for Nilo-Saharan affiliation. Moreover, the relationship between the members is still uncertain and no overall 'tree' is yet agreed by scholars.

Although Greenberg considered Nilo-Saharan to be a wholly distinct phylum, there has been a succession of papers adducing evidence for a close link with Niger-Congo. Gregersen (1972) originally proposed a 'Kongo-Saharan' superfamily, and Blench (1995a) has presented a detailed case for a unification of Nilo-Saharan and Niger-Congo, suggesting that Niger-Congo is most closely related to the Central Sudanic languages. The proposed macrophylum would be named 'Niger-Saharan'.

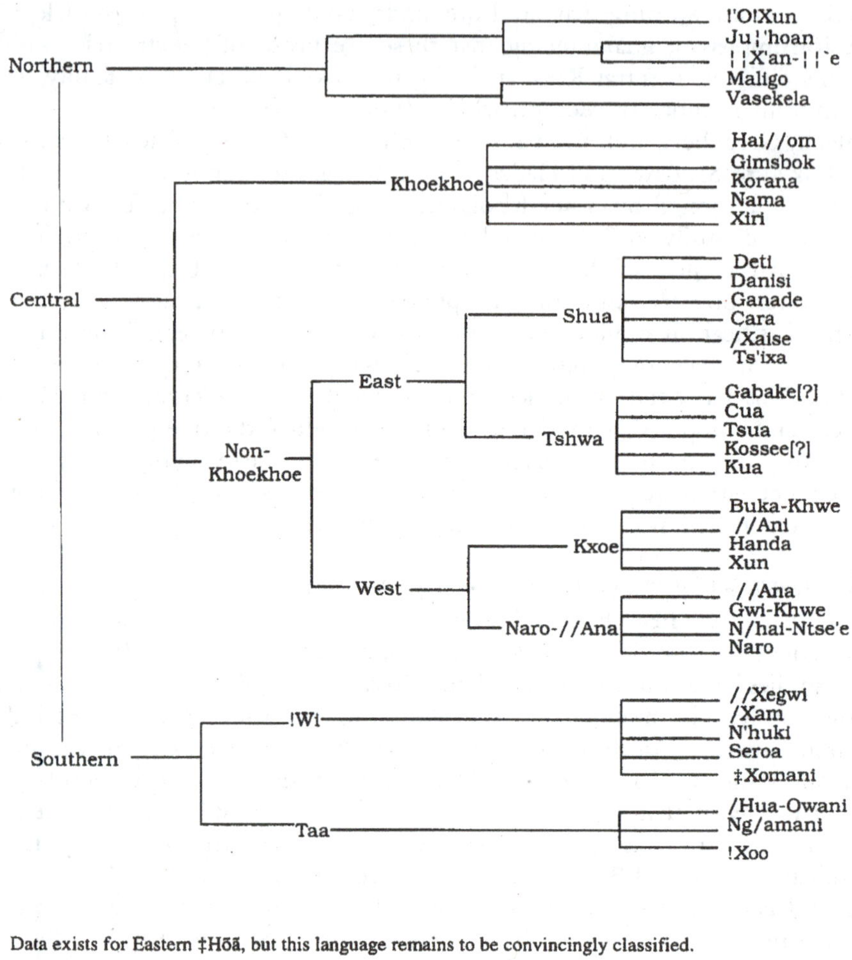

Figure 1.2 Classification of the Khoisan languages
Source: Rainer Voßen and Roger Blench

If a Niger-Saharan macrophylum is accepted, Niger-Congo then becomes one branch of it, rather like Bantu, despite its size, is simply one sub-branch of Niger-Congo. The challenge is then to see exactly where Niger-Congo branches from Nilo-Saharan. Blench (op. cit.) argues that Niger-Congo split off from Nilo-Saharan at the same time as Central Sudanic. Excluding the branches further away from Niger-Congo, a minimal 'tree' of Niger-Saharan can be constructed as shown in Figure 1.4.

If this model is substantiated by further work, then a clear conclusion can be drawn from it: that Nilo-Saharan is far older than Niger-Congo. The deep divisions within Nilo-Saharan have led previous writers to this conclusion.

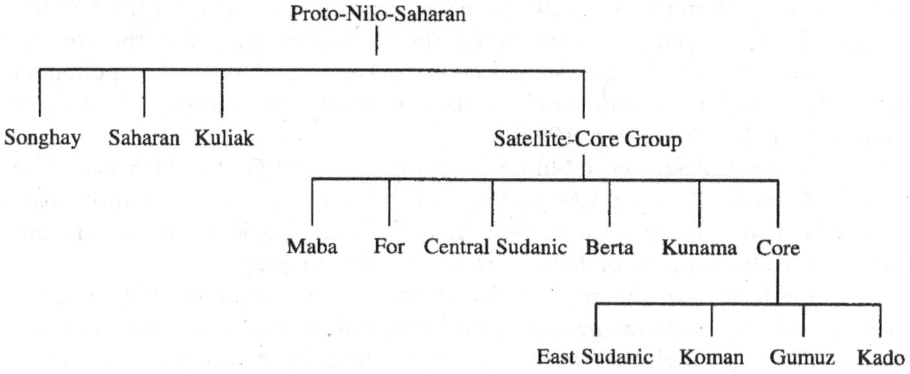

Figure 1.3 Nilo-Saharan in the classification of Bender (1996b)
Source: Bender 1996b

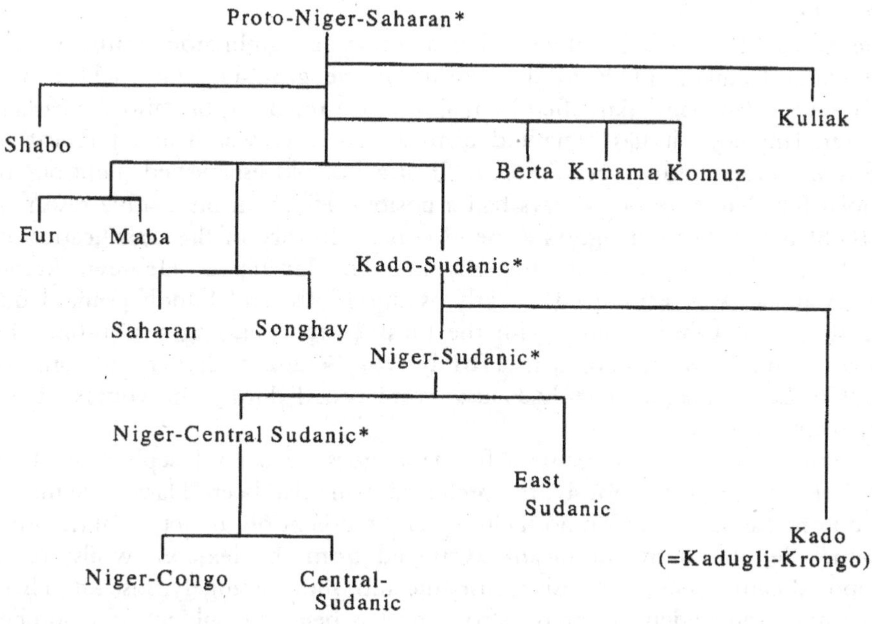

Figure 1.4 Proposed Niger-Saharan 'tree': minimal hypothesis
Source: Blench 1995a

The challenge is to attempt somehow to link this with processes evident in the archaeological record (see MacDonald, Volume II, Ch. 1).

Niger-Congo

Niger-Congo is the most widespread of Africa's language phyla and includes the majority of its languages – in excess of 1,500. The concept of Niger-Congo has its roots in Westermann's (1927) *Sudan-Sprachen*, and many of the families recognized today were first established there. Westermann was the first to illustrate the strong links between the Bantu languages and those spoken in West Africa, and his demonstrations have generally been accepted by later scholars (Blench 1989, 1992, 1993b).

The possible linkage with Nilo-Saharan is discussed above. However, the internal structure of Niger-Congo has itself been under review as more data becomes available for the various subgroups. Figure 1.5 is based on a recent study of the distribution of lexical items (Blench in prep):

A notable feature of this revised classification is the treatment of both Gur-Adamawa and Kwa-Benue groups as continua rather than as discrete language families (Kleinewillinghöfer 1996). The representation of continua with transverse double lines derives from a convention introduced by Ross (1988) and supposes that these originally represented dialect chains that coalesced into distinct language groupings.

Afroasiatic

The Afroasiatic language phylum has a somewhat ambiguous status among the major language phyla of the world. As the grouping that includes not only several languages sanctified by major world religions, but also the earliest written language, it has benefited from a massive research and publication effort in certain rather specific areas. It also has old-established traditions of scholarship that have not always had a positive effect on innovative research.

Ruhlen (1991: 87 ff.) gives a useful concise history of the classification of the languages that constitute the phylum. The kinship of Hebrew, Arabic and Aramaic was recognized as early as the 1530s, and Ludolf pointed out the affinity of Ethio-Semitic with the Near Eastern languages in 1702. The name 'Semitic' was proposed in 1781 by von Schlözer. Berber and some of the Chadic languages, notably Hausa, were added during the course of the nineteenth century.

A phylum under the name Afroasiatic goes back to Joseph Greenberg ([1950] rev. 1966). Previously, the preferred name had been 'Hamito-Semitic', an unfortunate conjunction both clumsy and redolent of suspect racial theories. Hamito-Semitic is by no means expunged from the lexicon, whilst other proposed names include Afrasian, Lisramic and, more strangely, Lislakh. These have not been widely adopted. Afroasiatic has been the subject of a number of overviews; historically, the most important of these have been Cohen (1947), Diakonoff (1988) and Perrot (1988). Reconstructions of Afroasiatic have been proposed by Ehret (1995b) and Orel and Stolbova (1995). The distribution of Afroasiatic languages is shown in Figure 1.6.

Significant developments in the classification of Afroasiatic have been:

1 the recognition that Greenberg's 'Western Cushitic' is quite distinct from other branches of Afro-Asiatic;
2 the break up of Southern Cushitic as traditionally constituted.

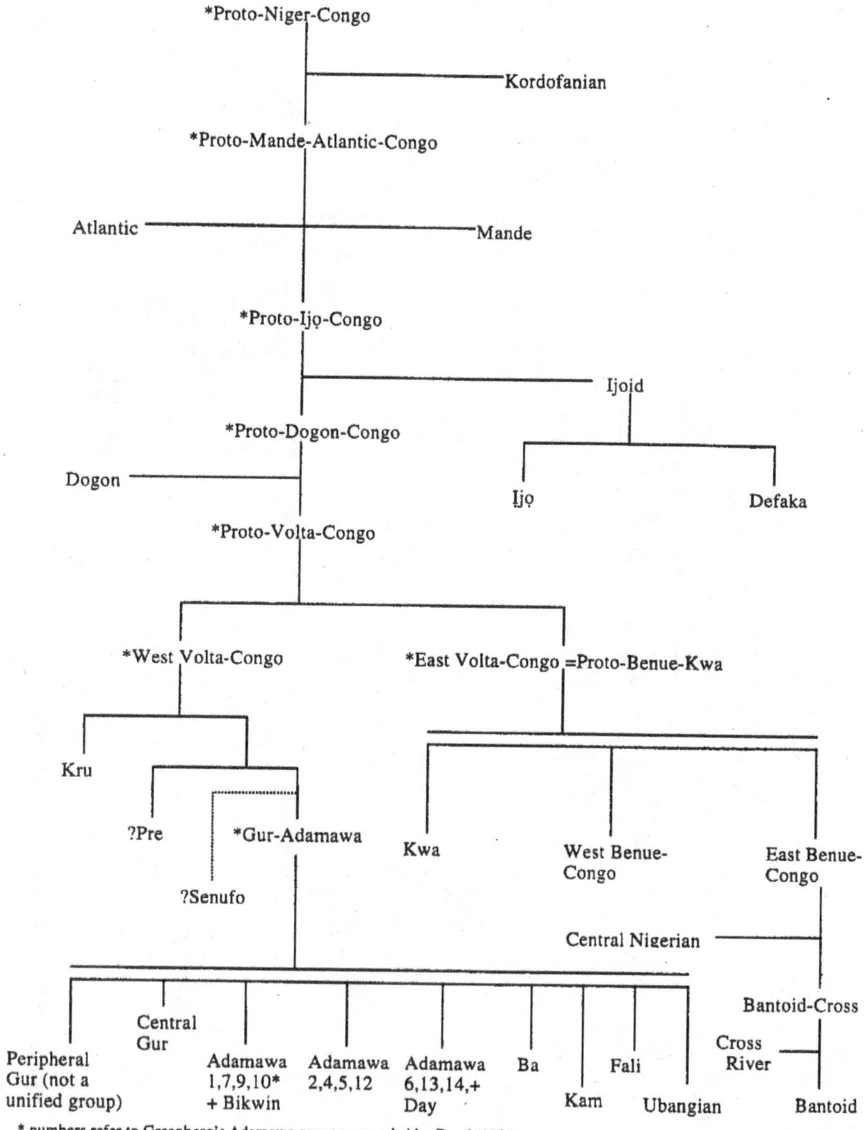

Figure 1.5 Niger-Congo: a reclassification
Source: Williamson and Blench 1999

Western Cushitic has been renamed Omotic (Bender 1975). Most scholars have accepted the coherence of Omotic as a group and agree on its assignment to Afroasiatic. Some researchers would prefer to retain Omotic within Cushitic, but these are now in a minority. In the case of the other branches of Cushitic, there has also been considerable discussion about whether it really

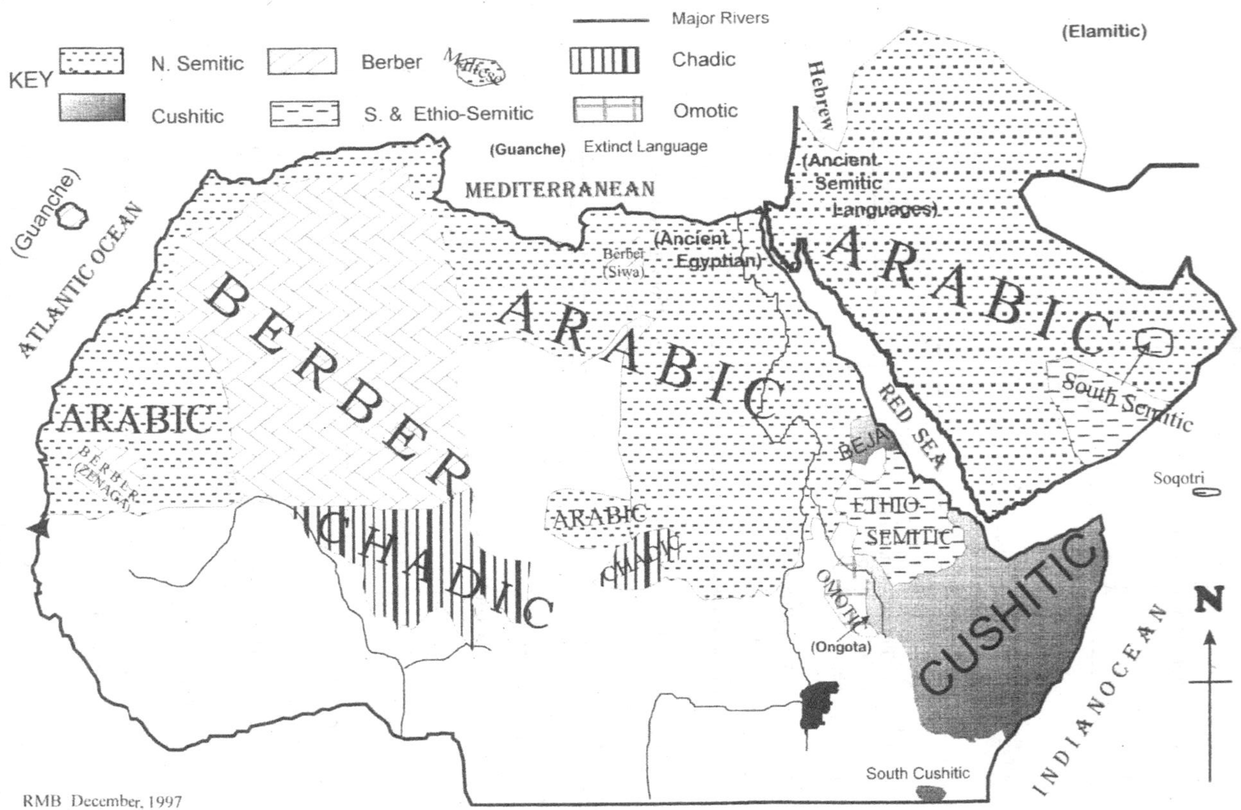

Figure 1.6 The distribution of Afroasiatic languages

Source: Blench

constitutes a family, and Beja, Ethiopian Cushitic and Southern Cushitic are often treated as distinct branches. Ehret (1987) has proposed a 'Proto-Cushitic', making explicit the hypothesis that these branches form a unity.

Ehret (1980) argued for a Southern Cushitic based on a membership of:

West Rift	Iraqw, Gorwa, Alagwa and Burunge
East Rift	Asa, Kwadza
Outside	Ma'a(=Mbugu), Dahalo

Recent work on Ma'a suggests that it is rather a 'register' of a Bantu language and that the Cushitic elements are a superstratum consisting of both eastern and southern elements (Mous, pers. comm.). Meanwhile, better data on Dahalo (Tosco 1991) suggests that it may either be a very aberrant member of Eastern Cushitic or form a branch of its own with Yaaku and Galaboid with which it shares some intriguing isoglosses. Ehret proposes the internal grouping of Cushitic shown in Figure 1.7.

Ehret's (1995b) schema of the internal structure for Afroasiatic is fairly similar to the models proposed formally or informally by other researchers. The major difference with Ehret's classification is that he does not see a special relation between Cushitic and Chadic, but does have Chadic branching off directly after Cushitic.

A radically different view is taken by Orel and Stolbova (1995), who consider that Cushitic and Omotic are not genetic groupings at all but an ancient *Sprachbund*. The outline proposed by Orel (p.c.) is shown in Figure 1.8.

Such a tree should be treated as a working hypothesis; it does not really address the relations of Berber-Egyptian and Semitic in detail, nor does it attempt to make the inter-relationships of Cushitic fully coherent.

Bender (1997) has also proposed a radically new structure for Afroasiatic ('upside-down Afrasian' in his terminology). His revised tree is shown in Figure 1.9. Bender proposes a homeland for Afroasiatic (the region where Chad, Sudan and Libya meet today) and a date (10,000 BP). Perhaps even more startlingly, he canvasses the possibility that Indo-European is somehow an offshoot of his 'Macro-Cushitic'. Whether these suggestions will be taken on board by the scholarly community will depend on the presentation of fuller evidence than is given in this short chapter.

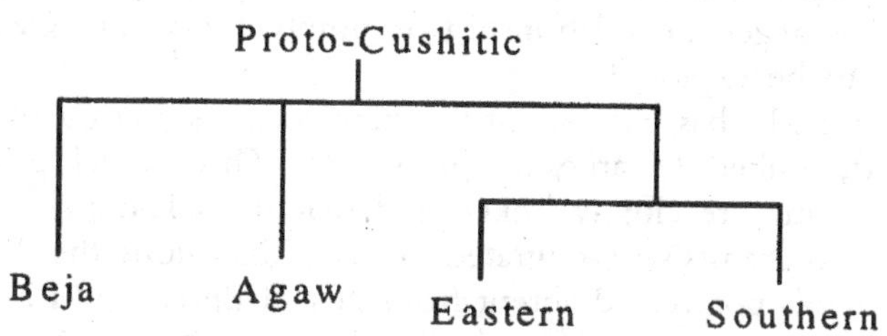

Figure 1.7 Internal structure of Cushitic
Source: Ehret 1995b

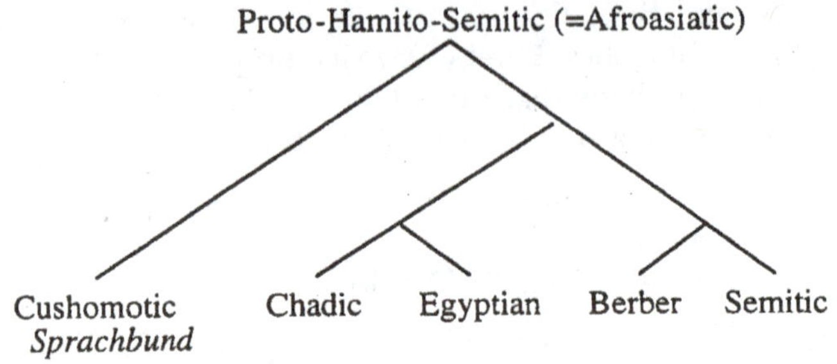

Figure 1.8 Internal structure of Afroasiatic (after Orel)
Source: Orel p.c.

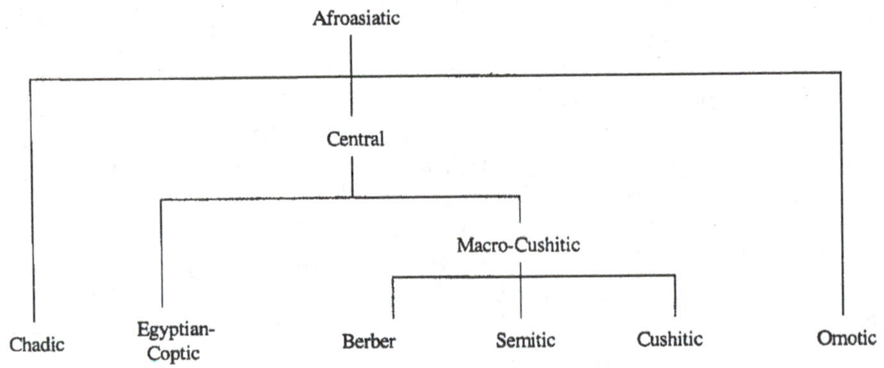

Figure 1.9 The internal structure of Afroasiatic (after Bender 1997)
Source: Bender 1997

Blažek (this volume) proposes that Elamite, an extinct language of the Ancient Near East, either constitutes a seventh branch of Afroasiatic or is co-ordinate with it. Elamite is usually classified with Dravidian, spoken in south India, but does show clear cognates with Afroasiatic. Blažek proposes a structure where Afroasiatic is related to Dravidian at a higher level and Elamite forms a bridge between the two. Whether the links between Elamite and Afroasiatic is a genetic relationship or simply a case of extensive loan-words remains to be explored.

Blench (in press, b) has put forward a view of the structure of Afroasiatic that is explicitly linked to archaeological data. This model proposes that Chadic and Cushitic are closely linked and that the Chadic speakers are in reality Cushitic pastoralists who migrated to Lake Chad along the Wadi Hawar. The resulting tree is not very different from that of Ehret, and I have adapted some of his proposed names for the nodes (e.g. North Afroasiatic and Erythraic). Figure 1.10 shows a composite view of Afroasiatic incorporating some of the recent proposals made concerning Elamitic, Ongota, etc.

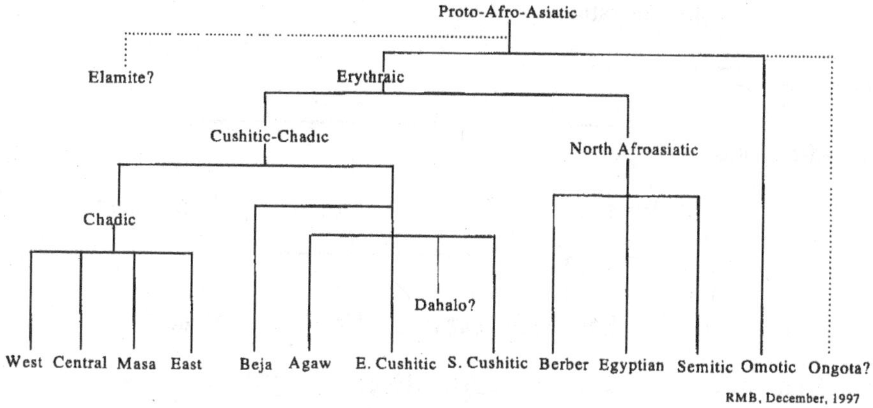

Figure 1.10 Proposed revised Afroasiatic classification
Source: Blench in press, b

EXTERNAL LINKS FOR AFRICAN LANGUAGE PHYLA

The Nostratic hypothesis

More speculative proposals have been advanced that are intended to explore 'deep-level' relationships of African language phyla. Afroasiatic is the African language phylum that has been most commonly proposed as related to other phyla of Eurasia. To enumerate all these proposals would be lengthy, but apart from Dravidian mentioned above, Afroasiatic is frequently connected to Indo-European and more broadly to 'Nostratic' (Bomhard 1994; Hegedüs 1997; Dolgopolsky 1998). Ruhlen (1991) provides a useful summary of these debates. Although there is definitely not a consensus in this area, there are two basic views:

1 that Afroasiatic (like Kartvelian and Dravidian) is co-ordinate with 'Eurasiatic' (Greenberg, Starostin);
2 that Afroasiatic is a member of Nostratic (Pedersen, Illitc-Svityč, Bomhard, Dolgopolsky).

A compromise view is represented in Figure 1.11.

These hypotheses are stimulating; they extend the debate on broad connections and similarities between languages at a very great time-depth. However, they remain linguistic hypotheses; their conclusions should not be extended to other disciplines, notably archaeology. Macrophylum classifications are much less well founded than low-level reconstructions. They are more tools to help linguists to think than representations of the past.

The lost language of the pygmies

The question of the origin of the pygmies of the African rainforest and their relative antiquity has remained controversial (Cavalli-Sforza 1986). Although

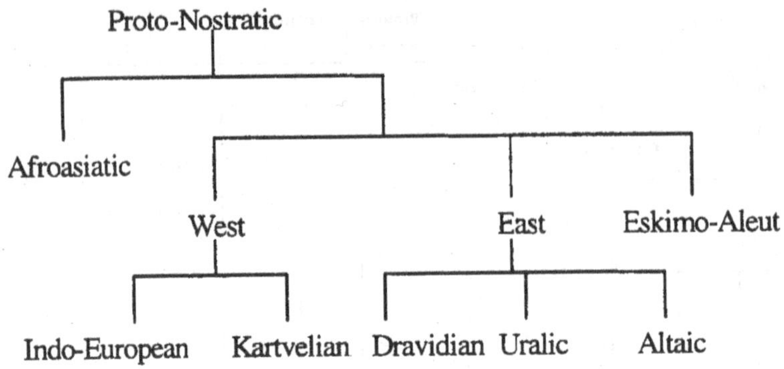

Figure 1.11 Afroasiatic and Nostratic (after Hegedüs, 1997)
Source: Hegedüs 1997

in many ways the pygmies appear to be the ancient inhabitants of the forest, partly displaced by the incoming Bantu, researchers have been disturbed by the complete absence of distinctive languages spoken by the pygmies and the ambiguous archaeological evidence.

Bahuchet (1992, 1993a, b) has recently presented an extremely challenging view of the history of the pygmy populations, in particular the Aka and the Baka. Despite speaking languages of quite different genetic affiliation, these groups prove to have common vocabulary, concerned especially with food gathering in the rainforest. If Bahuchet is right, then this vocabulary constitutes a trace of the lost language of the pygmies. Bahuchet further argues that the reduction in the rainforest at the end of the Pleistocene isolated pygmoid groups in relict forest. These groups diffused outwards when the forest began to expand again, eventually encountering the incoming Bantu cultivators. The results of this encounter, and in particular the evolution of client relationships, account for the ethnolinguistic pattern seen today.

An alternative view is presented in Blench (in press, a) which argues that the absence of a true pygmy language is no accident: that the pygmies are to be identified genetically with their cultivator neighbours. Instead, the fragmentary hunter-gatherer peoples and isolated languages that today form a ring around the rainforest represent the remaining traces of a lost complex of non-pygmoid hunter-gatherer populations, speaking highly diverse language, who inhabited Africa prior to the spread of the major language phyla. This diversity would then have been largely eliminated in the regions where the major language phyla expanded. Nilo-Saharan languages may have been the first to expand, but Niger-Congo was substantially more successful in terms of geographic range. If this is so, the situation would then resemble more closely regions of the world where hunting and gathering dominated the economies until recently and where no large-scale state systems evolved to impose uniformity. The main regions where this is the case are the New World, Australia, Papua and Siberia.[4]

CONCLUSIONS

The gradual increase in availability of data has led to major disagreements between scholars as to the internal classification of many of Africa's language phyla. As with Indo-European, it is easier to discern a large number of discrete groups than to fit them together into a hierarchical tree structure. Paradoxically, a slightly clearer image of African prehistory is beginning to emerge from the present synthesis. The four major phyla may be reducible to three if the Niger-Saharan hypothesis is accepted, but they are unlikely to split into more phyla. Even if the macrophyla proposals gain acceptance, they are unlikely to change estimates of the relative internal diversity of existing phyla and therefore estimates of their relative antiquity.

A scatter of 'remnant languages' across the continent may be unclassifiable because they have an extremely complex language history or because they are genuine survivors of otherwise vanished phyla. Such languages correspond to the position of Basque in Europe or Burushaski in Asia. These languages may be the only remaining traces of the earlier inhabitants of the continent. Originally spoken by mobile hunter-gatherers, they have remained in traces in isolated populations. Influenced by all the neighbouring languages, they usually show a complex texture of loanwords as well as a core of words of unknown etymology.

The Khoisan languages probably fit together but are so deeply internally divided that this question is likely to remain controversial. A useful parallel here is the situation in Papuan or Australian. Striking phonological similarities suggest genetic unity. There are good historical/archaeological reasons for supposing that we are dealing with essentially homogeneous populations in these cases. However, their languages have been ramifying for so long that the lexical connections between them have reduced to a point of near invisibility.

Khoisan is, therefore, probably the oldest phylum and the Khoisan languages that remain are just a fraction of those that were once spread over eastern and southern Africa. Whether the Niger-Saharan hypothesis is accepted or not, Nilo-Saharan is likely to be older than Niger-Congo and Afroasiatic. However, despite various claims, there are no convincing reconstructions of lexemes for agricultural items in the proto-languages of any of the African language phyla. In other words, the processes that led to their expansion took place in a pre-agricultural phase.

The likely correlate of these expansions is micro-climate fluctuations, but the doubt hanging over the internal classification of these phyla suggests that attempts to draw up large-scale models is presently a fruitless task. In the case of macrophyla proposals, it should be emphasized that these are very much linguists' constructs. Many of the methods used to argue for the very existence of these groupings are disputed by other linguists. Moreover, macrophylum proposals, especially in the case of language isolates such as Basque or Ainu, have a habit of producing multiplexes of different solutions.

More problematic is how the archaeologist should respond to the disagreements over the internal classification of the major language phyla. In principle, trying to model the dispersal of an entire phylum is a fruitless task at our present stage of knowledge. Even Indo-European, the most-researched phylum, has yet to be analysed as a convincing tree-structure, and there is considerable disagreement about the link between archaeology and present-day linguistic geography.

The task then should probably be to work with much smaller-scale groupings. Every phylum is divided into manageable units, most of which are generally accepted by the linguistic community. Examples of these would be Mande, Bantu, Nilotic, Omotic, etc. Although a satisfactory model of the expansion of these families is yet to be worked out, they are neither so ancient nor so vast that such a task is in principle unachievable.

NOTES

1 I am grateful to Matthias Brenzinger for inviting me to the Round Table on Endangered Languages in Africa at Leipzig, August 1997.
2 In addition, the representation of clicks is not as accurate as it might be, due to limitations in the drawing programme.
3 I am grateful to Anbessa Teferra for data on the Shabo language.
4 Smaller subregions such as the Andamans could also be included.

REFERENCES

Bahuchet, Serge. 1992. *Histoire d'une civilisation forestière. I: dans la forêt d'Afrique centrale, les Pygmées Aka et Baka.* Louvain: Peeters.

Bahuchet, S. 1993a. *Histoire d'une civilisation forestière. II: La rencontre des agriculteurs. Les Pygmées parmi les peuples d'Afrique Centrale.* Louvain: Peeters.

Bahuchet, S. 1993b. History of the inhabitants of the Central African Rain Forest: perspectives from comparative linguistics. In *Tropical Forest, People and Food,* C. Hladik, *et al.* (eds), 37–54. Paris: UNESCO.

Bender, M.L. 1975. *Omotic: A new Afroasiatic language family.* Carbondale: University Museum, Southern Illinois University.

Bender, M.L. (ed.) 1983a. *Nilo-Saharan Language Studies.* East Lansing: Michigan State University Press.

Bender, M.L. 1983b. Introduction. In *Nilo-Saharan Language Studies,* M.L. Bender (ed.), 1–10. East Lansing: Michigan State University Press.

Bender, M.L. 1988. Proto-Omotic: phonology and lexicon. In *Cushitic-Omotic: papers from the International Symposium on Cushitic and Omotic Languages,* M. Bechhaus-Gerst and F. Serzisko (eds), 121–162. Hamburg: Buske Verlag.

Bender, M.L. (ed.) 1989. *Topics in Nilo-Saharan Linguistics.* Hamburg: Buske Verlag.

Bender, M.L. (ed.) 1991a. *Proceedings of the Fourth Nilo-Saharan Conference. Bayreuth, 1989.* Hamburg: Buske Verlag.

Bender, M.L. 1991b. Subclassification of Nilo-Saharan. In *Proceedings of the Fourth Nilo-Saharan Conference. Bayreuth, 1989,* M.L. Bender (ed.), 1–36. Hamburg: Buske Verlag.

Bender, M.L. 1996a. *The Nilo-Saharan Languages: a comparative essay.* Munich: Lincom Europa.

Bender, M.L. 1996b. Nilo-Saharan '95. Proceedings of the 6th Nilo-Saharan Linguistics Conference, 1995. *Afrikanische Arbeitspapiere* 45, 1–26.

Bender, M.L. 1997. Upside-down Afrasian. *Afrikanische Arbeitspapiere* 50, 19–34.

Blažek, V. 1991. *Comparative analysis of the Ongota lexicon*. Paper to the 21st Colloquium on African Languages, Leiden.

Blench, R. 1989. New Benue-Congo: A definition and proposed internal classification. *Afrikanische Arbeitspapiere* 17, 115–147.

Blench, R.M. 1992. Recent research in the prehistory of Bantu languages. In *Datation et chronologie dans le bassin du Lac Tchad*, D. Barreteau (ed.), 147–160. Paris: ORSTOM.

Blench, R.M. 1993a. Recent developments in African language classification and their implications for prehistory. In *The Archaeology of Africa: food, metals and towns*, T. Shaw, P. Sinclair, B. Andah and A. Okpoko (eds), 126–138. London: Routledge.

Blench, R.M. 1993b. Is Kordofanian the Omotic of Niger-Congo? *Mother Tongue* 19, 33.

Blench, R.M. 1995a. Is Niger-Congo simply a branch of Nilo-Saharan? In *Proceedings of the Fifth Nilo-Saharan Linguistics Colloquium, Nice, 1992*, R. Nicolai and F. Rottland (eds), 83–130. Köln: Rudiger Köppe.

Blench, R.M. 1997. Language studies in Africa. In *Encyclopaedia of Precolonial Africa*, J.O. Vogel (ed.), 90–100. Walnut Creek, London and New Delhi: Altamira.

Blench, R.M. in press, a. Are the African pygmies an ethnographic fiction? In *Hunter-gatherers of Equatorial Africa* K. Biesbrouyck, G. Rossel and S. Elders (eds), Leiden: Centre for Non-Western Studies.

Blench, R.M. in press, b. The westward wanderings of Cushitic pastoralists: explorations in the prehistory of Central Africa. In *L'Homme et l'animal dans le bassin du Lac Tchad*, J. Boutrais and C. Baroin (eds). Proceedings of the MegaTchad Colloquium, Orléans, October 15–17 1997. Paris: ORSTOM.

Blench, R.M. in prep. *Niger-Congo roots in Westermann's* Sudan-sprachen: *a reclassification*. ms.

Bomhard, A. 1994. *The Nostratic Macrofamily*. Berlin and New York: Mouton de Gruyter.

Boyeldieu, P. n.d. *Presentation du làà:l ou "Gori" (Moyen-Chari, Tchad)*. ms. Paris: CNRS.

Cavalli-Sforza, L.L. 1986. *Pygmies as a model of the Upper Palaeolithic*. In *African Pygmies*, L.L. Cavalli-Sforza (ed.), 422–426. Orlando: Academic Press.

Cavalli-Sforza, L.L., P. Menozzi and A. Piazza. 1994. *The History and Geography of Human Genes*. Princeton: Princeton University Press.

Cohen, M. 1947. *Essai comparatif sur le vocabulaire et la phonétique du Chamito-Sémitique*. Paris: Honoré Champion.

Diakonoff, I.M. 1988. *Afrasian Languages*. Moscow: Nauka.

Dolgopolsky, A. 1998. *The Nostratic Macrofamily and Linguistic Palaeontology*. Cambridge: Macdonald Institute.

Doornbos, P. and M.L. Bender 1983. Languages of Wadai-Darfur. In *Nilo-Saharan Language Studies*, M.L. Bender (ed.), 43–79. East Lansing: Michigan State University Press.

Ehret, C. 1980. *The Historical Reconstruction of Southern Cushitic Phonology and Vocabulary*. Berlin: Reimer.

Ehret, C. 1986. Proposals on Khoisan reconstruction. *Sprache und Geschichte in Afrika* 7(2), 105–130.

Ehret, C. 1987. Proto-Cushitic reconstruction. *Sprache und Geschichte in Afrika* 8, 7–180.

Ehret, C. 1989. Subclassification of Nilo-Saharan: a proposal. In *Topics in Nilo-Saharan Linguistics*, M.L. Bender (ed.), 35–50. Hamburg: Buske Verlag.

Ehret, C. 1993. Nilo-Saharans and the Saharo-Sudanese Neolithic. In *The Archaeology of Africa: food, metals and towns*, T. Shaw, P. Sinclair, B. Andah and A. Okpoko (eds), 104–125. London: Routledge.

Ehret, C. 1995a. Do Krongo and Shabo belong in Nilo-Saharan? In *Proceedings of the Fifth Nilo-Saharan Linguistics Colloquium, Nice, 1992*, R. Nicolai and F. Rottland (eds), 169–193. Köln: Rudiger Köppe.

Ehret, C. 1995b. *Reconstructing Proto-Afroasiatic: consonants, vowels, tone and vocabulary*. Berkeley: University of California Press.

Ehret, C. in press. *Historical Reconstruction of Nilo-Saharan*. Submitted to Michigan University Press.

Elderkin, E.D. 1983. Tanzanian and Ugandan isolates. In *Nilotic Studies*, R. Vossen and M. Bechhaus-Gerst, (eds), 499–522. Berlin: Reimer.

Excoffier, L., B. Pellegrini, A. Sanchez-Mazas, C. Simon and A. Langaney, 1987. Genetics and history in sub-Saharan Africa. *Yearbook of Physical Anthropology* 30, 151–194.

Fleming, H.C., A. Yilma, A. Mitiku, R. Hayward, Y. Miyawaki, P. Mikesh and J.M. Seelig 1992. Ongota or Birale: a moribund language of Gemu-Gofa (Ethiopia). *Journal of Afroasiatic Languages* 3(3), 181–225.

Greenberg, J.H. 1966. *The Languages of Africa*. The Hague: Mouton for Indiana University.

Greenberg, J.H. 1971. Nilo-Saharan and Meroitic. In *Current Trends in Linguistics Vol. 7, Sub-Saharan Africa*, J. Berry and J.H. Greenberg (eds), 421–442. The Hague: Mouton.

Gregersen, E.A. 1972. Kongo-Saharan. *Journal of African Languages* 11(1), 69–89.

Grimes, J.E. and B.F. Grimes, 1993. *Ethnologue Language Family Index*. (12th edn) Dallas: SIL.

Hegedüs, I. 1997. Principles for Palaeolinguistic Reconstruction. In *Archaeology and Language I: theoretical and methodological orientations*, R.M. Blench and M. Spriggs (eds), 65–73. London: Routledge.

Kleinewillinghöfer, U. 1996. Relationship between Adamawa and Gur: the case of Waja. *Gur Papers* 1, 25–46.

Kleinewillinghöfer, U. in press. The language of the Jalaa: a disappearing language isolate. *Sprache und Geschichte in Afrika* 16.

Köhler, O. 1981. Les langues Khoisan. In *Les langues de l'Afrique subsaharienne*, G. Manessy (ed.), 455–615. Paris: CNRS.

Militarev, A. 1990. Afrasian cultural terms (preliminary report). In *Proto-Languages and Proto-Cultures*, V. Shevoroshkin (ed.) 33–54. Bochum: Brockmeyer.

Orel, V.E. and O.V. Stolbora 1995. *Hamito-Semitic Etymological Dictionary*. Leiden: E.J. Brill.

Perrot, J. (ed.) 1988. *Les langues chamito-sémitiques*. Vol. 3 of *Les langues dans le monde ancien et moderne*. Paris: CNRS.

Renfrew, C. 1992. Archaeology, genetic and linguistic diversity. *Man* 27, 445–478.

Ross, M. 1988. *Proto-Oceanic and the Austronesian Languages of Western Melanesia*, Pacific Linguistics series C-98. Canberra: Australian National University.

Ruhlen, M. 1991. (rev. ed. from 1987) *A Guide to the World's Languages. Volume 1: Classification*. London: Edwin Arnold.

Ruhlen, M. 1994. *The Origin of Language*. New York: John Wiley and Sons.

Sands, B. 1998. *Eastern and Southern African Khoisan: evaluating claims of a distant linguistic relationship*. Quellen zur Khoisan-Forschung. 14. Köln: Rüdiger Köppe.

Traill, A. 1986. Do the Khoi have a place in the San? New data on Khoisan linguistic relationships. *Sprache und Geschichte in Afrika* 7 (1), 407–430.

Voßen, R. 1988. Khoe linguistic relationships reconsidered: the data. In *New Perspectives on the Study of Khoisan*, R. Voßen (ed.), 67–108. Hamburg: Buske Verlag.

Voßen, R. 1997. *Die Khoe-sprachen*. Köln: Rüdiger Köppe.

Westermann, D. 1927. *Die Westlichen Sudansprachen und ihre Beziehungen zum Bantu*. Berlin: De Gruyter.

Westphal, E.O.J. 1963. The linguistic prehistory of southern Africa: Bush, Kwadi, Hottentot and Bantu linguistic relationships. *Africa* 33, 237–265.

Westphal, E.O.J. 1971. The click languages of southern and eastern Africa. In *Current Trends in Linguistics Vol. 7, Sub-Saharan Africa*, J. Berry and J.H. Greenberg (eds), 367–420. The Hague: Mouton.

Williamson, K. and R.M. Blench 1999 Niger-Congo. In *African Languages*, B. Heinze and D. Nurse (eds), Cambridge: Cambridge University Press.

Wilson, J.G. 1970. Preliminary observations on the Oropom people of Karamoja, their ethnic status, culture and postulated relation to the peoples of the late stone age. *Uganda Journal* 34(2), 125–145.

2 *Elam: a bridge between Ancient Near East and Dravidian India?*

VÁCLAV BLAŽEK

ABBREVIATIONS

AA Afroasiatic
Akk Akkadian
Arab Arabic
Aram Aramaic
Berb Berber
C Central
Ch Chadic
Cush Cushitic
Dr Dravidian
E East
Eg Egyptian (BD Book of the Dead, D 18/19 18/19 Dynasty, *plus*
 Gr Greek period,
 M/N/OK Middle/ New/Old Kingdom,
 Med Medical texts,
 Pyr Pyramid texts),
El Elamite
 A Achaemenid
 m middle
 n new
 o old
H Highland
Hbr Hebrew
IE Indo-European
L Lowland
N North
NP personal name
OSA Old South Arabian
Ph Phoenician
S South

Som	Somali
Sum	Sumerian
Syr	Syrian
Ug	Ugaritic
W	West

ELAMITE LANGUAGE AND SCRIPT

The first certain attestation of the Elamite language is from the twenty-third century BC. The so-called 'Treaty of Naram-Sin', written in cuneiform script, was concluded between Naram-Sin (2254–2218), a successor of Rimuš, the son of Sargon of Agade (2334–2279), and Hita, the ninth king of Awan, against their common enemies, the Qutians (Hinz 1964: 64; Steve 1992: 4). Hita's successor, Puzur (alias Kutik) – Inšušinak, the last of twelve kings of Awan (around 2200 BC), developed Linear Elamite (Proto-Elamite B = monumental) script, today known from nineteen inscriptions of the twenty-third century BC. The creation of the script can be explained as a reaction against the centuries-old cultural and occasionally political domination of Elam by Mesopotamia. The content of one of the inscriptions (A) is known thanks to a parallel Akkadian translation and represents a key to the decipherment of the script. Although the results and their application for the interpretation of other texts are not unambiguous, the language is certainly Old Elamite (Hinz 1969; Meriggi 1969a and 1971, 184–220).

The Linear Elamite script developed from the Proto-Elamite script, known from around 1,400 inscriptions relating to economic transactions found especially in Susa (3100–2900 BC). The Linear Elamite script, with 103 known, mostly syllabic, signs, represents a simplification of the older, pictographic, Proto-Elamite script, with at least 400 signs (Meriggi 1969b: 156 and 1971: 185, 193–205; Parpola 1994: 35).

The language of the Proto-Elamite script is not known, but there is no reason to suppose it is any language other than Elamitic. The oldest tablets with Proto-Elamite pictograms are from level 16 at Susa (3100 BC). Two 'numerical tablets' appear at level 18 (3300 BC) – contemporaneously with Uruk IV in Sumer, where the first discovery of writing was probably realised. This fundamental borrowing of the idea of writing (nothing other than the numerical symbols and some ten signs – see Vajman 1972; Meriggi 1969b) has been connected with so called 'First Conjuncture' (3300 BC) – the first wave of cultural expansion of the Sumerians.

In this period, three sites on the periphery of Mesopotamia were colonized:

1 Habuba Khabira on the Euphrates in Northern Syria;
2 Godin Tepe in the Zagros mountains of NW Iran;
3 Susa on the Mesopotamian alluvium in SW Iran (Lamberg-Karlovsky 1986a: 195).

In the 'Second Conjuncture' (3000–2900 BC), the Proto-Elamites repeat the pattern of the Sumerians during the First Conjuncture: they colonize foreign areas. Within the century after 3000 BC, the sites of Tepe Sialk, Tal-i-Malian, Tepe Yahya and, ca. 2900, Shahr-i Sokhta in Iranian Seistan are colonized by the Proto-Elamites from Susiana (Lamberg-Karlovsky 1986a: 197, 199; 1986b). The latter locality became a large urban complex on more than 100 hectares. It played an intermediary role connecting Elam, cultural centres in Turkmenistan (Geoksyur, Namazga III), Afghanistan (Mundigak) and the Indus valley (Parpola 1994: 17). The famous Proto-Indus script probably has its origin (or at least inspiration) in some later variety of the Proto-Elamite script (Fairsevis 1992: 228; Parpola 1994: 53; Meriggi 1977 on the inscription from Shahr-i Shokta).

The question of the genetic affiliation of Elamite remains unresolved. Many scholars have noted the remarkable similarities between Elamite and Dravidian, especially in morphology. The comprehensive study of McAlpin (1981) on this topic must now be completed and corrected (see Appendix 1). The relatively poor results of Elamite-Dravidian comparison (especially in the core lexicon) contrast with the more attractive comparison of Elamite to Afroasiatic (see Appendix 3). The hypothesis of a closer Elamite-Afroasiatic relationship can be supported at least indirectly by archaeological evidence. Before 3000 BC, there are only two periods when the material cultures of Khuzistan (Elam) and Sumer are closely comparable:

1 Late Uruk expanding in Susiana during the 'First Conjuncture' (see above);
2 Choga Mami Transitional (Iraq) expanding at the site of Choga Sefid (phase 5) in the Deh Luran plain (Iran) sometime in the sixth millennium BC.

The hypothesis of a cultural expansion is based on the introduction of certain plants and animals apparently not previously attested in Khuzistan: domesticated cattle and pig and various hybrid cereals, including hexaploid wheat, indicating that irrigation (attested in Choga Mami in the sixth millennium BC) was also introduced into Khuzistan at this time. It has been noted that a certain type of mud-brick also appears in Khuzistan at the same time. These simultaneous introductions have been interpreted as signifying an actual population movement into Khuzistan (Oates 1991: 24–25).

SUMERIAN LANGUAGE AND SCRIPT

The Sumerian language was spoken by the people who lived in the alluvial plains of the lower Euphrates and Tigris at least from Uruk III, especially the Jemdet Nasr period (3100–2900 BC) onwards, but very probably also in the Uruk IV period (3300–3100 BC) and even earlier (Parpola 1994: 30–31). During these periods, the first pictographic script was developed and the idea was exported to Elam and Egypt (3100 BC?). A carved flint knife from Upper

Egypt (Gebel el-ʿAraq) depicts on its handle a man in Sumerian dress conquering two lions, a common Mesopotamian motif, and (on the reverse) a naval battle in which Sumerian-type ships defeat Egyptian ships (Parpola 1994: 35–36; for the Late Uruk presence in Egypt see also Zarins 1992: 71).

It is usually held that the Sumerians were not indigenous to Mesopotamia. Höyrup cites some authoritative conclusions:

> The fundamental observation is that no Sumerian etymology for the names of the oldest cities can be constructed, and that a large number of words of cultural importance (tools, products and professions) seem not to fit the normal phonology of Sumerian. They are bisyllabic, which is rare for Sumerian roots, and often contain a consonantal cluster (Landsberger; Salonen). More about pre-Sumerian toponyms see in Appendix 2. (p. 63, ftn. 82) . . . The existence of entries in the Mesopotamian lexical texts with known syllabic values but with no corresponding logographic values indicates originally non-Sumerian words, which were perpetuated in the Sumerian writing, but not in the Sumerian language (Gelb).
>
> (Höyrup 1992 [94]: 60–61)

Speiser tried to identify the pre-Sumerian substrate language with Elamite, an idea that has some support. Among Elamite personal names the last two syllables are frequently repeated: Šilhaha, Kunene, Hilulu, Kinunu, Nabubu, etc. (Meriggi 1971: 182–183). These forms are interpreted as 'Kosenamen' by Hinz and Koch. A similar pattern is typical for some Sumerian divine names: ^{d}Bunene, ^{d}Zababa, ^{d}Kubaba, ^{d}Inana, ^{d}Igigi, ^{d}Aruru. Diakonoff (1981: 48) (his examples are quoted here) calls the source 'Banana-language'. The same pattern was a productive way of forming diminutives in Egyptian (*ḥfll.t* 'lizard', *ḥdqq* 'rat', *ḥwrr* 'divine calf', *ḫprr* 'scarabeus') and Berber, e.g. Shilh *asĕlmam* 'eel' vs. *aslĕm* 'fish' etc. (Vycichl 1961: 250).

Höyrup (1992 [94]: 34) presented the revolutionary hypothesis that Sumerian developed from a mid- or late fourth-millenium Uruk creole. The idea of a local melting-pot is doubtless fruitful and does not exclude an external origin for at least one component in this glottogenetic process. The preceding opinions agree with archaeological data indicating a large population growth in southern Mesopotamia during the Early Uruk period (3600 BC) – very probably as the result of immigration into this region (Lamberg-Karlovsky 1986a: 196).

New hypotheses concerning the genetic affiliation of Sumerian have been formulated more recently. Boisson (1989) has collected lexical parallels between Sumerian and Dravidian (see also Appendix 2). Blažek and Bengtson (1995) include Sumerian in a macrophylum they call 'Dene-Caucasian' together with North Caucasian, Yeniseian, Burushaski, Sino-Tibetan, etc., following Hüsing, Bouda, Braun, Christian. Militarev (1984a and p.c.) presents Sumerian-Afroasiatic lexical parallels that cannot be explained as Semitic borrowings.

AFROASIATIC, ELAMITE AND SUMERIAN AND THE QUESTION OF THE AFROASIATIC HOMELAND

The following language families are usually included in Afroasiatic (=Semito-Hamitic/Hamito-Semitic=Erythraic=Lisramic etc.): Semitic, Cushitic, Omotic, Egyptian, Berber and Chadic (see also Blench, this volume). Their common origin is generally accepted, but their internal classification and the site of their common homeland remains disputed. Two basic hypotheses for a localization of the Afroasiatic homeland have been presented: (1) northeast Africa and (2) west Asia. Diakonoff (1991: 12–13) provides a valuable overview of these alternatives. The main reason for rejecting an Asiatic homeland (besides an *a priori* rejection of Biblical tradition) is the fact that all branches of Afroasiatic except Semitic are/were spoken in Africa; but it is doubtful whether the question of a homeland can be solved mechanically in this way. There are many examples of a similar or even more disproportionate dispersal (Latin and Romance, Arabic, Indonesian, Swahili, English, Turkic). Whilst not rejecting *a priori* the African hypothesis, the following arguments appear to support an Asiatic homeland:

1. The neolithic character of the Proto-Afroasiatic cultural lexicon. The only area where the 'Neolithic Revolution' starts before the disintegration of Afroasiatic (ca. eleventh–tenth millennium BC) is its primary area: the Fertile Crescent in the Near East. Militarev and Šnirel'man (1984) and Militarev *et al.* (1988) identify the Proto-Afroasiatic ethnos with the early neolithic Natufian culture from the Syro-Palestinian region (eleventh–ninth millennium BC). This agrees with the fact that Egyptian cereals are of Asiatic origin (Diakonoff 1981: 45).
2. The zoological lexicon reconstructible for Afroasiatic reflects the wild fauna not of northeast Africa but of the Near East (e.g. elephant, hippo, but not giraffe or rhino – cf. Blažek 1994).
3. Very early mutual borrowings between Afroasiatic (not only Semitic) and northern Caucasian (Militarev and Starostin 1984, 1994).
4. The Afroasiatic stratum in Sumerian (§2), representing perhaps one originally independent dialect of Afroasiatic, later lost in the 'melting pot' of Sumerian glottogenesis (Diakonoff 1981: 66; Militarev 1984a, 1989; Kovalev and Militarev 1994).
5. An exclusive Cushitic – South Semitic/dialectal Arabic isoglosses reflecting probably a Cushitic substratum of Arabian peninsula (Militarev 1984b: 18–19; Belova 1989).

More controversially, the Nostratic hypothesis proposes a genetic relationship between many of the language phyla of the Old World (Afroasiatic, Kartvelian, Indo-European, Uralic and Yukaghir, Altaic, Dravidian, Elamite; probably also Chukchi-Kamchatkan, Nivkh and Eskaleut). The most natural 'epicentre' of such a primary disintegration would again be the Near East. Preliminary estimates of the time of divergence of Proto-Nostratic are not very different

Abbreviations

Ak	Akkadian
Ca	Caspian
CS	Central Semitic
Dr	Dravidian
Eb	Eblaite
EC	East Caucasian
Eg	Egyptian
El	Elamite
Ha	Hattic
Hi	Hittite
HU	Hurro Urarteian
II	Indo-Iranian
Ka	Kartvelian
Ks	Kassite
Ku	Kutian
Ll	Lulubean
Lu	Luivian
Su	Sumerian
WC	West Caucasian

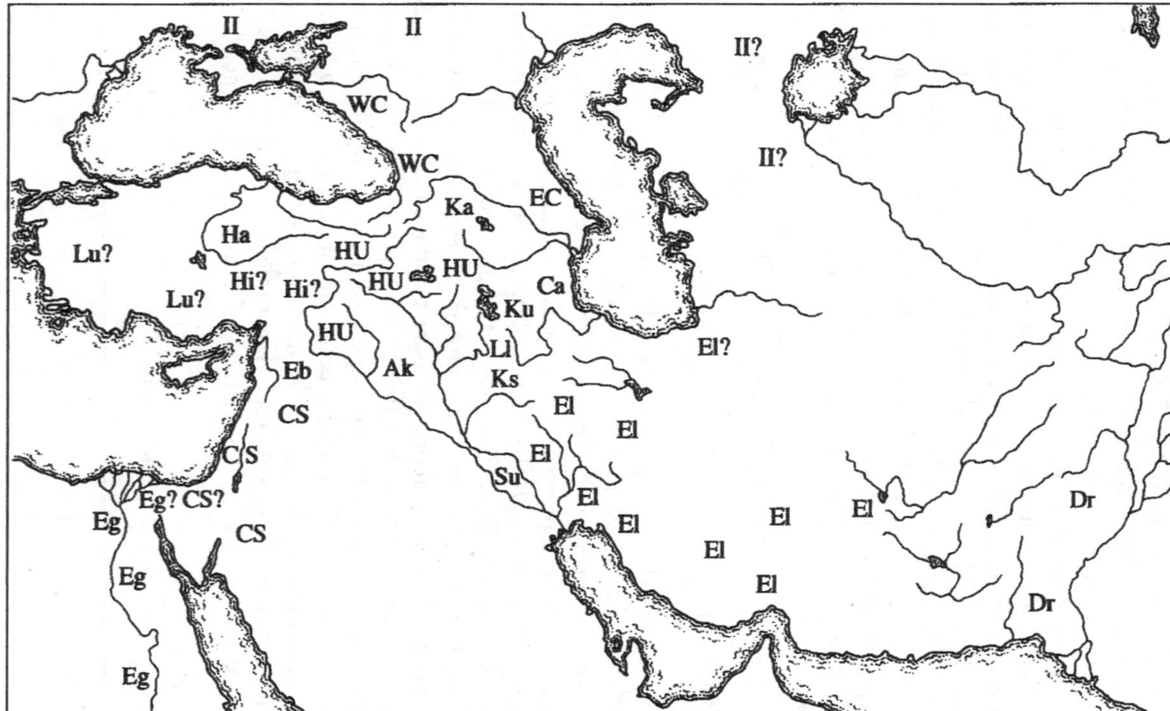

Figure 2.1 The distribution of known languages

Source: Diakonoff 1985: 39: Gragg *apud* Parpola 1994: 127

from the hypothetical time-depth of Afroasiatic (thirteenth millennium BP). Starostin – an author of this rather paradoxical result – explains this by a dichotomy of Afroasiatic vs. 'Micro-Nostratic' (i.e. Nostratic minus Afroasiatic). A modified version is presented by Greenberg, who postulates a Eurasiatic macrophylum consisting of the same language families as Nostratic minus Afroasiatic, Kartvelian, Dravidian and Elamite. Greenberg proposes that these languages have a closer relationship and suggests remote genetic links to his Eurasiatic. The principal authors of the Nostratic hypothesis, Illič-Svityč and Dolgopolsky, maintain a border between Western Nostratic phyla, characterized by apophony (Afroasiatic, Indo-European, Kartvelian) and Eastern Nostratic phyla with stable vocalism (for more detailed information see Blažek 1992b: 82–84).

CONCLUSION

Our present knowledge does not permit any definitive conclusions, but models can be formulated and tested in future research. Two possible schema are presented here to explain the cognates identified in the appendices. The position of Elamite could be represented as a bridge connecting Afroasiatic and Dravidian (Figure 2.2), although the Elamite-Afroasiatic relationship seems to be closer than the Elamite-Dravidian one (cf. Appendix 3 vs. 1).

An alternative, but not diametrically opposed, scheme is depicted in Figure 2.3. Elamite and pre-Sumerian represent here peripheral dialects of an Afroasiatic continuum comparable with Chadic or Omotic. The central position of Egyptian correlates with a relatively high rapidity in a development of its morphology (e.g. the loss of a prefixal conjugation), typical for a centre of any dialectal continuum in comparison with more conservative non-central dialects (Semitic, Cushitic, Berber). Refining these models depends crucially on further work on the internal structure of Afroasiatic, a topic that has

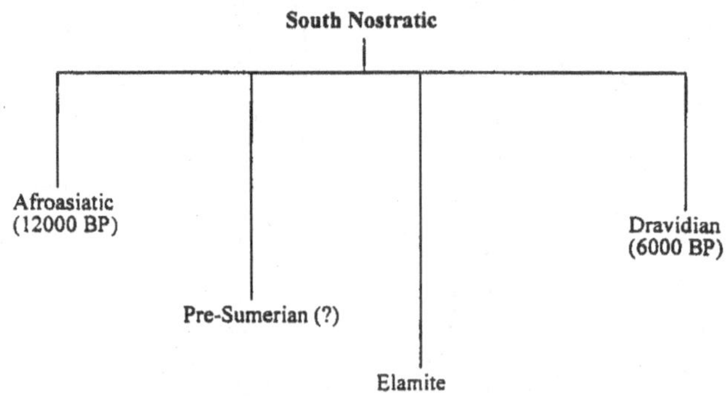

Figure 2.2 Tree model showing the place of Elamite
Source: Blažek

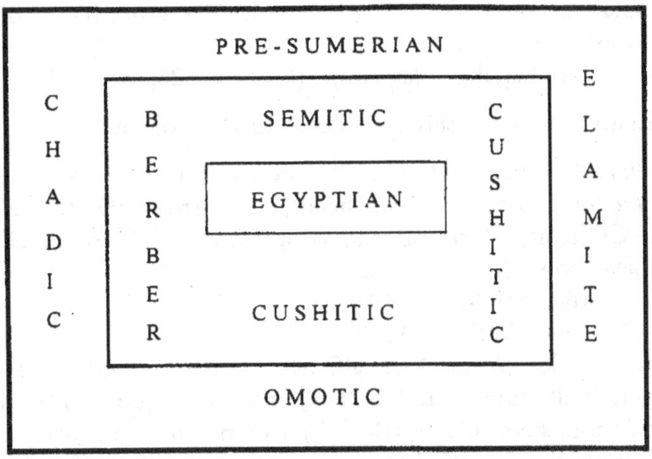

Figure 2.3 Continuum model of the place of Elamite
Source: Blažek

recently become highly controversial with the publication of two very different reconstructions (Ehret 1995; Orel and Stolbova 1995; see also Blench, this volume).

APPENDIX 1: ELAMITE AND DRAVIDIAN

A hypothesis of the Elamite–Dravidian relationship based especially on morphological comparisons has had numerous proponents (Norris, Caldwell, Hüsing, Trombetti, Bork, Diakonoff, Vacek, McAlpin). The most detailed study of Elamite–Dravidian connections was presented in a series of papers and summarized in a monograph by McAlpin (1981). Besides some promising cognates, he presents several semantically or phonetically questionable parallels, including evidently incorrect comparisons based on erroneous interpretations of Elamite words, such as:

a. El(m) *hun* 'water' (König 1965: 190; correctly probably 'light' – see HK 717)
Dr **uṇ* – 'to drink, eat a meal' (DEDR 600) – MA 145: El+Dr; Dr **uṇ* – has a promising cognate in ECush **ʿun-* 'to eat/drink' (He 1978: 100); or very probable borrowings from Sumerian or Akkadian:
b. El(m) *ukku* 'head, chief; on' (HK 1210; cf. Sum *ugu* 'head, skull, upper side; on' – IK 1104; perhaps Akk *ukkum* glossed SAG-SUHUR.SUHUR – see *AHw* 1405 – has the same origin)
Dr **uk(a)-* 'to ascend, rise, jump up' (DEDR 559; MA 95: El+Dr), but Kolami *cok-* 'to climb', Parji *cokk-* id., *cotip-* 'to raise' (DEDR 2828) signalize probably the initial **c-*;

c. El(m) *upat, upatta* 'brick' (HK 1240; cf. Akk *ur(u)bātu* 'coping stone' – *AHw* 1436)
Dr **uppar-* 'bricklaying, plastering' (DEDR 626, 628; MA 96: El+Dr).

The new Elamite lexicon (HK) suggests further cognates:

d. El(n) *ulkina* 'weapon' (HK 1218), 'reed arrow' ? (Bork)
Dr **alaku* 'blade of a weapon, head of an arrow' (DEDR 237)
cf. AA: CChadic: Mandara *əlka*, Gisiga *helek*, Mafa *leked'*, Glavda *laagha*, Margi *laga* 'bow' (Lukas 1970: 30);
e. El(o) *ik* 'votive gift' (HK 746)
Dr **ik-* 'to give' (DEDR 416);
f. El(n) *kutu* 'cattle', (A) *kiti* 'ox, calf, ass and cattle, foal of ass' (HK 489, 548)
Dr **kōṭ-ay* 'bull, cow' (DEDR 2199) and **kūṭ-ay* 'cow' (DEDR 1886); cf. Sum *gud* 'bull, steer, cattle' (IK 367) which can be a source of El words;
g. El(o) *kun(n)a* 'hair' (HK 513)
Dr **kūntal* 'hair' (DEDR 1892) – a compound; the second component is **tal-ay* 'head' (DEDR 3103); cf. also Dr **kunkaṭi* 'hair/crest of bird' (DEDR 1634);
h. El(A) *maka/i-* 'to consume, digest' (HK 861–2)
Dr **mookk-* 'to eat/drink' (DEDR 5127);
i. El(A) **nar-* in *naranda, narante/i, narada, nara(na)te* 'daily' (HK 991) besides *na(n)* 'day' (HK 967, 968), compared by MA 103 with Dr **nāḷ* 'day' (DEDR 3656) (having closer cognates in AA: ECush: Som *nal* 'light'/ EChadic: Ndam *nelnel* 'day')
Dr **ner-* 'sun, day, time' (DEDR 3774);
j. El(m) *nu* 'a sort of corn (barley ?)' (HK 1004)
Dr **nū* 'sesamum' (DEDR 3720) and/or **nuvaṇ-ay* 'Italian millet, panic seed' (DEDR 3712); cf. also Sum *nu(mun)* 'seed, offspring' (IK 771, 777);
k. El(n) *piti* 'vessel' (HK 224–5)
Dr **puṭṭi* '(earthen) vessel' (DEDR 4265A);
l. El(o) *ten* 'sweetness, kindness' (HK 305)
Dr **tēṉ*/**tīn* 'honey', cf. **tī* – 'sweet' (DEDR 3268);
m. El(A) *dud(d)u* 'foal' (HK 345), '(camel) calves' (H 102)
Dr **tūṭ-*/**tuṭṭ-* 'calf' (DEDR 3378).

APPENDIX 2: SUMERIAN AND DRAVIDIAN

Most Sumerian–Dravidian lexical parallels (such as those in Boisson 1989), including the oldest Sumerian toponyms with etymologies drawn from Dravidian (Fähnrich), can be supplemented by Afroasiatic data (Militarev sees in them an influence of pre-Semitic Afroasiatic substratum on Sumerian):

n. Sum *Buranun(a)*, Akk *Purattum* 'Euphrates' (IK 157; Edzard *et al.* 1977: 208)
El(n) *Pirin* 'river name, probably Karun in Susiana' (HK 209)

Dr **pur̤–ay* 'river' (DEDR 4318; Fähnrich 1981: 91: Sum < Dr); Note: Sum *Idigna*, Akk *Idiqlat* 'Tigris' has a hopeful etymology in the Sum compound **idi-gina* 'running river' (Albright and Lambdin 1970: 148).

o. Sum *Uri* 'a city from south Sumer', *uru* 'city' (IK 1137)
Dr **ūr* 'village, town, city' (DEDR 752; Fähnrich 1981: 91) or Dr **ur̤i* 'place, site, side' (DEDR 684)
? El **mur-* /**wur-?*/: (o) *muru* '(some)where', *murut* (g.) 'the earth', *murun* 'earth' (HK 952, 954, 964; MA 106: El+Dr)
? AA **war-*/**wur-* > ECush: Oromo *warra* 'family, kin', Arbore *warí* 'household'; Chadic: (W) Hausa *wúrli* 'place', (C) Gabin *wúnre* 'town', Muturua *urhai* 'Dorf', Makeri *wɔrɔ* 'village', (E) Dangla *wére*, Migama *wèré* 'place', Sumrai *wóram* 'kin'; ? Eg(OK) *w* (< **w3* ?) 'district, region' (EG I: 243; Takács p.c. Eg+El) and/or (Pyr) *ỉ3.t* (< **ỉu3.t* < **ỉurt* < **wur-t* ?) 'place' (EG I: 26); Note: Hattic *fur(i)* /*wuri* ?/ 'land' (Girbal 1986: 65, 69, 129, 150, 167) can represent the same term reflecting the beginning of Near Eastern urban civilization.

p. Sum *éri* 'city' (IK 278)
Dr **aṟ-ay* 'room of house' (DEDR 322)
AA **ᶜayr-*/**ᶜary-* > Sem: Ug *ᶜr* 'city', Hbr *ᶜīr* id., OSA *ᶜr* 'castle' (Segert 1984: 196; Aistleitner 1965: 241); ECush: Afar *ᶜàri* 'house, tent', Saho *ᶜarii* 'family, house, kin'; ? Eg(MK) *ᶜ.t* (< **ᶜ3.t*) 'chamber', (late) *ᶜ* (< **ᶜ3* ?) 'house' (EG I: 160, 159; Takács p.c.).

q. Sum *an* 'heaven; high, up' (IK 64) > Akk *Anu(m)* 'God of heaven' (*AHw* 55) // Dr **aṇ* 'upper part, above' (DEDR 110; Boisson 1989: 41: Sum+Dr)
? AA: Sem: Akk *an(a)* 'to, on' (*AHw* 47); HECush **hana* 'over, above' (Hudson 1989: 109).

r. Sum *é-ri-a* 'deserted country, steppe, pasture-land' (IK 254)
Dr **ere-* 'black soil' (DEDR 820).

s. Sum *gár* 'cream' > Akk *garūm* 'cream' (*AHw* 282), cf. Sum *ga* 'milk'
Dr **kaṟ-* 'to milk' (DEDR 1385; Boisson 1989: 43: Sum+Dr)
AA **kar-* > Sem: Syr *karᵉ-* 'beestings, colostrum, curdled milk'; Cush: (N) Beja *kar* 'butter'; (E) Rendille *keéra* 'fresh milk'; Berb: Ahaggar *a-kru* 'curdled milk, curds' (Militarev 1984a:#23: Sum+AA).

t. Sum *nundum/n* // Emesal *šumdum* 'lip' (Schretter 1990: 258)
Dr **noṇṭ* – 'to kiss, caress' (DEDR 3787) // **cuṇṭ* – 'bill, lip, mouth' (DEDR 2664; Boisson 1989: 42: Sum+Dr)
AA: Omotic: Koyra *nunaa*, Chara *noonaa*, Gimira *noon*, Anfilo *noonoo*, Mocha *noono* 'language, mouth, lip'.

u. Sum *sì* 'to give' (IK 866)
Dr **cī* – 'to give' (DEDR 2598; Boisson 1989: 17: Sum+Dr) // AA **say-* > Sem: Ug *šy*, Hbr *šay* 'gift' (Aistleitner 1965: 304); ECush: Som *sii*, Boni and Rendille *síi* 'to give' (He 1978: 95), Yaaku *-isɛʔɛ* id.; ? Eg(old) *ỉsw* 'compensation, salary, reward', Copt *asu* 'price' (EG I: 131; Vy 1983: 16; Takács p.c.: Eg+AA).

Some other cultural words appearing in Sumerian, Afroasiatic and Dravidian are discussed in Blažek and Boisson (1992). Separate Dravidian-Afroasiatic cognates are collected in Blažek (1992a).

APPENDIX 3: ELAMITE–AFROASIATIC COMPARISONS

Body parts and space orientation

1. El(m) *el(t)* 'eye' (HK 396, 394)
   AA **ʔil-(at-)* 'eye' (Greenberg 1963: 56) > Cush **ʔil-(t-)* (Do 1973: 144–145; Eh 1987: #326)
   Eg(Pyr) *ỉr.t* (EG I: 106)
   Berb: Shilh *tiṭṭ* < **ta-ʔil-t*, pl. *al(le)n* Chadic: (C) Hidkala *ílí*, Alataghwa *ilyia*, Vizik *iri/ili*, Buduma *yíl*, Mandague *ʔàl* (pl.)
   Dr **āli* 'pupil of eye, eye ball' (Zvelebil 1985a: 658).
2. El(m) *buni* 'heart' (HK 234)
   AA **b[u]n-* > ? Sem: Akk *abunnatu(m)* 'navel, umbilical cord' (*AHw* 9) // Eg(Med) *bn.tj* (du.) 'female breasts' (EG I: 457) // Chadic: (C) Gulfei *fɛnɛ*, Makari *fînɛ* 'breast'.
3. El(m) *kassu* 'horn' (HK 409)
   AA **ḳVsw/y-* 'horn' (Blažek 1989, #66) > Cush(N): Beja *koos* 'horn; tooth' Omot **ḳusim* 'horn' > Ubamer *qošma*, Dizi *usum*, etc.// Berb: Senhaja *a-qaššaw*, Matmata *qiš*, Harawa *kiišu* id. // Chadic: (C) Logone *káāśú* id.
4. El(m) *kir*, (A) *kur* 'hand' (HK 469, 523, 529)
   AA **ḳar-* 'arm, shoulder' > Cush(E): Som *qarqar* '(upper part of) shoulder' // Eg(MK)*qꜥḥ* > **q3ḥ* (?) 'arm, shoulder' (EG V: 19); *-ḥ* is probably a body parts suffix, cf. *ỉbḥ* 'tooth', *b3ḥ* 'penis', *ḏnḥ* 'wing, leg', *gmḥ.t* 'lock', *s3ḥ* 'toe', *sḏḥ* 'calf (with foot)', *spḥ.t* 'Rippenfleisch' – maybe identical with *ḥꜥ* 'body, flesh' (EG III: 37–38) // Berb: Shilh *ịgir*, pl. *ịgariun* 'shoulder', cf. *tagərurṭ*, pl. *tigoraḍ* 'shoulderblade'.
5. El(A) *mat*, *madda* 'with young' = 'trächtig' (HK 855)
   AA * *m[a]ṭ* – > Cush: (E) Afar *maḍaḍ* 'uterus, womb', cf. *Maḍ* – 'to copulate', Burji *maḍ -iss-* (caus.) 'to marry' (Sa 1982: 139)? // Berb: Ahaggar *temiṭ-* 'uterus, womb'.
6. El(m) *pat* 'foot; under' (HK 111)
   AA **pVd-* > ? Sem: Akk *padānu* 'way, path', Mehri *awōfəd* 'to look for a footprint, Arab *wafada* 'to come, travel' // Eg(Med) *p3d*, (D 18) *pd* 'knee; to run', Copt *pat* 'knee, foot, leg, thigh' (EG I: 500; Vy 1934: 165) // Berb: Mzab *fud*, Ghat *afud*, Zenaga *offuḏ* 'knee' // ? Chadic: (E) Mubi *fúudí* 'thigh'
   Dr **paṭ -am* 'palm (of hand)/sole (of foot)', **paṭi* 'step' (DEDR 3843, 3850). Note: A similar semantic dispersion is also known in the case of the Indo-European etymon **pōd-s*, g. **ped-és/-ós* 'foot', **pedo-m* 'bottom, place', **pedó-/ā́* 'sole, step, trace' (Pokorny 1959: 790), probably related on Nostratic level.

7. El(n) *pur* 'fingernail' or '(nail of) thumb' (HK 241)
AA **par-* or **far-* (Illič-Svityč 1984: 70–77, #362) > Cush(E) **far-* > Som *far* 'finger' (Do 1973: 41–42) // Omotic: Koyra *partaa* 'finger' // Chadic: (W) Hausa *farce* 'fingernail', Gwandara *apiraci*, Bolewa *paala* ; (C) Hina *mbraa*, Mandara *falidze*, Gidar *purzlumay*; (E) Mubi *féerí*, Jegu *pʰíllô*
? Dr **vir-al* 'finger, toe' (DEDR 5409), cf. Dr **pa ṟ- aṇṭu* and **vaṟ-aṇṭu* 'to scratch with fingernails' (DEDR 4023, 5322).

8. El(n) *san* 'blood' (HK 1053)
AA **ʒVn-(P-)* (Blažek 1989: #17) > Omotic: Zayse *zonne* 'pus', Hamer *zom(?)bi*, Karo *zunpi* 'blood' // Eg(Pyr) *znf* 'blood', Copt *snof* (EG III: 459; Vy 1983: 193) // Berb: Ifoghas *azeni*, Ghat *azəni*, Ayr *azni*, Ahaggar *ahĕni* id. // Chadic: (W) **zanyam* > Hausa *jíníi*, Montol *šiyìm*, Galambu *ǯàamá*, Kulere *zòm*; (C) Bata *ǯambɛ*, Bachama *zàmbäy*, Gudu *aǯin* id.

9. El **siha[n]*: (m) *sihha* 'tooth', (o) *sihhan* NP (HK 1071)
AA **siḥn-* 'tooth' (Do 1973: 91–92) > Sem **šinn-* (Ls 1938: 504) // Cush(S) **siḥn-* (Eh 1980: 180) // Berb: Ahaggar *esiin*, pl. *isiinen* // Chadic: (W) SBauchi **sin*, Ngizim *yaanau*; (C) Hurzo *tlahaaŋ*, Musgu *ši-*; (E) Jegu *sa ŋo* etc. id.

10. El(n) *siri* 'ear', cf. *siri* 'true, right' (HK 1089)
Cush(C): Waag *šər* 'to hear' // Eg(late) *sy3* 'to recognize, know' (Fa 212; EG IV: 30) // ? Chadic: (C) Zelgwa *tsạrạkạ* 'to hear'.

11. El(m) *šara* 'under' (HK 1132)
AA **sar-* 'back' (Co #269) > Sem: Arab *sarā* 'back', Soqotri *sar*, Mehri *sār* 'behind, after' // Cush: (N) Beja *saraat* 'back'; (C) Xamir *səra* id., Awngi *sər* 'lower part'; (E) Afar *sàrra* 'back, rear', Burji *saro* 'tail', Yaaku *sɛɛrɛy* 'below, down'; Dahalo *sàre* 'back'; (S) Burunge *sira* 'buttocks' // Eg(Pyr) *s3* 'back' (EG IV: 8).

12. El(A) *šimme* '(his) nose' (HK 1170); originally probably **šin-me* with the same suffix as *tit* and *tit-me* 'tongue' and the assimilation as *imme* 'not' < **in-me* (HK 342, 754, 757, 758)
AA **sin-/*sun-* > Cush: (C) **əsäŋ/*əsan-* 'nose'; (E) **sin-/*sun-/*san-* id. (Eh 1987: #476); Dahalo *sina* id. // Eg(Pyr) *sn*, *snsn* 'to smell' (EG IV: 153, 172, 277) // Chadic: (W) Hausa *sunsuna* id. Note: An alternative cognate can be seen in Sem **ś-m-m* > Arab *šamma* 'to smell', *ḫašm* 'nose'.

13. El(n) *tebba /teppa ?/* 'before, up' (HK 307)
AA: Eg(Pyr) *tp* 'head; on, upon' (EG 263, 273) // ? Cush: (E) Burji *tip-óo* 'skull' (Sa 1982: 177).

14. El(n) *tipi* 'neck' (HK 333)
AA **duby-* > Sem **d-b-r* 'to be hinder, back', Mandaic *dibra* 'back, tail', Arab *dubr* 'tail' // Cush: (E) **dib-/*dub/*dab-* 'tail, back' (Sa 1982: 57), cf. Oromo *duba* 'back, behind' // Omotic: Kullo *duupiya*, Karo *dibini*, Bako *doobanna* 'tail' // Chadic: (C) Gisiga, Mafa *dəba*, Gidar *dúbo* 'back'.

Human society

15. El(m) *ah(a)-pi* 'origin, kin, genealogy' (HK 15, 33, 35, 392); *-pi* = pl. suffix
AA **ʔaḫw-* > Sem **ʔaḫw-* 'brother', **ʔaḫw-at-* 'sister' (Cohen 1970: 15) ? Cush: (E) Arbore *ʔáw* 'maternal uncle' // Eg(BD,NK) *ḫ(w/y)* 'child' (Vy 1934: 258) // ? Chadic: (C) Boka *xwəyà*, Musgu *aḫíi* 'son'.
16. El(m) *bali* 'male' (HK 131–132)
AA **bal-/*bil-* > ? Sem **baʕl-* 'lord, husband' // Eg(OK) *by3* 'to be a powerful being', *b33w.t* 'virility' (EG I: 413, 417) // Chadic: (W) Sura *ɓál* 'strength, strong, powerful'; (C) Logone *bíle* 'man, male', Kuseri *bɛlom*, Gulfei *bələ* 'man'.
17. El **eri/*iri* 'uncle' (HK 401,774)
AA **ʔary-* 'kinsman' > Sem: Ug *ʔary* 'son' or 'brother' // Eg(Pyr) *ỉry* 'companion' (Ward 1961: 32; EG I: 105) // Cush: (N) Beja *'aar* 'female relatives'; (C) Awngi *ŋärá* 'her husband' < **ŋi-ärä*; (HE) **aroʔo* 'husband'; (S) Mbugu *m'áro* 'neighbor; kind, related thing', Asa *ʔarato* 'twins' // Chadic: (W) Kulere *ʔyer* 'brother', Bokkos *re*, pl. *ʔarya* 'man'.
18. El(m) *hiš* 'name', *hiša* 'praise, glory' (HK 662, 669)
AA **ḫ/ḥaS-* > Sem: Akk *ḫasāsu* 'to remember', *ḫasīsu(m)* 'ear, wisdom', Ug *ḫss* 'to feel'; Arab *ḥassa* id., *ḥiss* 'voice' // Eg(OK) *ḥsy* 'to sing' (EG III: 164–165) // Cush: (E) **haašaw-* 'to chat' (Sa 1982: 88) > Rendille *xawes* 'tale' etc.; ? (S) Asa *has-* 'to hear'
19. El(o) *hit* 'troops', (m) *hitra* 'warrior' (HK 665–666)
AA **ʕad-/*ʕid-* > Sem: Hbr *ʕɛdy-ek̲* 'the prime', Soqotri *ʕed(e)* 'vie, esprit', Geez *ʕəd* 'viri, masculi, mariti, viri fortes', Tigre *ʕad* 'tribe, family, people' (Ls 1938: 56) // Cush: (E) Som *ʕed*, *ʕid* 'people', Rendille *et* 'person, man', Arbore *ʔedan* 'people' // Omotic: Ometo **ad(d)e* 'man, male, husband', Aroid ** e(e)d* 'man' // Berb: Shilh *id* 'people', Ksur *idu* 'kin' // ? Chadic: (C) Musugeu *hiddi* 'man'. Note: The analogical semantic dispersion appears, e.g. in Indo-European: Hittite *tuzzi-* 'army, camp' vs. West IE **teutā* 'people'.
20. El(o) *igi* 'brother' (HK 743)
AA **ʔagy-* > Cush: (C) Bilin *ʔäg*, Kemant, Awngi *ag*, Xamir *ig*, Kunfäl *yaga* 'uncle'; (S) Asa *ʔagok* 'mother's brother' // Omotic: Ubamer *agi* 'aunt'.
21. El(o) *iza* 'cousin'
AA **ʔiS-* > Cush: (C) Awngi *iṣyaa* 'brother'; (E) Tambaro *izoa*, Kambatta *hizoo* 'brother'; ? Boni *éésə* 'mother's brother', Oromo *eessuma* 'maternal uncle' // Omotic: Basketo *išaa*, Wolaita *iša* 'brother', Koyra *iččaa* 'id., paternal uncle', Yemsa *ištaa* 'uncle'; Hamer *išma* 'brother'.
22. El(o) *liba* 'servant, groom' (HK 818–819)
AA ? **lyab-* (Do 1973: 164, 229) > Cush: ?(N) Beja *rába* 'male; capable, able'; (E) **leb-* 'male; strong' (Sa 1979: 22, 1982: 131); (S) Iraqw *láwaalee* 'slaves' // Omotic: Dizi *yabu*, Na'o *iab*, Sheko *yaab* 'man'.
23. El(o) **ma(a)n-* 'might, power' (HK 846)

AA *manw/y- 'man' (Illič-Svityč 1976: 58, #292) > Cush: (LE) Som *mun* 'male', (HE) *manna 'man (people)' // Omotic: Wolaita *minoo* 'warrior', Kachama *mono* 'strong' // Berb: Zenaga *miin*, pl. *måån* 'man', *(u)man* 'kin', Zwawa *iman* 'person, life' // ? Eg(Pyr) *mn* 'someone', Copt *man* 'a certain person / thing' (EG II: 64–65; Vy 1983: 114) // Chadic: (W) *mani 'man, husband, people' (St 232: #801); (C) Logone *meeni* 'man'
? Sum *emen > *ewen > *en*, Emesal *umun* 'lord' (Schretter 1990: 263)
Dr *maṉ 'king, lord, warrior' (DEDR 4774).

24. El(A) *mal /wal ?/* 'child, baby' (HK 903)
AA *wayl-/*waly- 'child' > Cush: (E) Sam *weil 'child', *wàláál 'brother'; Elmolo *wéil*, Dasenech *vèèl* 'child' // Berb: Libyan *w* 'son' vs. *wl.t* 'daughter' // Chadic: (C) Buduma *wuli*; (E) Sumrai *wiil* 'child'.

25. El(m) *mu(h)ti* 'woman, wife' (HK 948, 961–962)
AA *maṭ-/*matH- 'woman, wife' > ? Sem: Arab *m-t-t* 'to be related with somebody through marriage' or *m-ṭ-ʔ/w* 'cohabiter avec une femme' (Vycichl, AION 50, 1990, 80) // Cush: (E) Sidamo *matē* 'wife' // Omotic: Shinasha *maton* and/or Kachama *maaṭo* 'woman' // Berb: Ahaggar *taməṭ* 'woman', *məṭ* 'femme sans aucun valeur', Djerba *tamáttoṯ* 'woman' // Chadic: (W) *mata 'woman, wife' (St 232, #796); (C) Bachama *mata* 'woman', Wadi *miitti* 'Weib'.

26. El(o) *nab* or *nap* 'god' (HK 966, 970–971)
AA *na(ya)b- 'lord' > Sem: Arab *nāb*, pl. *ʔanyāb* 'tribal chief' (Ember, 1917: 83: Arab+Eg); Mehri *nōb* f. 'grand' // Eg(Pyr) *nbw* 'lord', Copt *nēb* (EG II: 227; Vy 1983: 138) // ? Cush: (E) Afar *naba* 'to be big', *nabam* 'very, much', Arbore *níib* 'greatly, very'.

27. El(A) *puhu* 'boy', cf. *punna* 'young' (HK 230, 238, 240)
AA *p/fu[ġ]- > Sem: Ug *p'y* 'boy', *p't* 'girl' (Segert 1984: 198) // Chadic: (W) Bokkos *fú*, Sha *foy*, Kulere *fwè*, *fo* 'boy, child'
Dr *poy 'girl' (DEDR 4532).

28. El(A) *ruh* 'man', (o) *ruhu* 'offspring' (HK 836, 1044–1046, 1049)
AA *rVḥ/ḫ[w]- > ? Sem: Akk *raḫū*, *reḫū(m)* 'to beget, pair' (*AHw* 969) // Eg(OK) *rḥ.w* 'people' (EG II: 441) or (Pyr) *rḫy.t* 'men' (EG II: 447), cf. *rḫ* 'to copulate' ? (Fa 152).

29. El(m) *šak* 'male offspring, son' (HK 1110)
AA *Sak/k^(w)- > Cush: (E) Oromo *sookiyyaa* 'adolescent' // Berb: Ahaggar *ašaġu*, pl. *šaġet* 'young man' <*aa-saaġuh/*sāġuh (Prasse 1974: 62); Guanche *suka* 'son' vs. *sukaha* 'daughter' (Wo 408) // Chadic: (W) Hausa *saako* 'a younger brother' vs. *saakuwaa* 'a younger sister'.

30. El(A) *zin* 'baby, suckling' (HK 1291)
AA *ʒin-/*ʒun- (?) > Chadic: (W) NBauchi *ʒin- 'child'; (C) Gisiga *zuŋ*, Bachama *nze* 'son, boy'
Dr *ciṉṉa 'small', cf. Brahui *cunaa* 'child' (DEDR 2594; MA 100: El+Dr).

Natural phenomena

31. El(m) *amni* 'mountains', (A) *amnu* 'mountain' ? (HK 55, 517)
AA **ʔabun-* 'stone' > Sem **ʔabun-* id. (Ls 1938: 4) // Eg(Med) *ỉbnw* 'mineral material, alun', Copt *obn*, *ōben* 'alun' (Vy 1934: 48–49) // Cush: (N) Beja *'awe* 'stone' <**ʔawen-*, cf. *siku-awn-eb* (acc.) 'Quartz' (Munzinger); (C) **ʔamb-* 'mountain' > **ʔabn-* // Berb **abūn* 'stone' > Sus *auwun* / *aggun* etc.; Guanche *t-abonas* (pl.) id. (Rössler 1964: 214) // Chadic: (W) **ʔabuni* 'millstone' (St 230, #781).
32. El(A) *bel* 'year' (HK 188)
AA **bVl-* > ? Sem: Ph *bl*, Hbr *bul* 'name of a month' (Cohen 1970: 51) // Cush: (E) Sam **bil-* 'month' (He 1978: 76), Sidamo *bululo* 'year'; (S) Qwadza *balaʔeto* 'year', ? Alagwa *balalu* 'days' // ? Chadic: (W) Fyer *wél*, Sha *wúl*, etc. 'year'.
33. El(o) *hal* 'land, bottom, region, city' (HK 574, 594)
AA **hal-* 'place' > Cush: (E) Som *hal* 'place' // Berb: Mzab *al* 'place', Zenaga *al* 'id., country'.
34. El(A) *har* 'Stein' (HK 623)
AA **har-* 'mountain, rock' > Sem **harar-* 'mountain' > Hbr *har*, *hererī*, Ph *hr* id. (Klein 1987: 167) // ? Cush: (E) Yaaku *hɛ́ɛ́rɔ'*, pl. *hɛrɔ́r* '(big) rock' // Berb: Ahaggar *ahor* 'accumulation of rocks'
Dr **aṟ-ay* 'stone, rock' (DEDR 321).
35. El(o) *hun* 'light' (HK 697, 717, 719–720); cf. (o) *nahi[n]ti* 'God of sun', (m) *nahhunte* 'sun' = **naN* 'Tag' and *hunti* 'Beleuchter' (HK 979–980)
AA **[h]Vn-* > ? Cush: (E) Burji *hin'-ícco* 'sun' // Chadic: (C) Zelgwa *həne*, Paduko *həni*, Hurzo *hənde*, Mandara *hạr* 'day (twenty-four hours).
36. El(m) *ki-el* 'region, district', (n) *ku-el* 'region' (HK 463, 501)
AA **kal[w]-* > Cush: (E) Oromo *kaloo* 'pasture land' // Berb: Adghaq *akal*, Zwawa *akkal*, Ntifa *akäl* etc, 'earth' // Chadic: (W) Tangale *kálaw* id.
37. El(o) *lali* 'source' (HK 813)
AA **lay-(l[ay-])* > Cush: (N) Beja *lil* 'to be wet, damp, moist'; (E) Afar *lay*, pl. *laayl* 'water', *layhintii* 'source'; ? Oromo *lolaa* 'flood' // Berb: Libyan *lilu* 'water' (Hesychios); Matmata *ilil* 'sea', Zenaga *ẹll* 'id., big river'. Note: Hittite *luli-* 'lake, pond, source, well' resembles rather El *lali* than Sum *túl* 'source' connected with Hittite by Puhvel (1976: 27).
38. El(o) *sud-/šut-me* 'night' (HK 1018, 1170, 1193–1194)
AA **sud-/*sut-* > Sem: Arab *swd* 'to be black', OSA *s(w)d* Cush: (N) Beja *sootay*, *suutay*, *sooday* 'of dark colour, dark-brown, -grey // Omotic: Dime *suut-u*, Galila *šoyt-i*, Ari *soyt-i*, Hamer *soyt-i*, *soot-i* 'night'.
39. El(n) *tep /deb ?/* 'rain' (HK 311)
AA **dib-/*dub-* > Cush: (E) Rendille *dubbat* 'cloud', Hadiya *duuba* id. // Omotic: Dizi *diɛb* 'to rain'; Kafa *dup* id.; Dime *deeb*, Ari *doob* 'rain' // Chadic: (W) Jimbin *dabuna* 'rainy season'; (C) Daba *dəbav□ya* 'rainy season', Gidar *dúbbya* id.; (E) Kera *dubueni* 'rain'.
40. El(n) *uhi* 'stone, rock' (HK 1202)

AA *ʔuḫay- > Berb: Menacer *uqi*, Iznacen *awqi* 'stone' // Chadic: (W) Montol *oho* 'rock'; Sha *wàḥày* 'mountain' or *hàw* 'stone', Daffo-Butura *hayaay* pl. id., Fyer *hoó* 'mountain'.

Dwelling, agriculture, tools and weapons, transport

41. El(o) *aapi-* 'to plough' (HK 15)
AA *hVb- > ? Sem: Arab *habba* 'to cut' // Eg(OK) *hb* 'plough' (EG II: 485), Copt *hebbe, hebi* (Vy 1934: 288)
Sum *apin* 'plough' (Blažek and Boisson 1992: 22).
42. El(A) *bardu* 'street' (HK 147)
AA *bVr[d]- > ? Cush: (E) Konso *pora* 'road' (*p-* < *b- regularly) // Omotic: Nao *buřun*, Gimira *bod* 'road' // Berb: Ahaggar *abarīd*, Ayr *abər*, Augila *tabarur* 'road' // Chadic: (W) Buli *bədàna*; (C) Hwona *banda*; (E) Mubi *bə̀də̀l*, Migama *bótól* 'way, road'.
43. El(A) *basram* 'hammer' (HK 126, 395)
AA *bVrVs- > Cush: (E) Oromo *burrisa*, Konso *purriša*; Dobase *purruša* 'heft'.
44. El(A) *elpi* 'saw' ? (HK 395)
? AA *ʔalb- > Cush: (E) Oromo *albee* 'knife'; Gollango *albeni* 'sickle'.
45. El(o) *halki* 'sweet', (A) *hal(?)-la(?)-ki* 'honey' (HK 599–600)
AA *ḥVl- > Sem *ḥluw > Arab *ḥalā* 'to be sweet, pleasant', *ḥulw* 'sweet', Syr *ḥ^{e}lí* 'to be sweet' // Eg(D 19) *ḥ3hrg* /*ḥlg*/ 'to be glad, to rejoice', (Gr) *ḥrg*, Demotic *ḫlk* 'sweet', Copt *hloj* 'to be sweet' (EG III: 34; Vy 1934: 298) // ? Berb: Tamasheq *sulleg^{y}et* 'to be sweet' (caus.) (Vy 1934: 85).
46. El(m) *hwel-/*hyel-* 'portal, gate; yard' (HK 391, 393, 657, 666, 683, 1201)
AA *^{c}/ġul- > ? Sem: Aramaic *ʿll*, Arab *ġa lla* 'to enter' // Eg(Pyr) *ʿ3* '(leaf of) door', (D 20) *ʿry.t* 'Türbalken', (Pyr) *ʿr(r)w.t* 'gate' (EG I: 164, 209–211) // Cush: (E) Oromo *ula* 'gate, portal' // ? Chadic: (W) Siri *hwuli* 'doorway'.
47. El(A) *hipis* 'ax', cf. *atti hipis* 'Spitzhacke' (HK 395, 668)
AA *ḫVb(-)Vs- > Eg(Pyr) *ḫbs* 'hacken' (cf. *ḫb3* id.), (BD) *ḫbsy.t* 'Hacke' (EG III: 256) // ? Berb: Ahaggar *eġwes* 'tailler, retrancher ce qu'il y a de trop'.
48. El(n) *menu-me* 'roof' ? (HK 915)
AA *min- > Cush: ? (N) Beja *mine* 'to create'; (C) *-ən- 'house'; (E) *min-/*man- id. (Sa 1982: 45), cf. Elmolo *míndu* 'roof'; (S) *min- 'house' (Eh 1987: #436) // Eg(Pyr) *mn(n)w* 'fortress' (EG II: 82; Takács p.c.) // Chadic: (W) Bole-Tangale *mina 'hut' (St 247)
Dr *maṉ-ay 'house' (DEDR 4776).
49. El(m) *mit[i]* 'needle' (HK 939)
AA *mut^{c}-/*mit^{c}- > Cush: (E) Elmolo *midi*, Dullay *mut(u^{c})^{c}o*, Gedeo *muta* 'needle'.
50. El(m) *ulhu* 'chamber', *ulhi* 'dwelling-place; Tempel-Cella'; (A) *ulhu* 'house, palace, yard' (HK 1216–1217)

AA **ʔuhl-* > Sem **ʔuhl-* > Akk *aalu(m)* 'village, city', Ug *ʔahl* 'tent, dwelling', Hbr *ʔohel* 'tent, shelter' etc. (Cohen 1970: 10) // Eg(D 19) *ỉh3y.t*, (D 18) *ỉhw* 'camp, stable' (EG I: 118) // Cush: (E) Oromo *oll-aa* 'village', Arbore *ʔollah* 'id., neighbors'.

Fauna

51. El(A) *bagimaš* 'halbwüchsig bei weiblichen Kleinvieh' (HK 118) = *bakemaš* 'intermediate (female) goat' (Hallock 1969: 673)
AA **bagg-* or **bagʿ-* (Co #390) > Cush: (N) Beja *bok* 'he-goat'; (C) **bäg(g)-* 'sheep' > Geez *baggəʿ* 'sheep, ram' // Berb: Ahaggar *abag*^{y}*ug*^{y} 'young ram', Iullemiden *abbegug* 'ram'. Note: El *bagimaš* can be a compound of a proper El word for 'goat' and Sum *maš*, *máš* 'he-goat, kid, gazelle' (IK 657, 660), cf. also *maš* 'son, boy' (IK 657).
52. El(A) *duma* 'wolf' (HK 356)
AA **duʔm-* or **dumm-* > Sem: Akk *dumaam-* 'gepard', Arab (Yemen) *dimm, dumm* 'cat' (Ls 1938: 136) // Cush: (S) **duʔuma* 'leopard' (Eh 1980: 347) // Omotic: Koyra *damaa* 'jackal' // Chadic: (W) **dami* 'leopard; hyena' (St 171, #240); (E) Bidiya *ḍèmḍèm* 'lynx'.
53. El(A) *it-ra-an-ku /dranku ?/* 'donkey' (HK 794)
AA ?: Cush: (C) **dəq*^{w}*ar-* 'donkey' // Chadic: (E) Mubi *ḍùrgúl*, Migama *ḍurkúl*, Dangla *ḍúrkur*, Bidiya *ḍurtikilo* id.
54. El(m) *hidu* 'sheep' (HK 656)
AA **ʿiid-/*ʿidd-* ? > Cush: (E) Saho *ʿeydo/ʿiido* 'sheep' (coll.), Asa-Lisan *ʿiddoo* pl. 'sheep', Elmolo *édi* 'goat'
Dr **iṭ* – 'to herd (esp. goats)'> Malayalam *iṭayan* 'a caste of shepherds and cowherds', Brahui *hiḍing* 'to gather, herd' (DEDR 450; MA 97: El+Dr).
55. El(m) *kumaš* 'he-goat' (HK 512); cf. *áš* 'cattle, herd' (HK 84)
AA **kVm-* > Cush: (C) **kəm-* 'cattle' // Chadic: (W) Bole-Tangale **k*^{w}*amV* 'cow' (St 246).
56. El(A) **kar(r)-/*kur(r)-* 'lamb' (HK 441, 442, 531)
AA **karr-* (Co #181) > Sem **karr-* '(male) lamb' // ? Cush: (E) Dasenech *kor-ac* 'male kid' // Berb: Qabyle *ikərri*, Ahaggar *ekrer* 'ram', Sus *ikru* 'goat' // Chadic: (W) Saya *kərò*, Wandai *karò* 'sheep'.
57. El(m) *lakpilan* 'horse' (HK 811); ? < **laki-*[*i*]*pilan*, cf. *laki-* 'to travel' (HK 806, 811), comparable with Beja *lagi* 'road'; Qwadza *lagalako* 'path, road' (Eh 1987: #316)
? AA: Sem **ʔib(i)l-* 'camel' (Cohen 1970: 3) // ? Eg *ỉb3w* 'Barbary sheep' (Fa 15; Takács p.c.: Eg+Sem)
Dr **ivuḷi* 'horse' (DEDR 500). Note: The domesticated horse (*Equus caballus*) was not introduced into south Asia until after 2000 BC. McAlpin (1981: 147) judges that Dr **ivuḷi* must refer to onager (*Equus hemionus*). On the other hand, the domesticated horse was introduced into Sumer just from West Iran / Elam beginning of 3rd mill. BC (Brentjes).
58. El(A) *putu* and *pitu* 'kid' (HK 226, 237)

AA *$pV\underline{h}(V)d$- > Sem: Akk *puḫadu* 'lamb, kid', Ug *pḫd* 'lamb' (Gordon 1965: 467) // Berb: Ahaggar *eifed* 'ram', Ayr *äɣfəḍ* id. (Prasse 1974: 21).

59. El(A) *tịla* 'calf' (HK 329)
AA *ṭaly- > Sem *ṭalay- 'young of sheep, goat, antelope' (Ls 1938: 590) // Cush: (E) *ḍal- 'to beget' (Sa 1982, 123), cf. Sidamo *ḍala* /*ṭala* 'she-donkey' // Berb: Iullemiden *āḍel* 'calf'. Note: Cf. also Hurrian *Tilla* 'a bull (of Teššub)' (Laroche 1977: 266).

60. El(A) *zamama* 'bird' = 'Geflügel' ? (HK 1280)
AA **cum-an-* > Sem: Akk *summatu* 'dove', Arab *summān* 'quail' (*AHw* 1058) // Eg(Pyr) *smn* 'goose' (EG IV: 136) // ? Berb: Ahaggar *a-jjam* 'sp. ostrich' // ? Chadic: (W) **ziman-* 'ostrich' (St 190).

61. El(A) *zibar-* 'camel' (HK 1288)
AA *[*z*]*VbVr-* > Cush: (C) Bilin *daabraa* 'bullock, Stier zum pflügen'; (E) Som *dubeer* 'decrepit pack-camel'. Note: The oldest discovery of domesticated camel (*Camel bactrianus*) is known from central Iran (Tepe Yahya, 4500–3800 BC) (Brentjes). Its spreading is attested from east Iran (Shahr-i-Sokhta, 2700 BC) and the Indus valley (2300 BC) (Banti 1993: 186). On the other hand, dromedary (*Camelus dromedarius*), originating probably from the Arabian peninsula, was depicted in Mesopotamia before 3000 BC and in Egypt in early 3rd mill. BC (Brentjes). The age of a presence of camel in Ethiopia and Somalia is discussed by Banti 1993: 193–199. The east African camel was imported from south Arabia. The similarity of Elamite and Cushitic words does not represent certainly a common heritage. If it is not an accident, it can be explained only as a result of a cultural diffusion.

Flora

62. El(n) *ahiš* 'pasture-land' (HK 34)
AA *ᶜ/ʔ*awis-* > Cush: (E) *ᶜ*awiš-* 'grass' (Sa 1979: 44, 45, 47) // Omotic: She *oš* 'cane, Bambusa abyssinica' // Chadic: (W): NBauchi **awasi* 'grass' (Skinner 1977: 24); (C) Ga'anda *ušìnna*, Masa *usna* id.; (E) Bidiya *ʔàwso*, Sokoro *ússii* id.

63. El(m) *par* 'seed, offspring' (HK 148)
AA **pVr-* (Co: #367) > Sem: Akk *pēru* 'fruit', Hbr *pᵉrī* id., *pārāh* 'to bear fruit', ? Arab *wafara* 'to be numerous, fruitful' Eg *pry* 'to give birth' (Ward 1961: 36–37: Sem + Eg) // Cush: (N) Beja *firi* 'to bear offspring, fruit', *faar* 'blossom, flower, seed, bud'; (C) **fər-* 'to flower, fruit; grain' (Eh 1987: #184).

64. El(o) *huk* 'wood' (HK 686, 689, 714)
AA **haq-/*ḥak-* ? > Cush: (E) Afar *hak*, Saho Irob *ḥak* 'branch', (HE) **haqqa* 'tree, wood' // Omotic: Koyra *akkaa* 'tree'; Ubamer *aqa*, Banna *haaqa*, Bako *(a)haka* etc. id. // ? Berb: Ahaggar *éké*, pl. *ikéwen* 'root'.

65. El(o) *husa* 'stem, stick, wood, tree, forest' (HK 702–703)
AA *ᶜ*iċ̣* – > Sem *ᶜ*iṣ̂* – 'tree, wood', cf. Akk *iṣu*, Arab Datina *ᶜaḍa*, *ᶜuḍah* (Ls 1987: 57) // Eg (Med,BD) *ḏᶜᶜ* 'branch' (EG V, 535) // Cush: (N)

Beja *'aḍa* 'pole, long stick'; ? (E) Afar *haḍaa* 'tree', *ḥàḍḍa* 'stick' // ? Chadic: (C) Mandara *háazlà* 'tree'.

66. El(m) *malu* 'wood' (HK 864)
AA **mal-* > Chadic: (W) Bolewa *mala* 'forest', Gera *màalà* 'bush' // Berb: Senhaja *amalu* 'oak'.

Adjectives

67. El(n) *hazza* – 'big' (HK 592–593, 653)
AA **ʿaʒ-* > Sem ** ʿ-z-z* 'to be strong, mighty' (*AHw* 269–270; Ls 1987: 81) ? Eg *ʿḏ* 'to be safe, vigorous, prosperous' (EG I: 237) // ? Cush: (S) Mbugu *-'ezá* 'long, tall' (Eh 1980: 275) // Omotic: Benchnon *ez-at-* 'to become big', *ez-ats-* 'to make big'.

68. El(A) *kara* 'old' (HK 437–438)
AA **gary-* > Sem: Arab *ǧārin-* 'to be worn out (clothes), be trained (beast)' // Cush: (E) **gerʿ-* 'old' (Black 1974: 20); (S) Alagwa *garmo*, pl. *gari* 'old man', *garaʔo* 'old woman' // Chadic: (W) Hausa *girme, girmaa* 'to be older than'
Dr **kiR-* 'old' (DEDR 1579).

69. El(m) *meli/u-* /*weli/u-* ?/ '(for) a long time' (HK 912, 918)
AA **w[a]ly-* > Eg(Pyr) *w3y* 'to be far', Copt *we(i)* id., (Pyr) *3wy* 'to be long', (MK) *w3ḥ* 'to be long (in time)' (EG I: 245, 255, 259; Vy 1983: 230) // Berb: Ahaggar *alu* 'to be large' (Co: #513 adds also Arab *waliya* 'to be near').

70. El(o) *mer* 'powerful' (HK 910)
AA **mVr-* > Sem **m-r-r* 'to strengthen' (Segert 1984: 193) // Eg *mr* 'strong' (Ward 1961: 36: Sem+Eg).

71. El(n) *purna* 'brown' (HK 242)
AA **buʔr-* > Cush: (E) **boʔr-* 'yellow, brown, red' (Sa 1982: 39); cf. Rendille *bóran* '(dark-) brown', Arbore *burrí* 'red' // ? Chadic: (E) Bidiya *baar* 'to become red', *barga* 'red'.

72. El(n) *riša-/ir(i)ša-* 'big' (HK 774, 779–80, 1041)
AA **riʔs-* > Sem **raʔiš-* 'head' – cf. Geez *r-ʔ-s* 'to rise above, become chief' (Ls 1987: 458) // Eg(Med) *3ys* 'brain' (EG I: 2).

73. El(n) *sir* 'heavy, rich' (HK 1087, 1089, 1090)
AA **s[u]r-* > Sem: Akk *ešēru*, Hbr *yašár* 'to be straight', Arab *sarā* (= *s-r-w*) 'to be brave, manly, noble, be firm' (Albright 1927: 212: Sem+Eg) // Eg(Pyr) *wsr* 'to be strong' (EG I: 860) // Cush: (E) **šor-* 'rich' (Sa 1979: 33) – add Boni **suur-* 'good' (He 1982: 110).

74. El(A) *teman-* 'evening' (HK 317)
AA **tVm-/*ṭVm-* > Sem: Arab *ʔaʔrama* 'devenir sombre' // Eg: Copt *thómtəm* 'to become dark' (Vy 1934: 43: Copt+CCush), derived perhaps from Eg *ḥtmtm* (Vy 1983: 316) // Cush: (C) **tem-* 'to be dark'; (HE) **t/ṭum-* 'darkness' // Omotic: Wolaita *tuumoo* id., Shinasha *tuumaa* 'night' (Do 1973: 53–54).

Adverbs, conjunctions and particles

75. El(m) *am* 'now' (Hallock 1969: 666; HK 14, 48, 51, 56)
AA **ʔam(m)-* > Sem **ʔam-/*ʔim-* 'if' (Cohen 1970: 22; Ls 1987: 22–23) // ? Eg(Pyr) *m(y)* 'how, if' (EG II: 1, 36; Vy 1983: 105) // Cush: (C) Bilin *emmáa, immáa* 'nun denn, also'; *e/imáanaa* 'time; earlier'; (E) **ʔamm-(an-)* 'time' (Do 1973: 132; Black 1974: 157; Sa 1979: 25) // NBerb **am* 'how' (Prasse 1972: 230: Eg+Berb).
76. El(n) *da* 'also, yet, then' (HK 245)
AA **dV* > Cush: (C) Bilin, Qwara *-dii* 'together with'; (E) Som *-daa* 'emphatic particle' // Chadic: (W) Angas *da* 'also' // Berb: Libyan *d* 'and, together with', Ahaggar *əd* 'with; and' (Prasse 1972: 225).
77. El(n) *hira* 'for' (HK 668)
AA: Eg(Pyr) *ḥr* 'for, (up)on, through', orig. 'face' (EG III: 132).
78. El(o) *in-* 'not' (HK 754, 757–758)
AA **ʔin-* > Sem: Akk *yānu* / *yaʔnu* 'isn't', Ph *ynny* id., Hbr *ʔayin, ʔeen*, Ug *in, yanu* 'there is not', Arab *ʔin*, Geez *ʔen* (Ls 1987: 27) // Eg(Pyr) *n*, (MK) *nn* 'not' (EG II: 195) // Cush: (E) Som *an* 'not', Oromo *en-* id., Afar *-inn* (in negative verbal constructions *mV*-verb-*inn*).
79. El(m) *sap* 'copy', (A) 'how' (HK 1054–1055; Hallock 1969: 751)
AA **ʒap-* > Sem: Arab *zaffat* 'once', *zafatāni* 'twice' (Ember 1913: 119: Arab+Eg) // Eg(OK) *zp* 'times' (= 'mal') (EG III: 435).

Numerals

80. El(o) *ki* 'one' (HK 459, 465, 468–469)
AA **kawy-* > Eg(Pyr) *kyy*, pl. *kwy* 'another' (EG V: 110), cf. *ky . . . ky* 'one . . . other' (Fa 285) // Cush: (N) Beja *kwo* 'unit'; ? (C) Bilin *kaayaa* 'empty; only, alone, solitary' or Qwara *kaw* 'to be in front, be first'; (E) **kaww-* 'one; alone' (Sa 1979: 44) // Omotic: Dizi *qòy*, Sheko *k(w)oy* 'one'; ? Gonga **ikk-* id.
81. El(n) *mar(i)* /=*wari* ?/ 'two' (HK 860, 876, 880)
AA **wary-* ? > Cush: (N) Beja *wari* 'other'; (C) **wäri* 'or' (Eh 1987: #578: N+CCush) – cf. Dahalo *watte* 'other' (Elderkin) vs. *watte* 'or' (Eh) // Chadic: (W) Hausa *waari* 'a pair'
Dr **waṟ-oṇṭi* 'next year' (DEDR 5375), cf. **onti* 'time, a turn' (DEDR 979) ? Note: Being El *m-* original, there is an alterative cognate in Dr **maṟu/i* 'another, following, next, again' (DEDR 4766). Al'bedil' 1986: 47 tries to prove a presence of this in the language of Proto-Indus script on the basis of a partial homonymity with Dr **maṟa-* 'hero' (DEDR 4764).
82. El(A) *ziti* 'three' (HK 1305)
AA ?: Sem **šidš-* and **šidt-* '6' < **šid+šid* = 3+3 ? – cf. Ug *ṯlṯt w ṯlṯt* '6' = '3+3', *ṯṯ tt* '12' = '6+6' (Gordon 1965: 503, 501) // Berb: **saḍīs* and **sūḍus* '6' (an old reduplication ?) // ? Chadic: (W&E) **sidu* '6'. Note: There are suggestive parallels in Nilo-Saharan: Berta *sittijini* '3'; Kunama *saate*, Ilit *satte*; Berti *soti* id. The position of ECush **s/šaz(zi)ḥ* '3' and **sizḥet* '8' is not clear, cf. also Mao (Omot) *t/siyaz-* and Tirma

(Surma) *sisi, dizi* '3'. A total puzzle is Soqotri (SSem) *ḍádeheh* '3' recorded by Bittner against the usual form *šile* / *šaʿte* m./f. by Johnstone.

83. El(n) *kut-* 'all' (HK 548, 565)
AA **gudd-/*gutt-/*gud-t-* ? (Greenberg 1963: 59) > Sem: Arab *ǧadda* 'to be great, rich, honoured' // Cush: (N) Beja *gud* 'to be much, many, full, big'; (C) Awngi *gud* 'good'; (E) **gudd-/*guud-* 'big', cf. Som *giddi* 'whole' (Eh 1987: #37) // Omotic: Wolaita *guute* 'much' // Chadic: (W) Kofyar *gʷɛɛt* 'many'; (C) Higi *gutàgəy* id. // Berb: Zwara *a-guda* id.

Verbs

84. El(m) *bakka-* 'to find' (HK 106)
AA **bVḳ-* > Sem **b-q-w* 'to seek, try' (Cohen 1976: 78) // Cush: (E) Som *beeq-* and *beeg-*, Oromo *bek* 'to know' // Omotic: Benchnon *beqᵃ* 'to see', Basketo *biq-*, Kafa *beg(g)-* id., caus. *beqq-* 'to know'.

85. El(n) *bera-* 'to read' (HK 185–186)
AA **ba[ʔ]r-* > Sem **b-ʔ-r* 'to explain' (Cohen 1976: 41) // Eg(MK) *sb3* 'to teach', (late) 'to learn', cf. (Pyr) *sb3.w* 'teacher' (EG IV: 84–85) with a frozen causative prefix *s-* ? // Cush: (E) **bar-*, cf. Afar *bar-is-* 'to teach', *bar-it-* 'to learn', Oromo Borana *bar-aḍ-* 'to understand' (Black 1974: 164); Dahalo *ɓar-* 'to know' (Eh 1980: 135)
Dr **paṟ-ay* 'to speak, say, utter' (DEDR 4031; MA 105: El +Dr) or **pēr-* 'speech' (DEDR 4439).

86. El(m) *da-/ta-* 'to lie, put' (HK 248, 254–255, 262)
AA **-d-ʿ* > Sem **(w-)d-ʿ* 'to put' (Ls 1938: 125) // Cush: (N) Beja *di'* 'to make, be ready, put', caus. *daa-s* // Chadic: (C) Musgu *da* 'to do, build', Logone *ḍə* 'to put, stand' (Do 1973: 186–187).

87. El(o) *du-* 'to get, take, keep' (HK 346–347, 356–357)
AA **-d-w/y* > ? Sem *ndy* > Akk *nadū* 'to throw (away), put down', Ug *ndy* 'to throw/drive away, remove', postbib. Hbr *ndy* 'to banish' (*AHw* 705; Segert 1984: 193) and Sem *wdy* > Ug *ydy*, Hbr yāḏā 'to throw', Arab *ʔawdā* 'to take away', Geez *wadaya* 'to put, add, lay, place, throw' (Ls 1987: 605) // Eg *ídy, wdy, ndy* 'to give, put, throw' (Ember 1930: 116: Sem+Eg) // Omotic: Kafa *dew-* '(ap)portare; consegnare, pagare; (ri)tornare'.

88. El(m) *duna/i-* 'to give' (HK 361–362)
AA **d[i]n-* > Sem: Akk *(i)din* 'give!', *nadānu(m)* 'to give', *tadānu* 'to give (back)', Hbr *nādān* 'gift', Arab *dūna-ka* 'you have here, take!', *dyn* 'to give on credit' // Eg(Pyr) *wdn* 'to make sacrifice', Copt *wōten* id. (EG I: 391; Vy 1983: 239; Ember 1930: 115: Akk+Eg) // ? Chadic: (W) Ron: Sha *ndi* 'to give'. Note: There is a voiceless variant in WSem: Hbr, OAram *ntn*, Ph, Ug *ytn* 'to give, pay' (Aistleitner 1965: 139–140).

89. El(m) *halpu/i-* 'to beat, kill', *halba* 'died' (HK 595–596, 605–607)
AA **ḫ-b-l* > Sem **ḫ-b-l* 'to ruin, destroy' (*AHw* 302; Ember 1930: 81: Sem+Eg) // Eg(Pyr) *ḫb3* 'to destroy' (EG III: 253).

90. El(o) *hani-* '(to) love' (HK 616–618)

AA *ḥ[a]n- > Sem *ḥnn 'to grant, favour, long for' (Aistleitner 1965: 105) // Eg(Pyr) ḥn 'to grant, favour', Demotic ḥn, Copt hne-, hna- 'to want' (EG III: 101; Vy 1983: 519; Ember 1913: 119: Sem+Eg) // ? Cush: (E) Konso heen- 'to want', heenaa 'love'
Dr *aṉ-/*aṇ- 'love, friendship' (DEDR 330; MA 97: El+Dr).

91. El(o) *hapu* 'to hear' (HK 578–579, 589–590, 622)
AA *hub- > Cush: (E) *hub- 'to know, be sure', cf. Afar -ob- 'to hear' (Sa 1979: 38, 40, 41); Dahalo *huɓ-aṯ-* 'to know' (Eh 1980: 336).

92. El(m/n) *hil-/hul-* 'to rob, loot' (HK 660, 673, 691)
AA *ᶜul- > Sem *ᶜ-w-l and *ġ-w-l < Hbr *ᶜalwā* 'disobedience', Arab *ᶜāla* (ᶜ-w-l) 'to deviate from the right course', Geez *ᶜalawa* 'to rebel, distort, reject, pervert' and Ug, OSA *ġlyt* 'wrath', Arab *ġ-w-l* 'to take unexpectedly, destroy' (Ls 1987: 78) // Eg(OK) *ᶜw3y* 'to rob, steal; robber; one robbed' and (MK) *ᶜwn* 'to rob, deceive' // Cush: (N) Beja *ol, ul* 'to strike'; (E) *ᶜol- 'war' (Do 1973: 162; Black 1974: 243; He 1978: 99; Eh 1987: #492)
Dr *ula- 'to become diminished, terminated, die, perish' (DEDR 671)
Sum *hul* 'bad, evil; to ruin, destroy; enemy' (IK 446–447). Note: There are hopeful cognates in IE: Hittite *halla-/ hallu-* 'to lay waste, ruin, savage', Greek *óllūmi* 'I destroy', Lat *ab-oleō* id. (Puhvel 1991: 13–14, 49–50 reconstructs IE *$A_2{}^{w}$l-n-*, rejecting the connection with Hittite *hulla-* 'to smash, quash, defeat' – p.368).

93. El(m) *huma-* 'to take, rob' (HK 691–694)
AA *ḫVm- > Sem *ḫ-m-y > Akk *ḫamū* 'to immobilize, paralyze', Geez *ḫamaya* 'to tie, shackle, chain' (Ls 1987: 262–263) // Eg(Pyr) *ḥmᶜ* 'to seize' (Ember 1930: 36: Sem+Eg) and/or Eg(Pyr) *ḫmy* 'to reach, touch, grasp' (EG III: 281–282).

94. El(o) *hutta-* 'to work, make', (linear script) *hut* 'work'
AA: Sem *ḫ-t-ʔ < Akk *ḫat¨* 'to vanquish', Ug *ḫtʔ* 'to disappear', Arab *ḫataʔa, ḫa tā, ḫaṭṭa* 'to be carried away' (Segert 1984: 187).

95. El *kani*: (n) *kanira* 'friend', (A) *kani* 'I would like' (HK 431–432)
AA *k-h-n > ? Sem *kāhin 'priest, prophet, augur' (Ls 1987: 278) // Cush: (N) Beja *kehan* 'to love, honor, venerate'; (C) Xamir *(i)ekan* 'to love, want', Awngi *ənkan-* id.; (E) Afar-Saho *kahan-* 'to love'
? Dr *kaṉi- 'to ripen grow tender', cf. Tamil *kaṉivu* 'ripeness, love, compassion' (DEDR 1408). Note: The semantic dispersion is plausible, cf. Sem *m-n-y 'to love, desire, wish' and 'to count' (Ls 1987: 352–353).

96. El(o) *kat* 'place, throne', (A) *kata/u-* 'to live' (HK 410, 452–454)
AA *kVt- > Cush: (N) Beja *keti* 'to seat, put together'; (S) Alagwa, Burunge *kiti* 'settlement' // Omotic: Chara *kot-it-*, Kafa *kot(e)-*, Mocha *kota-* 'to sit' (Do 1973: 246).

97. El(o) *kul(l)a-* 'to ask, call' (HK 508, 560–562)
AA *q-w-l > Sem *q-w-l 'to say, speak' (Ls 1987: 426) // Cush: (N) Beja *kwali* 'singing'; (E) Som *qayli* 'to cry, shout', Yaaku *-qɛɛl-* 'to sing'; (S) Qwadza *ḳwaʔaliko* 'voice'; ? Mbugu *-kaláʔe* 'to shout' (Eh 1980:

268, 1987: #513) // Chadic: (E) Jegu *kol-* 'to name, call', ? Gabin *guaal* 'to speak'.

98. El(n) *kuni-/kini-* 'to become, realize' (HK 477–478, 515, 564)
AA **k-w-n* (Co: #196) > Sem **k-w-n* 'to be, become' (Ls 1987: 299–300) // Cush: (C) Bilin *kʷín* 'to be, exist'; (E) Afar-Saho *kii(n)* id. // Berb: Ahaggar *eken* 'to do, arrange'.

99. El(o) *kura-* 'to burn, roast' (HK 518–519)
AA **kawr-* > Sem **kawr-* 'stove, furnace' (Ls 1987: 300) // Cush: (LE) **kar-* 'to boil' (Do 1983: 134: Sem+ECush).

100. El(o) *kusi-/kuši-* 'to build, bear (children)' (HK 538–539, 541)
AA **k[u]s-* > Cush: (N) Beja *kwsi* 'to mean; make, create', *kwása* 'heritage' // Berb: Shilh imper. *kkas*, fact. *yəkkus*, Ahaggar *kusāt* : *yəkkus* 'to inherit' (Rössler 1964: 206: Beja+Berb).

101. El(n) *kuti-* 'to carry, bring' (HK 505, 546–547)
AA **guty-* ? > Cush: (C) Xamir *gʷit-* 'to pull'; (LE) **giit-* id. (He 1978: 83; Do 1973: 245).

102. El(o) *li-* 'to give; gift' (HK 818, 820–821, 826–828)
AA **li-* > ? Sem: Arab (Taᶜiizz) *mā ʔallōs* 'there is not', Amhara *ʔall-* 'to be' (Co: #20) // Cush: (C) Qwara *lee* 'to give'; (E) **leḥ-* 'having' (Sa 1979: 41; Do 1973: 164–165) < **li-hay* 'to be by' ?; cf. Afar-Saho *-ell-* 'to come to have, possess', Elmolo *li* 'to possess'; (S) Qwadza *loʔ-* 'to give' (Eh 1980: 388) // Chadic: (C) Logone *lii* 'to be'; (E) Mokilko *ʔél-* 'to give' // Berb: Ahaggar *äl* : *yəla* (**l-ʔ-γ*) 'to have, possess' (Rössler 1964: 207: Som+Berb).

103. El(o) *muri-* 'to grasp', (A) *ma rri-/*m[o]rri-* 'to seize, hold, occupy' (Hallock 1969: 726; HK 885, 905, 953)
AA **mVr-* > Cush: (N) Beja *meri* 'to take, get, find, seize', *maray* 'to take, rob'; (HE) **moor-* 'to steal' (Hudson 1989: 143); ? (S) Mbugu *mmarú* 'load' (Eh 1980: 154).

104. El(m) *mirri-* 'to smear' (HK 923, 935)
AA **mVr-* > Sem: Arab *m-r-ḥ* 'to smear' // ? Eg(OK) *mrḥ.t* 'fat' (EG II: 111), if it is not derived from *wrḥ* 'to smear' (EG I: 334) // Cush: (E) **moor-* 'fat, sealing-wax' (Sa 1982: 147) // Chadic: (W) **ma/iwra* 'fat, butter' (St 233)
Dr **mer-* 'to smear, rub' (DEDR 4709) and/or **meṟuk-* 'to smear, plaster; wax' (DEDR 5082).

105. El(m) *na-* 'to say' (HK 975, 981, 990)
AA **nV* > Chadic: (W) Fyer *ne*, Bokkos *ni* 'to say'; Sura *nεε*; Bolewa *ni na*, Tangale *nεε*; SBauchi: Burrum *ne*, Kir *no* id. (St 235).

106. El(m) *ni-* 'to be' (HK 1000–1100; Hallock 1969: 738)
AA ?: Sem: Arab *ʔinn, ʔanniya* 'l'être', Amhara *na-* 'copula', Gafat *yän-* 'to be' (Co: #445) // Cush: (C) Bilin, Qwara *en*, Dembea *in* 'to be'; (E) Afar-Saho *na* id. // Chadic: (W) Hausa *na, ne* 'is, are, was, were'.

107. El(n) *para/i-* 'to go; arrive, come; draw, pull' (HK 146, 149)
AA **s-p-r* (with the causative prefix **s-* ?) > Sem: Akk *šapāru* 'to send',

Arab *sāfara* 'to travel' (Albright 1927: 228: Sem+Eg) // Eg(Pyr) *spr* 'to come, arrive, reach' (EG IV: 102)
Dr **pari-* 'to run, go out, move' (DEDR 3963; MA 104: El+ Dr).

108. El(m) *suku-* 'to destroy, exterminate' (HK 1102)
AA **suḳ-* or **sawḳ* – ? > Cush: (C) Bilin *suuk-*, Xamir *sooq-/sawq-* 'to kill (cattle)'; (E) **šoq-* 'to beat, hit' (Sa 1979: 33) // Omotic: Zala, Chara, Yemsa *šuk-*, Kafa *šuk(k)-* 'to kill (cattle)' (Do 1973: 115).

Pronouns

The correspondences between the sets of Elamite and Dravidian pronouns and nominal and verbal personal endings belong to the most convincing in McAlpin's demonstration of their genetic relationship. Let's confront these systems with the Afroasiatic one (see Tables 2.1, 2.2 and 2.3). The empty cells represent forms which are unreconstructable for more than one branch of Afroasiatic.

It is evident that some pronominal stems and even fragments of complete paradigms correspond. These cognates can be completed by other pronominal roots:

109. El(m) *akka* 'that, which' (rel.) (HK 37)
AA **ʔak(k)-* > Sem: Akk *akkāʔi*, Hbr *ʔēk*, Aram *ʔakam* 'how', *ʔaka* 'why', Ug *ik*, Mehri *ūkō* id. // Cush: (E) Oromo *aka* 'like', *akka* 'that, in order to; like' // Omotic: Yemsa *akka* 'thus, how ?' // Chadic: (W) Ngamo *aka* 'how'; SBauchi: Guruntum *akwaa* 'who', *akaa* 'what', Geji *yèk* id.; (C) Ngala *yaku* 'who'.

110. El(A) *-be*: *hu-be* 'that' ('jenes, das') where *hu-* corresponds to mEl *hu/i* 'this, dies' (HK 654, 676, 681)
AA **bV* > Cush: (N) Beja nom. *bɛ -n*, acc. *bɛ-b* 'that' // Omotic: Shinasha *bi/bo*, Kafa *bi/bonoosi* 'sg./pl. of demonstr. stem', Yemsa *baas/bar/baasɔ* m./f./pl. id.

Table 2.1 Middle Elamite pronouns and nominal and verbal personal endings

	nom.-dat.	*accusative*	*genitive*	*possessive*	*verbal*	*nominal*	*A=Achaem.*
sg 1	u	un	(A) unina/i	(A) u-ri	-h	-k	-ki / -ka
		(A) unan(-ku)	(A) u				
2	(o) ni / nu	nun		(A) -ni	-t	-t	-ti
3	ir	ir		-e	-š	-r	-ra
	(A) hi	(A) ir / in		(A) -e(-ri)			
dat.	(A) ha-						
dat.	kaš						
pl 1	nuku / nika		(A) nukami	-nika	-hu	(?) -unka	-un
2	num / nun				-hti		
3	api	apin / apun		-api-e	-hši	-p	-pi / -pa
				(A) -pini			

Source: McAlpin (1981) and Grillot-Susini (1987)

Table 2.2 Dravidian (*) / Brahui pronouns and nominal and verbal personal endings

	nominative	*oblique*	*possessive*	*appellative*	*verbal*	*Brahui*
sg 1	*yān̲ / í	*yan̲ / kan	*y- / -ka	*-en	*-ku	-v, -r, -ṭ
2	*nī(n̲) / nī	*nin̲ / nē	*ñ- / -ne	*-i & *-ay	*-ti	-s
3	*tān̲ / tēn	*tan̲ / tēn	*t- / -te	m. *-an̲rə n. *-(a)t	*-an̲rə	-k, -e, -s
pl 1 in	*nām / nam	*nam / nam		*-aṭ	*-t-aṭ	-n
1 ex	*yām	*yam		*-em	*-t-um	
2	*ním / num	*nim / num		*-ir	*-t-ir	-r
3	*tām / tēn	*tam / tēn	/ -tā	mf. *-ar n. *-av	*-ar *-ap	-r, -s, -õ

Source: McAlpin (1981) and Andronov (1980)

Table 2.3 Afroasiatic pronouns and nominal and verbal personal endings

	subject	*absolutive*	*dative*	*accusative*	*perfect*	*imperfect*
sg 1	*(ʔan-)ʔaku	*ya /*yi /*yu	*yiwāši	*yiwāti	*-ku	*ʔa-
2m	*(ʔan-)ta	*ku	*kuwāši	*kuwāti	*-ta	*ti-
2f	*(ʔan-)ti	*ki	*kiyāši	*kiyāti	*-ti	*ti-
3m	*šuwa	*šu	*šuwāši	*šuwāti	*-a	*yi-
3f	*šiya	*ši	*šiyāši	*šiyāti	*-at	*yi- / *ti
pl 1inclusive	*(ʔan-)muni	*na /*ni /*nu			*-na	*ni-
1exclusive	*(ʔan-)ḫina/u					
2m	*(ʔan-)tunwa	*kunwa			*-tunwu	*ti- ... -ū
2f	*(ʔan-)tinya	*kinya			*-tinya	*ti- ... -na
3m	*šunwa	*šunwa				
3f	*šinya	*šinya				

Source: Blažek (1995)

111. El(m) *-ka /i* 'I am (now)' (HK 459, 464–465)
AA **ʔaku* and **ʔan-ʔaku* 'I', **-ku* '1sg perf.'
Dr **-ku* '1sg of verbal conjugation', cf. Brahui *kan* 'me' and *-ka* 'my' (Tables 2.1, 2.2, 2.3).
112. El(A) *kaš* 'him' (dat.) (HK 418, 450)
AA **kV* > Cush: (C) **-y^w*: **la-y^w* m. vs. **la-ti* f. 'one'; (E) **ku* (subj.), **ka* (acc.) 'this' (Sa 1982: 111) // Omotic: Ari *koona* 'this': *koona-see* 'that'. Note: El *-š* can be a relic of old dative appearing in such forms as **yiwaaši*, **kuwaaši* (Sem: Akk and Eblaic; C+HECush – see Blažek 1991).
113. El(o/m) *ni/nu* 'thou' (HK 996, 1004, 1006)
AA: NOmot **ni(-ni)* (subj.), **ni(-na)* (obj.) 'thou'
Dr **nī(n̲)* 'thou' etc. (see Table 2.3).
114. El(o/m) *nika / nuku* 'we, us' (HK 1000, 1003, 1008, 1011), where *-ka/u* can correspond to *-ka/i* 'I (am)' or with *-ku* in (A) *unan-ku* 'me here'; (A) *-un* 'ending of 1pl of nominal conjugation'

AA **na* /**ni* /**nu* 'we, us, our' etc. (see Table 2.3)
Dr **nām* 'we', cf. Brahui *-n* 'verbal ending of 1pl'.

115. El(A) *hi-su* 'he self' (HK 669), cf. *hi* 'this'
AA: Cush: (C) Qwara *išuu*, Bilin, Xamir *šuu* 'self'; (E) **ʔis-* 'self' (Sa 1979: 34, 35, 1982: 107).

116. El(m) *-š* 'ending of 3sg of verbal conjugation'
AA **šuwa* 'he', **šiya* 'she' etc. (see above)
Dr: Brahui *-s* 'verbal ending of 3sg'.

Table 2.4 Preliminary phonetic correspondences

Afroasiatic	*Elamite*	*Numbers of entries*
**b*	*b*	2, 16, 22, 26, 42, 43, 51, 61, 84, 85, 110
	p	14, (26), 39, 41, 44, 47, 57, 71, 89, 91
**p*	*p*	6, 7, 27, 58, 63, 107
**d*	*d*	52, 54, 76, 86 (d/t), 87, 88
	t	6, 14, 19, 39, 53, 58
**t*	*t*	13, 74, 96, 117
**ṭ*	*t* / *-tt-*	59, (74) / 5
**ǯ*	*z*	30
**č*		
**č̣*		
**dʒ'*	*s* / *-z-*	8, 79 / 67
**c*	*z*	60 ?
**ç̣*		
**ĉ*		
**ĉ̣*	*s*	65
**s*	*s*	9, 38 (s / š), 73, 100 ?, 108
	š	11, 12, 62, 72
**š*	*sˇ*	112, 116
**ŝ*		
**g*	*g*	20, 51
	k	47, 68, 83
**k*	*k*	36, 55, 56, 80, 95, 96, 98, 99, 100, 108, 111
**ḳ*	*k* / *-kk-*	3, 4, 97 / 84
**ġ*	*h*	27, 92 ?
**ḫ*	*h*	15, 18, 28, 40, 45, 47, 89, 93, 94
**ʿ*	*h* / -Ø -	19, 46, 54, 62, 65, 67, 92 / 86
**ḥ*	*h*	9, 77, 90
**h*	*h*	33, 34, 50, 91
**ʔ*	Ø	1, 15, 17, 20, 21, 31, 40, 44, 50, 75, 78, 109
**y*		
**w*	*m*	24, 69, 81 ?
**m*	*m*	5, 23, 25, 48, 52, 55, 66, 70, 74, 75, 93, 103, 104
**n*	*n*	2, 8, 23, 26, 31, 35, 48, 78, 88, 90, 95, 98, 106, 113, 114
**l*	*l*	1, 16, 22, 24, 32, 33, 36, 37, 45, 46, 57, 59, 66, 69, 92, 97, 102
**r*	*r*	4, 7, 11, 17, 28, 34, 42, 56, 63, 68, 70, 72, 73, 85, 99, 103, 104

Note: The borrowed cuneiform orthography does not differentiate voiced and voiceless stops (cf. Hittite).

117. El(m) *-t* 'ending of 2sg of verbal and nominal conjugation'
AA **ti* and **ta* 'pronoun and ending of 2sg' (see Table 2.3)
Dr **-t* 'ending of 2sg of verbal conjugation'.

118. El(o) *u*, (A) *hu* 'I, me' (HK 676, 1195)
AA **[ʔ]yu* 'personal pronoun of 1sg' > Sem (Akk and Eblaic) **y[iw]ā-si/ti* 'dat./acc. of indirect case of a pronoun of 1sg' // Eg *□w*, (later) *wy* 'I' (dependent series) // Cush **yi/*yu* 'object case of a pronoun of 1sg' // Chadic: (W) Hausa *-wa* 'possessive pronoun of 1sg'; (C) Kotoko **nta-wu* (indep.), **[ʔ]wu* (obj.), **-wu* (poss.), Gidar *-wu* (poss.), Musgu **-u* id.; (E) Sokoro *-u*, Mokilko *-o* id. // Berb **íw* 'pronoun of 1sg of indirect object (simple)', **ū/w* (compound)
Dr: Brahui *î* 'I' and/or *-v* 'verbal ending of 1sg'.

ACKNOWLEDGEMENT

I had a chance to study the Elamite language and its relatives thanks to a grant from the Alexander von Humboldt Foundation during my stay in Bonn and Cologne in 1993–94.

REFERENCES

Within the appendices, some authors whose works are referred to repeatedly have had their names – or the name of their work – abbreviated for ease of reference as follows:

CO	Cohen 1947
DEDR	Burrow and Emeneau 1934
Do	Dolgopolsky
EG	Erman and Grapow 1971
Eh	Ehret
Fa	Faulkner 1981
He	Heine
HK	Hinz and Koch
IK	Hübner and Reizammer 1985
Ls	Leslau
MA	McAlpin 1981
Sa	Sasse
St	Stolbova 1987
Vy	Vycichl
Wo	Woelfel 1965

In addition, the following abbreviations have been used to refer to some publications:

AAL *Afroasiatic Linguistics*
AHw *Akkadisches Handwörterbuch, I-III,* W. von Soden (ed.). Wiesbaden: Harrassowitz 1965–81.
AION *Annali di Istituto orientale di Napoli.*
IF *Indogermanische Forschungen.*
JAOS *Journal of the American Oriental Society*
JNES *Journal of Near East Studies*
LRDIV *Lingvističeskaja rekonstrukcija i drevnejšaja istorija Vostoka.* Moscow: Institut vostokovedenija/Nauka.
MeE *Mesopotamie et Elam. Actes de la XXXVIème Rencontre assyriologique internationale* (Ghent 1989). University of Ghent 1991.
RHA *Revue hittite et asiatique.*
ZÄ *Zeitschrift für ägyptische Sprache und Altertumskunde.*
ZDMG *Zeitschrift für Deutschen Morgenländischen Gesellschaft.*
ZES *Zeitschrift für Eingeborenen Sprachen.*

Aistleitner, J. 1965. Wörterbuch der ugaritischen Sprache. Berlin: Akademie-Verlag.

Al'bedil', M. F. 1986. Tipy prootoindijskix nadpisej. In *Drevnie sistemy pis'ma. Étničeskaja semiotika,* J.V. Knorozov (ed.), 36–68. Moscow: Nauka.

Albright, W.F. 1927. Notes on Egypto-Semitic Etymology, III. *Journal of the American Oriental Society* 47, 198–237.

Albright, W.F. and T.O. Lambdin 1970. The evidence of language. In *The Cambridge Ancient History, Vol. I, Part 1,* 122–155. Cambridge: Cambridge University Press.

Andronov, M.S. (1980) *The Brahui Language.* Moscow: Nauka.

Banti, G. 1993. Ancora sull'origine del cammello nel corno d'Africa: osservazioni di un linguista. In *Ethnos, lingua e cultura,* A. Belardi (ed.), 183–223. Rome: Calamo.

Belova, A.G. 1989. *Isoglosses yéménites-couchitiques.* Paper presented at II International Congress on Cushitic and Omotic Languages. Turin (Nov 1989).

Black, P. 1974. Lowland East Cushitic: subgrouping and reconstruction. Unpublished Ph.D. dissertation, Yale University.

Blažek, V. 1989. *Omotic lexicon in Afroasiatic perspective: body parts.* Paper presented at II International symposium on Cushitic and Omotic languages. Turin (Nov 1989).

Blažek V. 1992a. The new Dravidian-Afroasiatic lexical parallels. In *Nostratic, Sino-Caucasian, Austric and Amerind,* V. Shevoroshkin (ed.), 150–165. Bochum: Brockmayer.

Blažek V. 1992b. (Review article of) Ruhlen, M. 1987. *A Guide to the World's Languages.* Stanford University Press. *Asian and African Studies* 1, 76–92.

Blažek V. 1994. Elephant, hippopotamus and others: some ecological aspects of the Afroasiatic homeland. *Asian and African Studies* 3, 196–212.

Blažek V. 1995. The microsystems of personal pronouns in Chadic, compared with Afroasiatic. In *Studia Chadica et Hamito-Semitica (Frankfurt, May 1991),* D. Ibriszimow and R. Leger (eds), 36–57. Cologne: Köppe.

Blažek, V. and Bengtson J. 1995. Lexica Dene-Caucasica. *Central Asiatic Journal* 39/1, 11–50 and 39/2, 161–164 .

Blažek, V. and Boisson C. 1992. The diffusion of agricultural terms from Mesopotamia. *Archív orientální* 60, 16–37.

Boisson, C. 1989. Sumerian/Nostratic/Sino-Caucasian isoglosses. Unpublished ms.

Brentjes, B. 1973. *Die Erfindung des Haustieres.* Leipzig, Jena and Berlin: Urania.

Burrow, T. and M.B. Emeneau 1984. *A Dravidian Etymological Dictionary.* Oxford: Clarendon Press.

Cohen, D. 1970, 1976. *Dictionnaire des racines sémitiques, Vols I, II.* Paris-La Haye: Mouton.

Cohen, M. 1947. *Essai comparatif sur le vocabulaire et la phonétique du chamito-sémitique.* Paris: Champion.
Diakonoff, I.M. 1981. Earliest Semites in Asia. Agriculture and animal husbandry according to linguistic data (VIIIth-IVth millennia BC). *Altorientalistische Forschungen* 8, 23–74.
Diakonoff, I.M. 1985: Elam & Media. In *The Cambridge History of Iran*, Vol. 2: The Median and Achaemenian Periods, ed. I. Gershevitch. Cambridge: University Press, T-24, 36–148.
Diakonoff, I.M. 1991. Afrazijskie jazyki. In *Jazyki Azii i Afriki IV, 1: Semitskie jazyki*, (ed. volume), 5–69. Moscow: Nauka.
Dolgopolsky, A. 1973. *Sravnitel'no-istoričeskaja fonetika kušitskix jazykov.* Moscow: Nauka.
Dolgopolsky, A. 1983. Semitic and Cushitic sound correspondences and cognate sets. In *Ethiopic Studies*, S. Segert and A. Bodrogligeti (eds), pp. 123–142. Wiesbaden: Harrassowitz.
Edzard, D.O., G. Farber and E. Sollberger 1977. Die orts- und Gewässernamen der präsargonischen und sargonischen Zeit. Wiesbaden: Reichert.
Ehret, C. 1980. *The Historical Reconstruction of Southern Cushitic Phonology and Vocabulary.* Berlin: Reimer.
Ehret, C. 1987. Proto-Cushitic reconstruction. *Sprache und Geschichte in Afrika* 8, 7–180.
Ehret, C. 1995. *Reconstructing Proto-Afroasiatic: consonants, vowels, tone and vocabulary.* Berkeley: University of California Press.
Ember, A. 1913. Kindred Semito-Egyptian words. *Zeitschrift für Ägyptische Sprache und Altertumskunde* 51, 110–121.
Ember, A. 1917. Kindred Semito-Egyptian words (cont.). *Zeitschrift für Ägyptische Sprache und Altertumskunde* 53, 83–90.
Ember A. 1930. *Egypto-Semitic Studies.* Leipzig: Asia Major Verlag.
Erman, A. and W. Grapow 1971. *Wörterbuch der ägyptischen Sprache, Vols I-VI.* Berlin: Akademie-Verlag.
Fähnrich, H. 1981. Das Sumerisch und die Kartwelsprachen. *Georgica* 4, 89–101.
Fairsevis, W.A. 1992. *The Harappan Civilization and its Writing: a model for the decipherment of the Indus script.* Oxford: IBH Publishing Co.
Faulkner, R.O. 1981. *A Concise Dictionary of Middle Egyptian.* Oxford: Griffith Institute.
Gordon, C.H. 1965. *Ugaritic Textbook.* Rome: Pontificium Institutum Biblicum.
Greenberg, J.H. 1963. *The Languages of Africa.* Bloomington: Indiana University Press.
Grillot-Susini, F. 1987. *Éléments de grammaire élamite.* Paris: Recherche sur les Civilisations.
Hallock, R.T. 1969. *Persepolis Fortification Tablets.* Chicago: Oriental Institute Pubs. 92.
Heine, B. 1978. The Sam languages. The history of Rendille, Boni and Somali. *AAL* 6/2, 23–115.
Heine, B. 1982. *Boni Dialects: the non-Bantu dialects of Kenya X.* Berlin: Reimer.
Hiebert, F.T. and C.C. Lamberg-Karlovsky 1992. Central Asia and the Indo-Iranian Borderlands. *Iran* 30, 1–16.
Hinz, W. 1964. *Das Reich Elam.* Stuttgart: Kohlhammer.
Hinz, W. 1969. *Altiranische Funde und Forschungen.* Berlin: de Gruyter.
Hinz, W. and H. Koch *Elamisches Wörterbuch.* Berlin: Reimer.
Höyrup, J. 1992 [94]. Sumerian: The descendant of a proto-historical creole? AIΩN. Annali del Dipartimento di Studi del Mondo Classico e del Mediterraneo Antico. *Sez. ling.* 14, 21–72.
Hübner, B. and A. Reizammer 1985. *INIM KIENGI II. Sumerisch Deutsches Glossar.* Marktredwitz: Selbstverlag.
Hudson, G. 1989. *Highland East Cushitic Dictionary.* Hamburg: Buske.
Hüsing, G. 1916. Völkerschichten in Iran. *Mitteilungen der anthropologischen Gesellschaft in Wien* 44, 199–250.

Illič-Svityč, V.M. 1971, 1976, 1984. *Opyt sravnenija nostratičeskix jazykov, I–III.* Moscow: Nauka.

Klein, E. 1987. *A Comparative Etymological Dictionary of the Hebrew Language.* New York and London: McMillan.

König, F.W. 1965. *Die elamischen Königinschriften.* Graz: Archiv für Orientforschung, Beiheft 16.

Kovalev, A.A. and A. Ju. Militarev 1994. Sumerians and Semites: an encounter of the equipotential cultures. Paper presented at 6th International Congress on Hamito-Semitic Languages. Moscow April 1994.

Lamberg-Karlovsky, C.C. 1986a. Third millennium structure and process: from the Euphrates to the Indus and from the Oxus to the Indian ocean. *Oriens Antiquus* 25, 189–219.

Lamberg-Karlovsky, C.C. 1986b. Further traces on the earliest history of the Iranian Plateau, In *Drevnie civilizacii Vostoka. Materialy II Sovetsko-amerikanskogo simpoziuma.* B.M. Masson (ed.), 166–171. Taškent: FAN.

Laroche, M. 1977. Glossaire de langue Houmite. *Revue hittite et asianique* 35, 163–323.

Leslau, W. 1938. *Lexique soqotri.* Paris: Klincksieck.

Leslau, W. 1987. *Comparative Dictionary of Ge'ez (Classical Ethiopic).* Wiesbaden: Harrassowitz.

Lukas, J. 1970. *Studien zur Sprache der Gisiga.* Glückstadt: J.J. Augustin.

McAlpin, D.W. 1981. *Proto-Elamo-Dravidian: the evidence and its implications.* Philadelphia: Transactions of American Philosophical Society 71/3.

Meriggi, P. 1969a. Zur Lesung und Deutung der proto-elamischen Königinschriften. *Bibliotheca Orientalis* 26, 176–177.

Meriggi, P. 1969b. Altsumerische und Proto-elamische Bilderschrift. In *XVII. Deutscher Orientalistentag (Würzburg, Juli 1968).* (*ZDMG*, Suppl. I), 156–163. Wiesbaden: Steiner.

Meriggi, P. 1971. *La scrittura proto-elamica, I: La scrittura e il contenuto dei testi.* Rome: Accademia Nazionale dei Lincei.

Meriggi, P. 1977. Iscrizioni proto-elamiche dell'Iran orientale. *Kadmos* 16, 1–4.

Militarev, A. Ju. 1984a. Afrasian–Sumerian parallels. Unpublished ms. [A shortened Russian version 'Afrazijsko-šumerskie leksičeskie paralleli' was published in *LRDIV* 1, 58–61]

Militarev, A. Ju. 1984b. Sovremennye sravnitel'no-istoričeskoe afrazijskoe jazykoznanie: čto ono možet dat' istoričeskoj nauke? *LRDIV* 3, 3–26.

Militarev, A. Ju. 1989. *Asian home for Cushitic?* Paper presented at II International Congress on Cushitic and Omotic Languages. Turin, Nov. 1989.

Militarev, A. Ju. and V.A. Šnirel'man 1984. K probleme lokalizacii drevnejšix afrazijcev (opyt lingvoarxeologičeskoj rekonstrukcii). *LRDIV* 2, 35–53.

Militarev, A. Ju. and S.A. Starostin 1984. Obščaja afrazijsko- severnokavkazskaja kul'turnaja leksika. *LRDIV* 3, 34–43.

Militarev, A. Ju. and S.A. Starostin 1994. *Names of body parts in Afro-Asiatic and Sino-Caucasian.* Paper presented at 6th International Congress on Hamito-Semitic Languages, Moscow, April 1994.

Militarev, A. Ju., I.I. Pejros and V.A. Šnirel'man 1988. Metodičeskie problemy lingvoarxeologičeskix rekonstrukcij ftnogeneza (1. Problema proisxoždenija afrazijcev. 2. Nekotorye problemy ftnogeneza dravidov). *Sovetskaja ėftnografija* 1988/4, 24–38.

Oates, J. 1991. Babylonia and Elam in Prehistory. *MeE* 23–26.

Orel, V.E. and O.V. Stolbova 1995. *Hamito-Semitic Etymological Dictionary.* Leiden: E.J. Brill.

Parpola, A. 1994. *Deciphering the Indus Script.* Cambridge: Cambridge University Press.

Pokorny, J. 1959. *Indogermanisches etymologisches Wörterbuch.* Bern and München: Francke.

Prasse, K.-G. 1972–4. *Manuel de grammaire touarege (tăhăggart) Vols I-III, IV-V, VI-VII.* Copenhagen: Akademisk Forlag.

Puhvel, J. 1991. *Hittite Etymological Dictionary, Vol. 3.* Berlin and New York: Mouton de Gruyter.
Puhvel, J. 1996. Finger in Greek, Latin and Hittite. *Indogermanische Forschungen* 81, 25.
Rashidi, R. 1983. The Kushite origins of Sumer and Elam. *Ufahamu* 12/3, 215–233.
Sasse, H.-J. 1979. The consonant phonemes of Proto-East-Cushitic (PEC): a first approximation. *Afroasiatic Linguistics* 7/1, 1–67.
Sasse, H.-J. 1982. *An Etymological Dictionary of Burji.* Hamburg: Buske.
Schmandt-Besserat, D. 1986. Tokens at Susa. *Oriens Antiquus* 25, 93–117.
Schretter, M. 1990. *Emesal-Studien.* Innsbruck: IBKW 69.
Segert, S. 1984. *A Basic Grammar of the Ugaritic Language.* Berkeley and London: University of California Press.
Selz, G.J. 1991. 'Elam' und 'Sumer' – Skizze einer Nachbarschaft nach inschriftlichen Quellen der vorsargonischen Zeit. *MeE* 27–43.
Skinner, N. 1977. The Northern Bauchi Chadic languages. *AAL* 4/1, 1–49.
Steve, M.-J. 1992. *Syllabaire elamite. Histoire et paleographie.* Neuchatel and Paris: Recherches et Publications.
Stolbova, O.V. 1987. Sravnitel'no-istoričeskaja fonetika i slovar'. In *Afrikanskoe istoričeskoe jazykoznanie.* 30–268. Moscow: Nauka.
Tosi, M. 1986. The archaeology of early states in Middle Asia. *Oriens Antiquus* 25, 153–187.
Trokay, M. 1991. Les origines du dieu elamite au serpent. *MeE* 153–161.
Vajman, A.A. 1972. O svjazi protoėflamskoj pis'mennosti s protošumerskoj. *Vestnik drevnej istorii* 1972, 124–133.
Vallat, F. *et al.* 1993. *Les noms géographiques des sources suz-élamites.* Wiesbaden: Reichert.
Vycichl, W. 1934. Hausa und Ägyptisch. *Mitteilungen des Seminars für Orientalische Sprachen* 37/3, 36–116.
Vycichl, W. 1961. Diminutiv und Augmentativ im Berberischen. *ZDMG* 111, 243–253.
Vycichl, W. 1983. *Dictionnaire étymologique de la langue copte.* Leuven and Paris: Peeters.
Ward, W.A. 1961. Comparative studies in Egyptian and Ugaritic. *Journal of Near Eastern Studies* 20/1, 31–40.
Woelfel, D. 1965. *Monumenta Linguae Canariae.* Graz: Akademische Druck.
Zarins, J. 1992. The early settlement of southern Mesopotamia: a review of recent historical, geological, and archaeological research. *JAOS* 112, 55–77.
Zvelebil, K.V. 1985. Review of McAlpin 1981. *JAOS* 105, 364–372.
Zvelebil, K.V. 1985a: The body in Nilgiri tribal languages. A contribution to areal linguistic studies. *JAOS* 105, 653–674.
Zvelebil, K.V. 1990. *Dravidian Linguistics: an introduction.* Pondicherry: Institute of Linguistics and Culture.

3 *Language diversification in the Akoko area of Western Nigeria*

CHINYERE OHIRI-ANICHE

ABBREVIATIONS

AAP	Afrikanistische Arbeitspapiere
CR	Cross River
C	Consonant
JOLAN	Journal of the Linguistic Association of Nigeria
LGA	Local Government Area
PBC	Proto-Benue-Congo
PIYE	Proto-lgboid-Yoruboid-Edoid
PYIG	Proto-Yoruba-Itsẹkiri-Igala
PYOR	Proto-Yoruba
STD	Standard
SY	Standard Yoruba
V-	a mid-back unrounded vowel (in Arigidi)
u_	a high front rounded vowel (in Arigidi)
!	downstep
'C	a lenis consonant
V	Vowel

INTRODUCTION

Akoko area dialects and languages

Since the late 1960s, scholars have drawn attention to the existence of enclaves of non-Yoruba languages amidst dominant Yoruba speakers in the Akoko area presently falling within Ondo State, one of the five western Nigerian states (Williamson 1970; Okẹ 1972). In this area, known as Àkókó, Ondo State today shares borders with Ẹdọ and Kogi States, whilst its population of 400,000 (1991 estimate) forms part of the total population of about 3.9 million of Ondo State (Adalemo and Baba 1993). Following subsequent researches, a consensus has emerged that the speech forms (or lects) of the Ondo State Akoko area fall into five main groups (see Figure 3.1).

THE YORUBA GROUP

Monolingual Yoruba speakers form the majority of the population of the Akoko area. These speakers live in the towns and communities of Ukare (Ikare), Irun, Ọ̀kà, Orà, Ifira, Afọ, Idoani (Amusigbo and Isure quarters), Imeri, Ikun, Ugbe (Igbe), Boropa (Ìbòròkpá), Supare (Ʃúkpáre), Ipesi (Ìkpèsì), Àkùngbá, Ọ́bà, etc. (Williamson 1970; Okẹ 1970; Akinkugbe 1978; Capo 1989; Ohiri-Aniche 1995).

THE EDOID GROUP

This group comprises such lects as Ùkpè (Ipe), Ekpinmi (Epinmi), the Iyayu quarters of Ido-Ani, Isua, etc. (Williamson 1970; Okẹ 1970; Elugbe 1973, 1989).

THE UKAAN GROUP

Ukaan is a dialect cluster spoken in such towns as Ìkákụ̀mọ́-kéji (formerly Ikakụmọ-Àwòrò, Àúga and Ìshẹ̀ on the Ondo State side of the Akoko border. Ukaan is also spoken in two other towns, Kákùmọ́-Akoko and Àŋyàráŋ on the Edo side of the border (Jungraithmayr 1973; Williamson 1989; Abiodun 1989). Another Ukaan-speaking village, Oreju (Òrèdʒù), was sited in the present Kogi State but has now been lost 'since the war', presumably the Biafran war. As previous writers have mentioned, the term Ukaan is not acceptable to all these towns. It will, however, be used in this study, until a more acceptable term emerges.

THE AKPES GROUP

Akpes is a dialect cluster spoken in the towns of Akunnu, Asẹ, Ikaram, Ibaram, Iyani, Gedegede, Daja and Esuku (Ibrahim-Arirabiyi 1989a; Williamson 1989; Crozier and Blench 1992). Again, the term Akpes to describe all the speakers is not acceptable to some, but is retained here pending the emergence of a more acceptable one.

THE AKOKOID GROUP

This is made up of such lects as Arigidi, Oyin, Uro (Iro), Igasi, Erusu, Ahan and Oke-Agbe, the latter comprising four subdialects – Ùdò (Ìdò), Oge, Afa and Aje (ẹʃẹ). Two languages, Ayere and Ahan, are related to one another and may also be related to Akokoid, although information on these two communities is presently too meagre to make any definitive statement (Williamson 1989; Hoffmann 1974; Akinkugbe 1978; Capo 1989; Crozier and Blench 1992).

All these language groups are generally considered to belong to Benue-Congo. On the detailed affiliations of these groups, those of Yoruba and Edoid are well established, whilst Akokoid remains controversial; none is yet

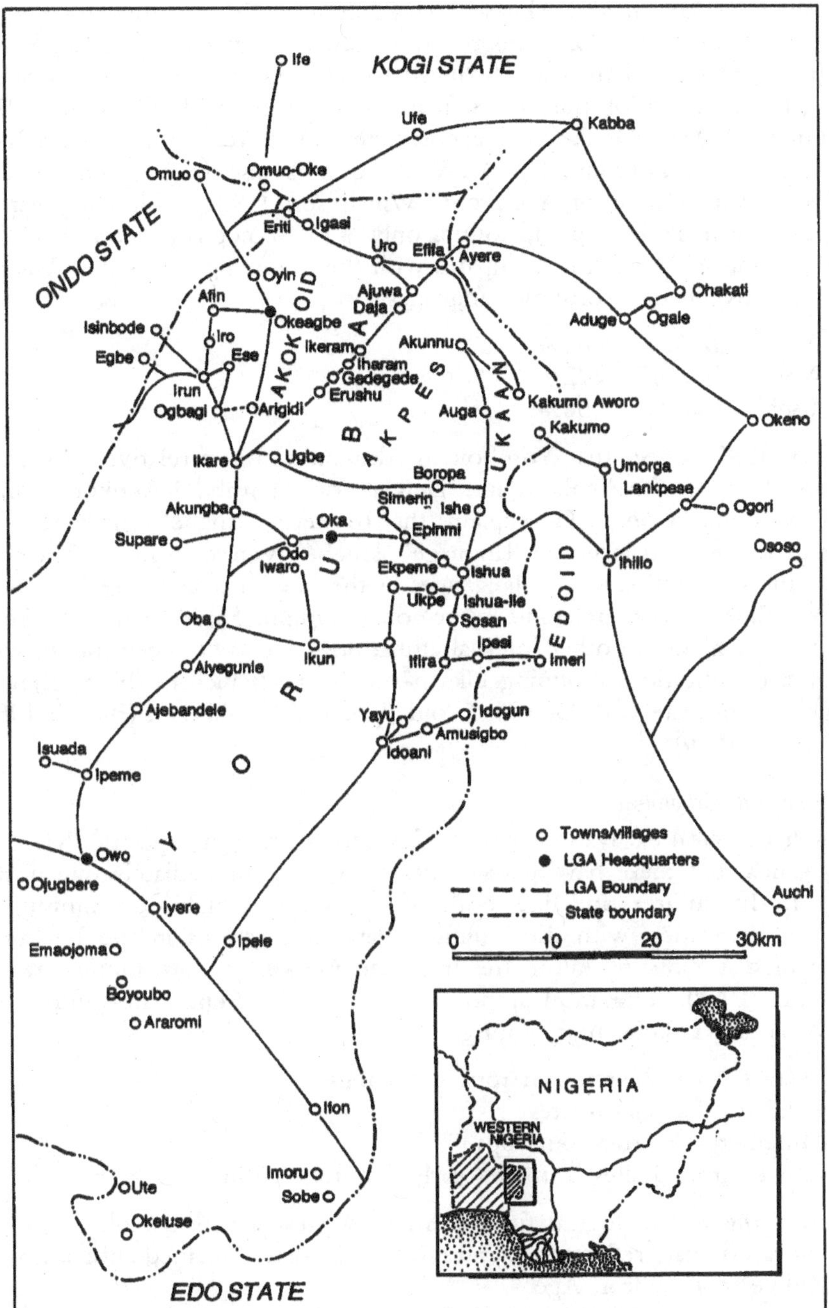

Figure 3.1 Speech forms (or lects) of the Ondo State, Akoko area
Source: Ogundele and Okoro

established for Ukaan and Akpes. The Edoid lects are classified by Elugbe (1973, 1989) under the Ossẹ subgroup of northwestern Edoid languages. For Yoruba, its closest relatives are believed to be Itsẹkiri (now in Delta State) and Igala, southeast of the confluence (Williamson 1973; Omamor 1976; Akinkugbe 1978). Akinkugbe reconstructed a Proto-Yoruba-Itsẹkiri-Igala (PYIG) or Yoruboid group. The Akokoid group has also, so far, been regarded as the nearest relative of Yoruboid. Williamson (1982), following earlier proposals from Fresco (p.c.), found only a 42% cognate score between Yoruboid and Akokoid; this compares with the following scores she obtained in 1973 between the Yoruboid languages:

Yoruba/Igala	66%
Yoruba/Itsẹkiri	72%
Itsẹkiri/Igala	56%

Capo (1989) agrees that Yoruboid is Akokoid's closest relative, proposing the term 'Ā̱mgbɛ́' for Akokoid and 'Defoid' for a Yoruboid-Akokoid branch of (New) Benue-Congo. He explains that the term 'Ā̱mgbɛ́' reflects the fact that all the Akokoid lects call 'language' 'Ā̱mgbɛ́', whilst 'Ifoid' (=Defoid) is due to the fact that almost all languages of the branch refer to Ifẹ (in central western Nigeria) as their cradle or region of origin. For Ukaan and Akpes, on the other hand, no other internal affiliations have as yet been found, with the latest classifications grouping Ukaan-Akpes as a branch of (New) Benue-Congo, co-ordinate with Defoid, Edoid, Igboid, Nupoid, etc. (Blench 1989; Williamson 1989).

Archaeological background

No archaeological excavations have as yet been carried out specifically in the Akoko area, but there have been some excavations immediately due north, in the Ife-Ijumu area, and due south, at Iwo-Eleru, which are promising in terms of correlation with the linguistic data. Excavations in the Ife-Ijumu area, which is about 60 kilometres from the Akoko area, are summarized in Obayemi (1982). The most important sites are at Akpaa, Ife-Ijumu, Ife-Olukotun and Lokoja, representing:

1 Caves, rock-shelters and rock overhangs
2 Hill-top habitation sites
3 Inselberg-like rock outcrops
4 Level ground sites from the early historical period.

Although the archaeological finds seem to suggest a considerable antiquity, no radiocarbon date reflects this, and most of the dates obtained cluster around the Middle to Late Iron Age.

An excavation at Itaakpa, which is also in the immediate vicinity of Ife-Ijumu, has produced a slightly older date of 2210 ± 80 BP on a burnt palm kernel (Allsworth-Jones and Oyelaran 1991). The material culture inventory suggests that this may be representative of the ceramic phase of the Late

Stone Age in Nigeria. Nonetheless, these are far more recent than the sixth-millennium BC dates reported for Dutsen Kongba near Jos (York 1978).

Excavations at Iwo Eleru, some 120 kilometres from the Akoko area and 60 kilometres south of the northern forest boundary, have produced considerably greater time-depths (Shaw 1980; Shaw and Daniels 1984). Iwo Eleru (literally 'cave of ashes' in Yoruba) is described as a large rock-shelter with good evidence of late stone age occupation. Finds at the excavation site include microliths, ground stone axes and the remains of an ancient hominid. Carbon dates obtained at Iwo Eleru go back to 11,200 ± 200 BP, or 9250 BC. This has generated much interest since it is the first time that evidence for a stone age occupation has been found in the present forest belt in Nigeria (see, however, MacDonald, Volume II, for evidence of considerably greater time-depths).

Oral tradition

Oral traditions were not collected systematically, but in the course of this study, narratives of the existence of caves, and of underground tunnels that served as dwelling places in times past and as hiding places during wars, such as the lost Ikakụmọ village of Oreju, were frequently mentioned. Ikakụmọ speakers insisted that they were of Edoid stock, adding that during their most famous festival, the Ikaya festival, the person possessed by the ẹ̀gbá spirit of fire went into a trance and spoke in a tongue that is not Ikakụmọ, but which is still spoken in a village situated in Edo State. A Yoruba oral tradition mentions the ruins of Áhóró Iká, said to lie about 64 kilometres northwest of present-day Ọyọ́ town. The word áhóró means 'a desolate place' in Ikakụmọ.

THIS STUDY

Objectives and methods

This study re-examines the existing subgroupings and classifications of some lects of the Akoko area, with a view to exploring how they could be correlated with archaeological data. Four lects, Ọkà (Yoruba), Arigidi (Akokoid), Akunnu (Akpes) and Ikakụmọ (Ukaan), are compared with Standard Yoruba (SY), the dominant language in western Nigeria, and with neighbouring languages such as Ẹdo, Igbo and Nupe. In addition, Ikakụmọ, which is a noun-class language, is compared to two other Edoid noun-class languages, Degema and Ibilo, whilst data from some Kainji-Platoid languages, spoken north of this area, is also considered. The findings from the linguistic comparisons are then briefly compared with oral traditions regarding the origins and past migrations of the groups.

Results: lexicostatistics

The Swadesh 100 wordlist, as modified by the University of Ibadan, was used to calculate cognate scores, the results of which are displayed in Table 3.1.

Table 3.1 Cognate scores

Std Yoruba	*Ẹdo*	*Arigidi*	*Igbo*	*Akunnu*	*Ikakụmọ*	*Nupe*	
95	50	50	44	37	36	23	Ọkà
	56	55	51	35	36	24	Std Yoruba
		41	50	40	34	24	Ẹdo
			45	36	34	24	Arigidi
				32	31	25	Igbo
					45	20	Akunnu
						19	Ikakụmọ

The cognate scores relating to Standard Yoruba (SY)/Ẹdo, SY/Igbo and Ẹdo/Igbo are higher than those obtained in earlier studies involving these languages, such as Schadeberg (1986) and Williamson (1988). The higher scores arise partly from the fact that reconstructed material (from Ohiri-Aniche 1991) has been used, which tends to reveal some hitherto hard to recognize cognates. Another reason has to do with an inherent weakness of lexicostatistics which is the subjectivity in deciding what to count as cognates.

SOME INFERENCES ON LANGUAGE AFFILIATIONS

Ọkà (Yoruba)

A 95% cognate score between Ọkà and Standard Yoruba (SY) confirms the former is a dialect of Yoruba. Ọkà may be closer to Proto-Yoruboid than any of the dialects used in Akinkugbe's 1978 reconstruction of Proto-Yoruba (Ohiri-Aniche 1995). Ọkà has generally maintained consonants of stronger stricture than those of other dialects. For instance, Ọkà has phonemic dental plosives t̪, d̪ which correspond to affricates and fricatives in other dialects; in some words, also, the voiced palatal affricate /dʒ/ and the voiced labialized velar /gʷ/ of Ọkà correspond to the palatal and labialized velar approximants y, w respectively in other dialects (see Table 3.2).

Ohiri-Aniche (1995) hypothesized that Ọkà's dental plosives t̪, d̪ developed from earlier PYOR palatal plosives *c,*ɟ, which became fronted in Ọkà whilst weakening to affricates in other dialects. If this is so, then Ọkà is one of the primary lects to have branched out from Proto-Yoruboid.

Table 3.2 Yoruboid correspondences

PYOR	*SY*	*Ọkà*	*Gloss*
*tʃe	ʃe	t̪e	'do, make'
*dʒó	dʒó	d̪ó	'dance'
*yá	yá	dʒá	'hasten'
*àwú	ɛ̀wù	àgwù	'gown, robe'

In addition, Ọkà does not share some lexical innovations reported for PYOR. An example is PYOR *ɔ̀-cụ̀kpá 'moon', which compounds the Proto-Yoruba-Itsẹkiri-Igala (PYIG) stem for 'month' *ocụ̀ with -kpá; Ọkà has /otụ̀/ for both 'moon' and 'month', just as Igala has /otʃụ̀/ for both words. Another example is PYOR *bì 'vomit', which Ọkà does not share; rather, Ọkà's ʃị 'vomit' is a preservation of the original PYIG stem *ʃɛ̀, also reflected in Itsẹkiri /ʃɛ̀/ and Igala /rɛ̀.

Some earlier writers have suggested that central and southeastern Yoruba dialects, such as Ifẹ, Ijesha and Ekiti, are closer to PYOR than dialects of other Yoruba areas (Armstrong 1964; Adetugbọ 1973; Oyelaran 1977). This chapter suggests that the Ọkà dialect is even closer to PYOR, and that a further study of some other Yoruba dialects of the Akoko area might also shed more light on the nature of Proto-Yoruba.

Arigidi (Akokoid)

Although Arigidi has the highest cognate score with SY (55%), this is far short of the range of between 70% and 86% that has been variously suggested by different authors for the recognition of dialects of the same language (Swadesh 1955; Williamson 1982, 1988; Crozier and Blench 1992). This study, therefore, upholds earlier findings that recognize the Akokoid languages as distinct from Yoruba (Williamson 1982, 1989; Capo 1989). On the basis of lexicostatistics, Akokoid comes out almost equidistant from Yoruboid, Edoid and Igboid; in other words, there is no marked relationship with Yoruboid. The SY/Arigidi score is marginally less than the SY/Ẹdo score of 56%, whilst the SY/Igbo score of 51% is not far behind. It is significant that after these scores, there is a sharp drop to the 30s level in scores between SY/Ẹdo /Arigidi/Igbo and Ukaan-Akpes, and another drop to the 20s level in scores between Nupe and the other languages. These scores suggest that the languages compared fall into three different subgroups. It is, therefore, proposed that a Yoruboid-Edoid-Akokoid-Igboid (YEAI) branch be recognized. This position is close to that expressed by Blench (1993, 1997), who posits that (YEAI) belong together in a branch that is then co-ordinate with each of the other separate branches of (new) Benue-Congo, namely Ọkà (Ogori), Nupoid, Cross River, Kainji-Platoid and Bantoid.

After Yoruba, the next closest relative of Arigidi is Igbo, with which it scores 45%, followed by Ẹdo, with which it scores 41%. The Arigidi/Igbo closeness is also reflected in shared lexical items, which are not found in SY or in Ẹdo; in some other lexical items that occur also in SY and Ẹdo, the Igbo forms appear most similar to the Arigidi ones. These are shown in Table 3.3.

Ikakụmọ (Ukaan)

Among the languages compared, Ikakụmọ scores highest (45%) with Akunnu (Akpes). Whilst this makes them the closest relatives of each other, the score is too low for them to be considered one language, as suggested in Ibrahim-Arirabiyi (n.d). Another argument, though weaker, for considering them as

Table 3.3 Lexical items suggesting Arigidi/Igbo closeness

Item	*PIYE*	*Ẹdo*	*SY*	*Arigidi*	*Igbo (Owere)*
kolanut		(ɛ̀ʊɛ́ɛ̀)	(obì)	éʃó	ɔ́dʒí
poison/medicine				ńʃí	ńʃí
say				hɛ̰	hʊ́ (archaic)
oil				òògò	ɔ̀gà (palm oil)
war				oolo	lʊ̀ (verb)
child	*V-mʊa	ɔmɔ̰	ɔmɔ̰	ʊ́'ɳʷɔ̰́	ɳ́ɳʷá̰
tongue	*CV-'də'mi	ala̰ʊɛ̰	(ahɔ̰́/awɔ̰́)	ɛ́rɛ́	íré
vagina	*V-ʃɪ̰ʊ̰	ùhé	(òbò)	ɛ́h:ɔ̰	ɔ́hʊ̰́
an elder	*V-dḭə̰	ɔdyɔ̰́	(àgbà̰	ùdá̰	ǹdá'á
black	*cɪ̰ʊ̰	xwixwi	du	dʒḭ́	dʒḭ́
walk	*cɪ̰V̰	xḭa̰	(rḭ̀)	dʒɛ̰	dʒḛ́
fish	*V-cə'ni	éhɛ̀ɛ̀	ɛdʒa	esɔ	ázʊ̰̀
back	*V-cɪ̰ʊ̰	(ìyèke)	(ɛ̀yḭ̀/ɛ̀hḭ̀)	ɔsa̰	àʒʊ̰́
pepper	*V-tḭə̰̀	e-hḭɛ̰	ata	ɛɛsɛ	ósḛ̀
sand	*'dɪ-'cɪ̰V	exae	ìyɔ̰̀rḭ̀	isa	á'dʒá̰

different languages is that Ikakụmọ is a full noun-class language with both prefixed nominal class morphemes and a well-established concord system (Abiodun 1989). Akunnu, on the other hand, has only a residual class system involving the use of prefix alternations to indicate singular/plural in the person category. Since both Abiodun and the Ikakụmọ informant mention Edoid links, the language was also compared to two Edoid noun-class languages, Degema and Ibilo, with the following results:

Ikakụmọ/Ibilo	32%
Ikakụmọ/Degema	32%

These scores are hardly different from those between Ikakụmọ/SY (36%), Ikakụmọ/Ẹdo 34%, Ikakụmọ/Arigidi 37% and Ikakụmọ/Igbo 31%. This suggests that Ikakụmọ is as related to Edoid as it is to Yoruboid, Akokoid and Igboid.

A look further afield to other noun-class languages in Nigeria, however, yielded some interesting vocabulary agreements between Ikakụmọ and Kainji-Platoid languages, notably with Basa-Benue (formerly Basa-Komo). The Kainji-Platoid forms were extracted from Williamson and Shimizu (1968, 1973) and from Kropp Dakubu (1980) (see Table 3.4). For the most part, the words in the table represent retentions from Benue-Congo or even from Niger-Congo. It is, however, significant that, as far as is known, these items do not occur in languages of western Benue-Congo as outlined in Blench (1993). Blench suggests a two-way division of Benue-Congo, with the western half comprising Ọkọ, Nupoid, Idomoid, Yoruboid-Akokoid, Edoid, Igboid and Akpes-Ukaan; the eastern half, on the other hand, would comprise Platoid, Kainji, Bantoid and Cross River. The striking lexical agreements mentioned above prompt the suggestion that Ukaan-Akpes might represent an important link between the western and eastern halves of Benue-Congo.

Table 3.4 Lexical items suggesting Ukaan/Kainji-Platoid closeness

Gloss	*Ikakụmọ (Ukaan)*	*Kainji-Platoid*		*cf: others*	
1. six	rhàdá	Rindre	andra	Bantu	taɳda-tu
2. ten	òpú	Basa	ópo:a	Tiv	pue
3. belly	ìmí/à-	Basa	a:me	CR. Ukele	eme
4. black	ṛíṛí	Duka	rim	Magongo	oririm
5. dry	hɔ́hɔ́	Clela	hɔ̆	CR. Ikom	kòtʃi^
6. chest	ekuadʒi	Kahugu	katʃi/karɪʃja	Edo	kóko#du?
7. leg	ùkàị/à-	Reshe	ú-kaɳá	CR. Kohumono	kɛ́-háɳ
8. road	ɔ̀rɛ̂	Kambari Central	úrɛ́	CR. Gokana	èèrè
9. snake	ùfùrháà	Basa	óhŭa	PBC	ɳoko, i-'/i-
10. water	umɔ	Clela	mhɔ̀	CR: Bekwara	ù-mó
11. hear	kpí	Jukun	foji	Idoma	-poi
12. hunger	ému	Jukun	ambo		

Akunnu (Akpes)

Lexicostatistics suggest that, after Ikakụmọ, the next closest relative of Akunnu is Ẹdo. Again, as in the case of Ikakụmọ, a look further afield yielded some interesting lexical agreements between Akunnu and the Jukunoid sub-branch of Platoid (Table 3.5). Sources of the Jukunoid data were Williamson and Shimizu (1968, 1973) and Shimizu (1980). The observations regarding the lexical agreements between Ukaan and Kainji-Platoid are also valid for those between Akunnu (Akpes) and Jukunoid, which are mostly also retentions

Table 3.5 Lexical items suggesting Akpes/Jukunoid closeness

Items	*Akunnu(Akpes)*	*Jukunoid*		*cf: others*
1. head	ítʃṳ́	PJ	*ki (ri-/ki-)	
2. one	éètʃì	Mbarike	ń-tʃo/ń-dʒo	
		PJ	*ńdo (i-)	
3. pound	tʃói	PJ	*kim	
4. mountain	ɔ́ʃṳ́	PJ	*kùn(ri-/la-)	
5. smell/odour	ɔ́ɳtʃàì	PJ	*kùṇ (ku-/la-)	
6. beer	íńʃá	PJM	*kin	
7. walk	tʃíǹ	PJ	*kyàṇ	
8. iron	íɳkéị̃	PCJ	*kín (-u/-a)	
9. year	íyé	PJ	*gi (u-/i-)	
		Kutep	ì-yé	
10. eye	áyô	PJ	*gip (ri/-a)	
11. salt	iɳmayị̃	PJ	*ṇwa	
12. ten	ìyóf	PJ	*dub	CR Ufia dʒɔ̀f
13. fall	kú	PJ	*kò	
14. think	tam	PJ	*taṇ	
15. hear	kpáś̀	PJ	*pwog	
16. they	àbèɳ	Jukun	*abe	Igboid ɓɛ
17. rubbish heap	óʃúˈyà	Jibu	*sùɳ	Igboid (Mbieri) ǹʃúʃɔ́

rather than genuine innovations. Their occurrence, however, also reinforces the earlier suggestion that Ukaan-Akpes might be a link between western and eastern Benue-Congo. This suggestion is represented graphically in Figure 3.2.

SUMMARY AND CONCLUSIONS

By combining mass comparison and lexicostatistics, this study has shed more light on the possible genetic affiliations of the dialects and languages of the Akoko area. The Ọkà dialect retains some phonological features here proposed as characteristic of Proto-Yoruba. Akokoid (represented by Arigidi) is upheld as a group separate from Yoruboid, both independently joining Edoid and Igboid in one co-ordinate branch of (new) Benue-Congo. An Ukaan-Akpes branch (represented by Ikakụmọ and Akunnu respectively) is also upheld, but some striking vocabulary agreements with Kainji-Platoid suggest that it is an intermediate branch linking western and eastern Benue-Congo.

One of the exciting possibilities that emerges from the linguistic diversity of the Akoko area is that these residual languages represent those that remained *in situ* after the expansion of Benue-Congo. The Niger-Benue confluence area is often suggested as the probable homeland of Benue-Congo and the

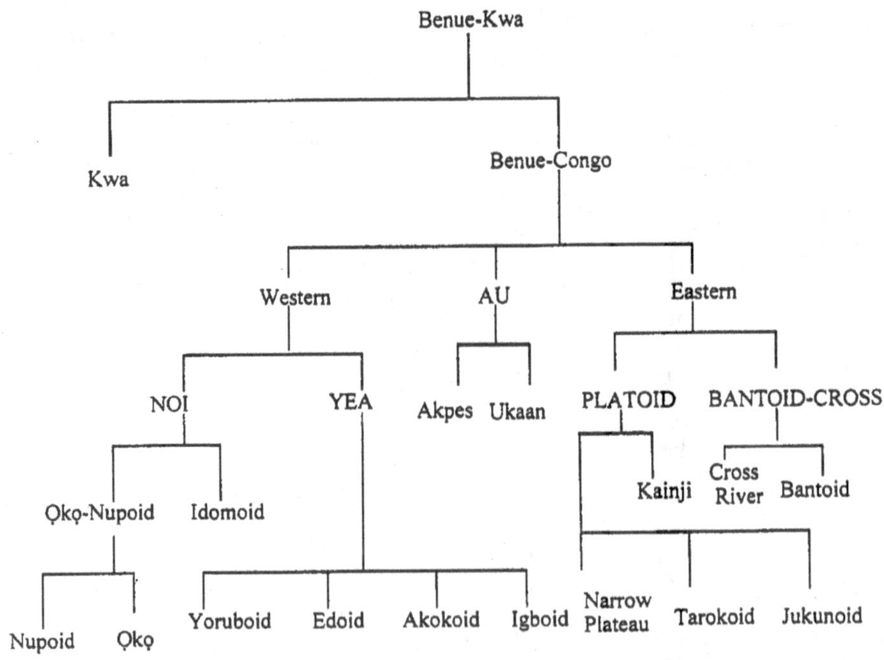

Figure 3.2 Proposed new internal classification of Benue-Congo
Source: Ohiri-Aniche

focal point from which the branch dispersed (Williamson 1989). The time of dispersal has also tentatively been put at around 6000 BP or earlier (Armstrong 1964; Horton 1995). I therefore suggest that some of the Akokoid languages have been distinct from other groups for at least 6,000 years, and most probably spoken in the same vicinity. This claim of antiquity is plausible if it is recalled that archaeological evidence from Iwo Eleru confirms that human beings have inhabited the general area for at least 11,000 years.

A date of 11,000 years would be to posit an antiquity for Benue-Congo far greater than most scholars would be willing to attribute to it. Indeed, this is earlier than Niger-Congo in most chronological schemes (MacDonald, Volume II; Blench 1997). The Iwo Eleru communities must therefore have represented earlier hunting-gathering populations, presumably the users of the microliths often reported from excavations in this area. Indeed, MacDonald (1997) has recently postulated that microlith-users may have persisted in many parts of West Africa as residual hunting-gathering populations well after the introduction of iron. The confluence area is, therefore, likely to have represented a mosaic of different populations with interlocking subsistence strategies, and this ethnolinguistic complexity has been retained up to the present.

Blench and Williamson (p.c.) hypothesize that north-central Nigeria was formerly occupied by a continuous band of Gur-Adamawa speakers. The subsequent expansion of Benue-Congo speakers outwards from the Niger-Congo confluence area divided and presumably assimilated the North Volta-Congo speakers. Those remaining in the west formed modern Gur (represented in Nigeria by Bariba) and those in the east became Adamawa-Ubangi (represented in Nigeria by such languages as Mumuye and Waja).

FUTURE RESEARCH

Much work remains to be done on the dialects and languages of the Akoko area, an immediate area of research being the fuller investigation of the Ukaan/Kainji and Akpes/Jukunoid relationship. The present survey used only sample languages from these groups, but the first requirement is a more comprehensive survey of each group. Ukaan, for example, consists of three lects, whose degree of relationship is yet to be determined. Languages such as Ayere-Ahan and the Ọkọ cluster need also to be brought more fully into the picture. Nonetheless, this preliminary survey suggests that the confluence region is crucial for understanding the evolution of the Benue-Congo languages, which include Bantu, and is therefore the single largest and most complex language family in Africa.

Archaeology and the collection of oral tradition in this region have to date been very unsystematic, and certainly not integrated with linguistic research. Although there are some indications of time-depth, actual sites in the region do not so far seem to have the antiquity suggested by the linguistic diversification. The priorities of linguists and archaeologists should therefore be:

1 collection of more complete lexical data, including cultural vocabulary;
2 more archaeological survey and excavation of stratified sites;
3 systematic recording of oral traditions.

ACKNOWLEDGEMENTS

My special thanks go to the following speakers from whom the wordlists used in this chapter have been collected:

Ọkà	Mr S.A. Aiyeyẹmi of Ubaka village, Ọkà
Arigidi	Mr S.O. Okunola of Ìlèkpá quarters, Arigidi
Akunnu	Mrs C.O. Pelemo of S'ẹkẹ quarters, Akunnu
Ikakụmọ	Mr Hussein Adjoto

Another Ikakụmọ list was collected from Mr Augustine Kolade, whilst Prince George Niyi proved extremely helpful in providing some background information on Ikakụmọ. I also received great co-operation from my colleagues at the University of Lagos, notably Professor Olugbemi (from Isua) and Dr Akinyeye (from Idanre) in Ondo State. My profound gratitude also goes to Prof. Kay Williamson of the University of Port Harcourt, Dr Roger Blench of the Overseas Development Institute and Dr Philip Allsworth-Jones of the University of the West Indies who made very useful comments on an earlier draft of this chapter. Finally, my thanks go to Mr Femi Ogundele and Mr Okoro of the Geography Department, of the Universities of Lagos and Port Harcourt respectively, who produced the map on the Akoko area, and to Miss Margaret Egbagiri, who worked hard to produce the typescript at very short notice. May God bless you all.

APPENDIX

Table 3.6 Comparative wordlist of four lects of the Akoko area of western Nigeria

	Ikakụmọ (Ukaan)	*Akunnu (Akpes)*	*Arigidi (Akokoid)*	*Ọkà (Yoruba)*
I	ìhyèdʒì	òɳì	ame	òmí
you (sg)	ìhyèrɔ̀	òsì	árɔ̰̂	ùwɔ
we	ìhyèbɔ̀	àbès	áò	àwa
you (pl)	ìhyèmɔ́	àbèɳ	ámɔ̂	ɛ̀ɳʷḭ̄
three	táárh	ís'ás	í'dá	mɛ́ta
four	ɳáíí	í'ɳíɳ	ìɳéé	ɛ̀rḭ
five	tóòǹ	í'ʃóṵ̀	útɔ̰̂	mą̰́ą̰́rʊ̰́
child	òʃú/à	ɔ́báǹ	ʊ́'ɳʷɔ̰̂	ɔma̰
navel	òkpódù/i-	ɔ̀ɳkɔ̀	ɛkpɔ̰	ùhɔ́
roast	tɔ́	mɔ́w̰ɔ́s̀	ʃɔ̰	sɔ̰
one	tʃíí-	éètʃì	ìkɔ̰́	uɳɛ̰́

	Ikakụmọ *(Ukaan)*	*Akunnu* *(Akpes)*	*Arigidi* *(Akokoid)*	*Ọkà* *(Yoruba)*
two	wáá	ídíáǹ	ìyí	èdʒì
big	ùgbá/dídíbɔ́k	édwúédú	dɔ̰̀	ńlá
long	ùdíǹ	ɲítɔì	gbàrà	ʃḭ́
small	àtḭ́/òwíyó	íɳkḭ́ɳkḭ́	kèmgbè	kḭ̄kḭ̄/gwɛ̄gwɛ̄
woman	òyéɳ/à	òɲóꜝɲó/à	èʃíéꜝré	obḭ̀rɪ
man	òyòhʷírh/à-	óɳó!óʃ/à-	ɔ́kɔ́ꜝrí	ɔkìrḭ
person	ɔ̀ɳɪ̀/à-	ɔ̀ɳí/á	ɛ̀ɳɛ́	eɳḭ
fish	éʃḭ́ʃḭ́/í	ɛ̀tɛ̀ɳ	éʃɔ́	ɛḍa
bird	ɛ̀kâ̰/ì-	ɛ́ꜝɳámà	ɔ́!rɔ̰́	ɛdʒɛ
dog	èɳʷṵ́/ì-	ébó	ofo	aḍá
goat	ɛ̀wí/ì-	ɛ́bí	àárɔ̰́	òwúɔ
tree	ɔ̀hʊ́/ì-	ɔ́hṵ́	ɔ́ꜝhɔ̰́	igḭ
seed	íhʷê	ɛ́mṵ́	ɔ́gɔ́ꜝrɔ́	eṯo/òʃó
leaf	ùfã́/à	áɳfã́	ímɛ̂	ewé
root	àíyimɔ̀hʊ́	ókúmɔ̀	irḭ̀	irḭ̀
housefly	ètʃó/ì	íɳtʃí	iʃḭʃḭ	iʃḭʃḭ
skin	ɛ̀hʷɛ́ꜝrhɛ́	ɛ́hʷá̰	àwɔ̀	ahɔ
meat	ɛ̀ɳá/–	éɳám	arɔ̰	ɛra̰
blood	ɔ̀dìà/–	ìkɔ̀ǹ	ɛ̀dʒɛ̰̀	ɛ̀ḍɛ̀
bone	òhʷó/ì	óhṵ́mɔ̀	éꜝkpḭ́	ugṵgṵ
fat	àyɔ̰́rʊ́/-	íyɔ́í	éꜝhɛ́	ɛ̀rà
egg	íʃḛ̂/á-	áɳtʃi^	ɛ́dʒáꜝhá	ɛhḭ
horn	ìkɔ̀ǹ/à-	àkùɳù	ɛ̀hɔ̰́	ùho
tail	òrṵ̀m/ì-	òmɔ̀	ùyà	ùrṵ̀
feather	àkpòr/-	ófiɔ̰́	èyɔ̀	ìdʒìdʒɛ́
hair	éwúrû/í-	ètìf	ìʃírí	irɔ̰
head	ìtʃṵ́	ítʃṵ́	ègírí	orí
ear	úrhóg̀/á-	áɳʃṵ̂	ōtō	etḭ́
eye	ídʒí/á-	áyô	ódʒû	oḍú
nose	ɔ̀kɔ́rɔ̰̀/à-	áhṵ̂	ódʒúɳʷɔ̰̀	imɔ̰́
mouth	òɳmó/à-	óɳú	orṵ	arṵ
tooth	óyḭ̀/á-	íɲû	éɲî	eɲḭ́
tongue	ɛ̀mú/í	íɳdâ	érê	iɳʷá̰
nail (finger)	ìɲɔ̰́/à-	ìheꜝɲɔ́bɔ́	ɛkɛ̰	àgbákíká̰
leg	úkàḭ̀/à-	òdù	ʊ́hɔ̂	ɔsɛ̀
knee	ìhṵ̀rṵ̀/à-	áhúɳòdù	éwóhɔ̀	igbogbo
hand	ùwɔ́/à-	ɔ̄bɔ̄	ɔ̄wɔ̄	ɔwɔ́
belly	ìmí/à-	áɳkú	ɔ̤gɔ̤	úkṵ̀
neck	ɔ̀hɔ́/ì	ɔ́hɔ̂	úgɔ̂	ɔfɔ̰
breast	íɳmâ/à-	ímbóùè	épɔ̂	ɔɲà
heart	èrùrùk	ɔ̄kɔ̰̀	ɔ́kɔ̰̀	ɔkà̰
swallow	mḭ̀	bɔ̀rɛ́	ʃérɔ̀mí	kpáɲì
drink	wʊ́	wɔ̰́	bɔ̤	mɔ̰
eat	yé	dʒɔ̰́	dʒə	ḍɛ
bite	yɔ̰́mɔ́	tʃè	rɔ̰	rɔ̰
see	há	yé	rí	rí
hear	kpì	kpás̀	ʃɛ	gbó
know	bá	ʃà	rɔ̰̀	mà̰
sleep	kʊ̀rà	míʃî	sɛ̰̀	sṵ̀
die	hʷó	hú	kṵ́	kú

Table 3.6 Continued

	Ikakụmọ (Ukaan)	*Akunnu (Akpes)*	*Arigidi (Akokoid)*	*Ọkà (Yoruba)*
kill	yú	wé	kpə́	kpa
bathe	kó	hɔ́	yé	gwɛ̀
jump	tárì	bɛ̄r	tà	fò
walk	ʃéní	tʃíǹ	dʒɛ̰̀	rḭ̀
come	wàk	bà	va/ʊa	wá
lie down	kùrà	ɳàɳíyéʃ	ʃíʃɛ̰̀	ʃṵ̀ɳálè
sit down	ɳmɛ́ɛ́	ɳání	ʃɛ́ʃḭ	dòkó
blow (wind)	fùɳù	mī	kee	fɛ́
give	rètʃ/mèdʒ	gbàtʃá	gbàɲɔ̰́	hṵ́
say (something)	hyɔ̰́ɳá	mwétʃí	hɛ̰	fɔ̀/gwí
sun	ìɲɔ́ɲɔ́	ɛ̄wû	ūhɔ̀	òorṵ̀
moon	òdʒó/-	àdɔ̀m	ērīdʒā	ot̰ṵ̀
star	ìʃḭ́ʃḭ́ꜜrḛ́ḭ/à	ìɳtʃḭtʃḭ	àláꜜɳʷḭ̀èdʒù	ìràhɔ̀
water	ʊ̀mɔ́	íɲí	ēdʒḭ̄	omḭ
steal	tʃédʒí	yútú	dē	dʒí
stone	èkpɔ̰̂/ì	íꜜɳʃá	ɛ́ɛ́ꜜtá	ɔta
sand	ɔtʃɛ̰̂	ìhṵ̀rɔ̀	ɲèɲè/íꜜʃá	ìɲà̰rḭ̀
ground	ʊ̀rhá/–	átɔ́	ēēʃḭ̄	alɛ̀
rope	òkó/à-	ɔ́ɳkú	òòkṵ̀	okṵ̀
smoke	èrhìd	ìmúńꜜdá	ūyū	àfi
fire	ìyá̰/–	íɳdá	éꜜʃɔ̰̣	uɳa̰
ashes	ìhṵhṵ́/–	íhṵ́ɳgó	ɔ́dɔ̰́	èrúrú
saliva	àdɔ̀rṵ́/–	ńʃúɔ̀	ūtɛ̰̄	itɔ́
road	ɔ́rɛ̂/í	ókpó	úwóꜜró	ɔ̀ɳà
hill (mountain)	èrhá/ì-	ɔ́ʃṵ́	édê	òkè
red	wá̰wá̰rh	éɲíɲá	ʃɔ̰̣́	kpa̰
give birth	dʒó	dʒóbáǹ	ʃúwɔ̰̀	bí
bury	wùgì	yṵ́	lyū/lü	ʃḭ/wó
white	há̰há̰rh	éfṵ́fɔ̃	hə́	fṵfṵ
black	rḭ́rḭ	éhḭ́tíɳa/éyítíɳá	dʒḭ́	dṵ́dṵ́
night	àráꜜhwó/–	íɳʃú	òdúꜜdó	àʃa̰rṵ
hot	túétùè	étʃítʃó	ʃɛ̰̀	gbó
cold	uria̰ꜜʃḭ́	éyɔ́ṵ́	tṵ́	tṵ
full	wɛ̂	mùà	hɔ̰́	kɔ̰
new	yá̰wá̰	éɳíɳá	tṵ́ɔ̰̀tṵ̀ɔ̰́	titɔ̰
good	fɔ́	tʃɔ́ǹ	ʃɔ̰̀	sa̰
fowl (chicken)	ɛ̀kɔ̀kɔ̀/ì-	ìkókó	ɛ́ꜜhɛ́	àkìkɔ́/ad̰ùrɛ
dry	hɔ́hɔ́	éhúhú	gə́	gbɛ
name	ìíɳí/-	ímúǹ	éɲî	orúkɔ

BIBLIOGRAPHY

Abímbọ̀lá, W. 1973. The literature of the Ifa cult. In *Sources of Yoruba History*, S.O. Biobaku (ed.), 41–62. Oxford: Clarendon Press.

Abiodun, M. 1989. *Noun class system in Ukaan*. Paper presented at the 10th annual conference of the Linguistic Association of Nigeria.

Adalemo, A. and J.M. Baba (eds) 1993. *Nigeria: giant in the tropics Vol. 1*. Lagos: Gabumo Publishing Co.

Adetugbọ, A. 1973. The Yoruba language in Yoruba history. In *Sources of Yoruba History*, S.O. Biobaku (ed.), 176–204. Oxford: Clarendon Press.

Akinkugbe, F. 1976a. An internal classification of the Yoruboid group (Yoruba, Itsẹkiri, Igala). *Journal of West African Languages* 11.1–2, 1–19.

Akinkugbe, F. 1976b. A comparative phonology of Yoruba dialects. Itsẹkiri and Igala. Ph.D. dissertation, University of Ibadan.

Allsworth-Jones, P. and P.A. Oyelaran 1991. Itaakpa Rockshelter, Kwara State: recent results. *West African Journal of Archaeology* 21, 208–213.

Armstrong, R.G. 1964. The use of linguistics and ethnographic data in the study of Idoma and Yoruba History. In *The Historian in Tropical Africa*, J. Vansina, R. Mauny and L.V. Thomas (eds), 127–144. London: Oxford University Press for the International Africa Institute.

Armstrong, R.G. 1967. *The Study of West African Languages*. Ibadan: University Press for the Institute of African Studies.

Blench, R.M. 1989. New Benue-Congo: a definition and proposed internal classification. *Afrikanische Arbeitspapiere* 17, 115–147.

Blench, R.M. 1993. Recent developments in African language classification and their implications for prehistory. In *The Archaeology of Africa: food, metals and towns*, T. Shaw, P. Sinclair, B. Andah and A. Okpoko (eds), 126–138. London: Routledge.

Blench, R.M. 1997. Language studies in Africa. In *Encyclopaedia of Precolonial Africa*, J.O. Vogel (ed.), 90–100. Walnut Creek, London and New Delhi: Altamira.

Capo, H.B.C. 1989. Defoid. In *The Niger-Congo languages*, J. Bendor-Samuel (ed.), 275–290. Lanham: The University Press of America.

Crozier, D.H. and R.M. Blench (eds) 1992. *An Index of Nigerian Languages*. Dallas: Summer Institute of Linguistics.

Elugbe, B. 1971. The Ẹdo languages of Akoko division. *Research Notes (University of Ibadan)* 4(1), 35–38.

Elugbe, B. 1973. A comparative Ẹdo phonology. Ph.D. thesis, University of Ibadan.

Elugbe, B. 1989. *Comparative Edoid. phonology and lexicon*. (Delta Series 6). Port Harcourt: University of Port Harcourt Press.

Hoffmann, C. 1974. *Nigerian Languages by Language Families*. Mimeograph. Ibadan: Department of Linguistics and Nigerian Languages, University of Ibadan.

Horton, R. 1995. The Niger-Congo diaspora: language, geography and history. In *Issues in African Languages and Linguistics: essays in honour of Kay Williamson*, E.N. Emenanjo and O.-M. Ndimele (eds), 306–339. Aba: National Institute for Nigerian Languages.

Ibrahim-Arirabiyi, Femi. 1989a. A comparative reconstruction of Akpes lects, Akoko North, Ondo State. M.A. dissertation. University of Port Harcourt.

Ibrahim-Arirabiyi, Femi. 1989b. *Aspect of ethno-linguistic study on part of Akoko-North, South, Owo and Ekiti East Local Government Areas of Ondo State*. Mimeograph. Prepared for the National Museum, Port Harcourt.

Kropp Dakubu, M.E. (ed.) 1980. *West African Data Sheets, Vol. 2*. Leiden: West African Linguistic Society.

Jungraithmayr, H.G. 1973. Notes on the Ishe dialect of Ukaan (Akoko, Western State, Nigeria). *Africana Marburgensia* 6(1), 38–52.

MacDonald, K.C. 1997. Kourounkorokalé revisited: the *pays Mande* and the West African microlithic technocomplex. *African Archaeological Review* 14(3), 161–200.

Obayemi, A.M. 1982. *The present state of archaeological work in Northern Yorubaland*. Paper presented at the Annual Conference of the Nigerian Archaeological Association, Zaria.

Ohiri-Aniche, C. 1991. A comparative phonology of consonants in the Igboid, Edoid and Yoruboid languages of Nigeria. Ph.D. thesis, University of Port Harcourt.

Ohiri-Aniche, C. 1995. Ọkà-Àkoko phonology and its implications for Proto-Yoruba consonant reconstructions. In *Issues in African Languages and Linguistics: essays in honour of Kay Williamson*, E.N. Emenanjo and O.-M. Ndimele (eds), 119–131. Aba: National Institute for Nigerian Languages.

Okẹ, D. 1970. *Language choice in the Yoruba-Edo ethnic border area of Nigeria.* Paper presented at the 9th West African Languages Congress, Freetown, March 1970.

Omamor, A.P.E. 1976. The place of Itsẹkiri in relation to Yoruba. *Research notes (Department of Linguistics) University of Ibadan* 1, 1–2.

Oyelaran, O. 1977. *Linguistic speculation and Yoruba history*. Seminar paper, Department of African Languages and Literatures, University of Ife.

Schadeberg, T.C. 1986. The lexicostatistic base of Bennett and Sterk's reclassification of Niger-Congo with particular reference to the cohesion of Bantu. *Studies in African Linguistics* 17, 69–83.

Shaw, T. 1980. Prehistory. In *Groundwork of Nigerian History*. O. Ikime (ed.), Ibadan: Heinemann Educational Books (Nigeria).

Shaw, T. and S.G.H. Daniels 1984. Excavations at Iwo Eleru, Ondo State, Nigeria. *West African Journal of Archaeology* 14, 1–269.

Shimizu, K. 1980. *Comparative Jukenoid.* (3 Vols) (Veröffentlichungen der Institut für Afrikanistik und Ägyptologie der Universität Wien.) Vienna: Afro-Pub.

Swadesh, M. 1955. Towards greater accuracy in lexicostatistic dating. *International Journal of American Linguistics* 21, 121–137.

Swadesh, M. 1959. The mesh principle in comparative linguistics. *Anthropological linguistics* 1(2), 7–14.

Williamson, K. 1970. *Northern Edo: a provisional classification with notes on manuscript material in the Department of Languages.* Mimeograph. Ibadan: University of Ibadan.

Williamson, K. 1973. *Lexicostatistical comparison of Itsẹkiri, Standard Yoruba, and Igala.* Mimeograph, Department of Linguistics and Nigerian Languages, University of Ibadan.

Williamson, K. 1982. The classification of East Niger-Congo. *Journal of the Linguistic Association of Nigeria* 1, 101–106.

Williamson, K. 1988. Linguistic evidence for the prehistory of the Niger-Delta. In *The Early History of the Niger-Delta,* E.J. Alagoa, F.N. Anozie and N. Nzewunwa (eds), 65–119. Hamburg: Helmut Buske Verlag in association with the University of Port Harcourt.

Williamson, K. 1989. Benue-Congo overview. In *The Niger-Congo Languages,* J. Bendor-Samuel (ed.), 247–274. Lanham: The University Press of America.

Williamson, K. and K. Shimizu (eds) 1968. *Benue-Congo Comparative Wordlist, Vol. 1.* Ibadan: West African Linguistic Society.

Williamson, K. and K. Shimizu (eds) 1973. *Benue-Congo Comparative Wordlist, Vol. 2.* Ibadan: West African Linguistic Society.

York, R.N. 1978. Excavations at Dutsen Kongba, Plateau State, Nigeria. *West African Journal of Archaeology* 8, 139–163.

4 *Revising Polynesian linguistic subgrouping and its culture history implications*

JEFF MARCK[1]

INTRODUCTION: ON ARCHAEOLOGY AND LINGUISTICS IN POLYNESIA

Interdisciplinary co-operation is a hallmark of cultural research in Polynesia. Anthropologists, linguists and archaeologists working in this region have tended to know each other and communicate their results to each other since about the middle of the present century, when work in the contemporary idioms of those disciplines began to emerge on the Polynesian culture area. We may think of research in Polynesia as extending back into the nineteenth and even eighteenth centuries, although this refers to general ethnographic work; subsurface archaeological studies and highly disciplined comparative linguistic studies did not begin to emerge until after about 1950.

Trained linguists were not working in any number around Polynesia until the 1960s, and even today there are very few carrying on active research. But the sounds of Polynesian languages are simple and clear, and there is a long history of competent description and dictionary making by anthropologists, archaeologists, natural historians, missionaries and colonial administrators. What the linguists lack in numbers they can often make up for in the abundance of reliable data. Sometimes archaeologists have not waited for linguists, and themselves do the comparative linguistic work on a topic of urgency from their point of view. Works by Kenneth Emory (1938, 1963) and Roger Green (1966, 1988) are examples.

It is sometimes said that interdisciplinary work is undisciplined work. There *are* synthetic statements about Polynesian prehistory that combine both archaeology and linguistics (cf. esp. Kirch and Green 1987), but critics of such work sometimes complain that it is not clear how the theory, method and data of the two disciplines are being integrated or that they are being integrated in a deficient manner (e.g. Dunnell 1987; Terrell 1987; Welsch 1987). But, more often in Polynesian studies, linguists and archaeologists make their own statements about the past and ask about the findings of the other group.

Synthesis is tempting, as there are three central areas of agreement between what the more conservative archaeological and linguistic models of Polynesian prehistory independently suggest:

1 Both models suggest a period of common development in Polynesia prior to internal diversification. This is called the Pre-Polynesian period by linguists, who believe it was a long period of time passed in the Tonga and Samoa area. The archaeological equivalent is the period from about 1100 BC to the early–mid first millennium AD, for which the archaeologists have many securely dated cultural sites from around Tonga and Samoa but none of that age from Eastern Polynesia or the Polynesian Outliers. There are no Outlier dates earlier than 2,000 years with distinctively Polynesian materials.
2 Both models suggest a period of common development in central Eastern Polynesia once it was settled. On the archaeological side, there is a somewhat distinctive Eastern Polynesian material culture, whilst the oldest universally accepted dates are from the centre. Linguistically, there is a highly distinctive Eastern Polynesian language subgroup.
3 Both models suggest a dispersal to the margins of Eastern Polynesia from its centre.

Thus, the linguists and archaeologists watch each other's results quite closely in questions of settlement, dispersal and borrowing, because there is currently the appearance that they are studying the same basic human events. In modern anthropology, however, answering questions about dispersals is no longer considered an appropriate end in itself. The questions must be related to problems of culture change or cultural process before either archaeologists or linguists have much of an audience amongst anthropologists. Archaeological studies and linguistic studies make quite distinct kinds of contributions to investigations of culture change and cultural processes. It is commonly the precise phylogeny of linguistics that so intrigues archaeologists, whilst the precise dating of the archaeologists similarly intrigues linguists.

Beyond the settlement question, archaeological evidence and linguistic evidence tend to offer rather different genres of opportunities for speculating about the past. There is, for instance, no linguistic evidence parallel to that developed by the archaeologists when they tell us that people inhabited an area, hunted out all the *moa* and seals, then abandoned the locality within about fifty years (Anderson and Smith 1996). Similarly, there is no archaeological evidence parallel to that developed by the linguists when they reconstruct terminologies having to do with social phenomena to one past language rather than another. Synthesis between archaeology and linguistics occurs when their independent lines of reasoning and argumentation have some opportunity to comment on the same issue.

POLYNESIAN LINGUISTIC SUBGROUPING

This chapter presents the standard theory of Polynesian language subgrouping and then considers the specific problem of the position of Eastern Polynesian within Nuclear Polynesian. Proto Nuclear Polynesian is the purported ancestor of all Polynesian languages other than Tongan and Niuean (the Tongic languages).

Although there was a great deal of data on Polynesian languages available even at the beginning of this century, trained linguists were not applying the comparative method of linguistics to the Polynesian problem until quite recently. The anthropologist turned linguist Andrew Pawley (1966, 1967) and the archaeologist Roger Green (1966) produced the first formal subgrouping of Polynesian in the idiom of comparative linguistics, as developed by the Indo-Europeanists and others (Figure 4.1). Based mainly upon shared innovations in morphology ('shape' and components of words) as well as earlier studies based on shared sound changes and lexicostatistics, their subgrouping has largely stood the test of time and has provided a basis upon which further insights into language relations within Polynesia have arisen.

Pawley and Green were able to set aside what they knew of the archaeology and anthropology of the area and apply subtle precepts of the comparative method of linguistics to obtain their result. Given this account

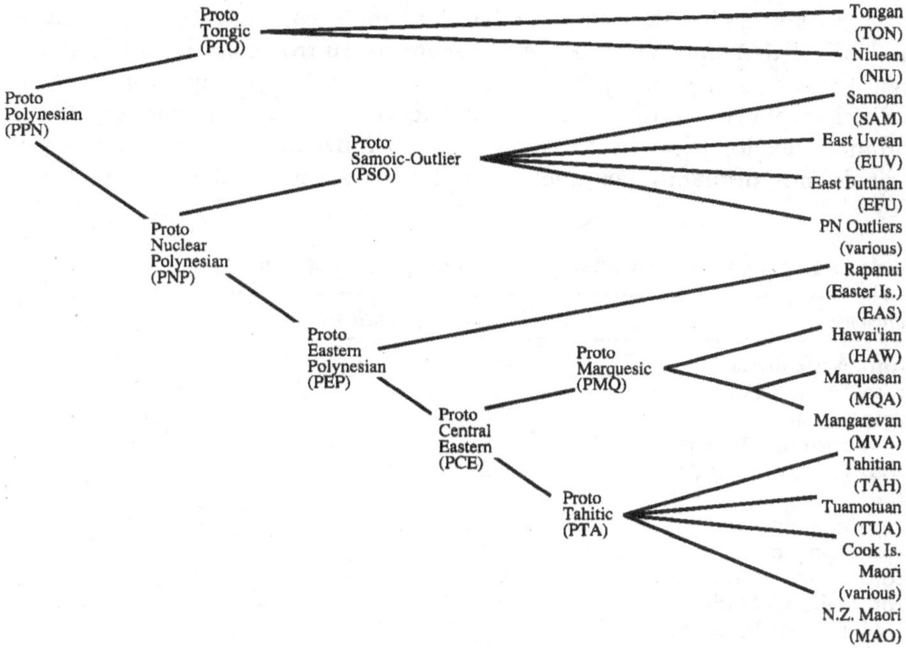

Figure 4.1 The standard subgrouping of Polynesian languages
Sources: after Pawley (1966, 1967), Green (1966, 1988) and Marck (1996)

of linguistic subgrouping in Polynesia, it is easy to imagine how the history of Polynesian archaeology and linguistics might seem blurred together to the outsider. But it was solely the comparative method of linguistics that was applied, not some interdisciplinary method, and it was specifically not one influenced by archaeology or anthropology.

'The standard theory of Polynesian subgrouping' (Figure 4.1) is recognized by its supporters and its detractors by that name. I present the current defence and some revisions of the standard theory elsewhere (Marck 1996, 1999) and elaborate here on Nuclear Polynesian only. In Figure 4.1, Mangarevan is reclassified as diverging from Marquesan after Hawai'ian (Marck 1996), but the suggestion that Hawai'ian may have come from the Southern Marquesan dialect remains defensible (Green 1966).

Bruce Biggs has developed a *Comparative Polynesian Lexicon* (*POLLEX*) that now contains over 4,000 reconstructions coded according to the subgrouping of Polynesia given in Figure 4.1. It is *POLLEX* data and reconstructions that are used here when unreferenced data is provided, updated versions were obtained annually in the course of this work (1992, 1993, 1994a). Biggs' (1994a) reconstructions, by proto language, are given in Table 4.1.

The convention in *POLLEX* is to label etymological groups according to the highest known proto language in which they occurred. Thus Proto Tahitic reconstructions are labelled '.TA', and the reconstruction is in the orthography of Proto Tahitic. When the highest known occurrence of the word is a language spoken earlier than Proto Polynesian, the etymological group is given the label of that language, e.g. '.AN' for Proto Austronesian, but the phonological and semantic reconstruction is made to the Proto Polynesian level.

Total vocabulary comparisons are not diagnostic measures of uniquely shared common developments in languages, nor can other measures be diagnostic if they do not distinguish uniquely shared innovations from uniquely shared

Table 4.1 Biggs' (1994a) reconstructions by proto language

Language	*No. of reconstructions*
Proto Austronesian	170
Proto Malayo-Polynesian	145
Proto Oceanic	175
Proto Eastern Oceanic	114
Proto Central Pacific	95
Proto Fijian	238
Proto Polynesian	1389
Proto Tongan	34
Proto Nuclear Polynesian	430
Proto Samoic-Outlier	108
Proto Eastern Polynesian	111
Proto Central Eastern Polynesian	450
Proto Marquesic	31
Proto Tahitic	141

retentions. Diagnostic evidence is best found in shared changes in the morphology of grammatical function words, sporadic sound changes and other developments where a uniquely shared retention is demonstrably *not* involved. The Polynesian subgroup is defined by numerous such developments, and one that occurs in several words is the reduction of an earlier **-ani* to Proto Polynesian **-ai*. Examples from Biggs (1994a) are shown in Table 4.2. Examples of such occasional changes between Proto Nuclear Polynesian and Proto Eastern Polynesian can be seen in Table 4.3 (from Biggs 1994a). Some examples of unexpected changes between Proto Eastern Polynesian and Proto Central Eastern Polynesian from Biggs (1994a) are shown in Table 4.4.

In each of these three tables we are able to demonstrate that the lower level language had a sporadic[2] difference from its next highest parent amongst these proto languages. The same is true for Proto Marquesic and Proto Tahitic in relation to Proto Central Eastern. However, an examination of Biggs (1992) revealed that Nuclear Polynesian and Samoic-Outlier differ from these cases in a basic way (Marck 1999). The daughter languages of all the other purported proto languages are marked by numerous sporadic sound changes limited only to those groups. This is not true, however, of Nuclear Polynesian much

Table 4.2 Comparisons between Proto Central Pacific and Proto Polynesian

Reconstruction level	*eat*	*sharp*	*ghost, spirit*
Proto Central Pacific[a]	*kani	*kani[b]	*qanitu
Proto Polynesian	*kai	*kai	*qaitu

Notes: a The language ancestral to Fijian, Rotuman and Polynesian
b This is my own reconstruction based on the PPN agreement with Proto Micronesian **kangi*.

Table 4.3 Comparisons between Proto Nuclear Polynesian and Proto Eastern Polynesian

Reconstruction level	*slate-pencil urchin*	*fleshy membrane*
Proto (Nuclear) Polynesian	*watuke	*lewelewe
Proto Eastern Polynesian	*fatuke[a]	*were[b]

Notes: a The expected and otherwise universal correspondence of PEP to PPN **w* is PEP **w*.
b Metathesis.

Table 4.4 Comparisons between Proto Eastern Polynesian and Proto Central Eastern Polynesian

Reconstruction level	*sharp*	*whale*
Proto Eastern Polynesian	*kai	*tafora(q)a
Proto Central Eastern Polynesian	*koi[a]	*toforaa[b]

Notes: a PEP **a* normally remains PCE **a*.
b The loss of the glottal stop is expected but PEP **a* normally remains PCE **a*.

beyond those cases established by Pawley (1966), and is not true at all of Samoic-Outlier. I will first consider the case of Nuclear Polynesian and then the case of Samoic-Outlier.

NUCLEAR POLYNESIAN

Figure 4.2 shows Western Polynesian islands with a circumference line extending around each island at the approximate distance of an overnight voyage (100 statute miles, specifically a voyage of twenty-four hours or less, most commonly beginning as the stars come out for the evening and ending before the last light of the following day). If a radius of such a voyage encounters another island, the circumference arcs are linked. In both Micronesia (Marck 1986) and Polynesia, the typical modern situation is one in which such linked arcs define areas linked by mutual intelligibility, whilst greater distances have tended to result in language or major dialect boundaries. Linguists want to establish how an early situation, where the language must have been unified, diversified into the distinct languages of today. In the present case, I will investigate the progress of dialect development within Nuclear Polynesian at the time of the divergence of Eastern Polynesian.

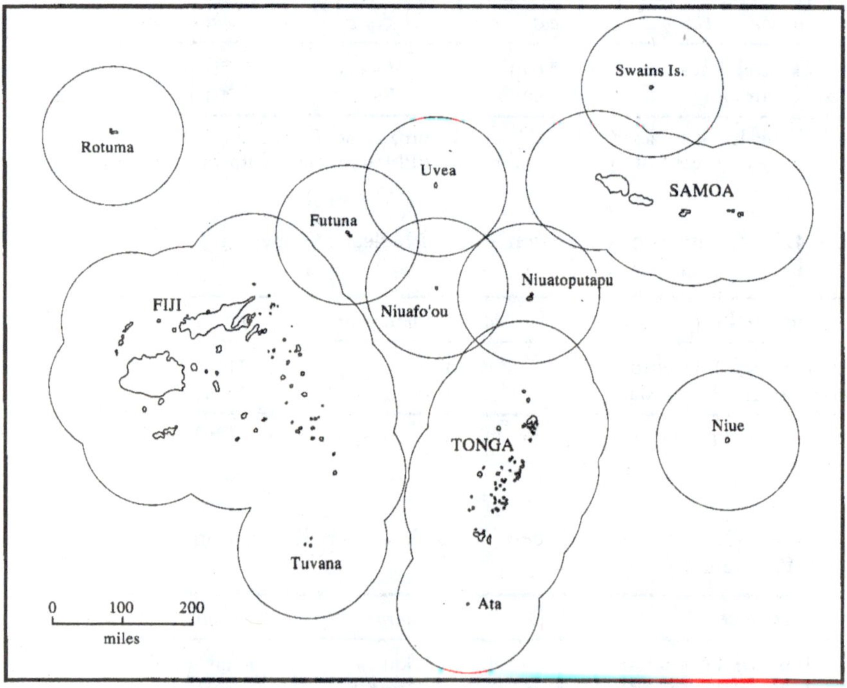

Figure 4.2 Island interlinked by voyages of twenty-four hours or less
Source: Marck

The number of reconstructions in Proto Nuclear Polynesian that presently have no known antecedents in Proto Polynesian is quite large (420 in Biggs 1992, 430 in Biggs 1994a). But these necessarily include many uniquely shared retentions from Proto Polynesian since each word had more chances of survival in Nuclear Polynesian; there are many Samoic-Outlier and Eastern Polynesian languages but just two Tongic languages. If we were to say that only two Samoic-Outlier languages will be used in the reconstruction of Proto Nuclear Polynesian, and compare this to the number of Tongic agreements with Eastern Polynesian, the results would be similar numbers. By this and other measures, there is no evidence for suggesting that the vocabulary of the Samoan area was highly distinct from the vocabulary of the Tongan area at the time of the divergence of Eastern Polynesian. The most we can be sure of is that pronunciations differed slightly and that some pronouns and demonstratives were different (Pawley 1966, 1967, 1996; Wilson 1982, 1985).

We can ask if total vocabulary comparison provides any striking points for discussion, but this should not be understood as diagnostic as large numbers of agreements do not necessarily involve *any* shared innovations. Table 4.5 gives the known cognate vocabulary from Biggs (1994a).

The magnitude of each number is a function of five main variables*:

1 how much is known about each language;
2 the genetic relations of the languages;
3 the conservative or innovative nature of vocabulary retention for each language;
4 the amount of work Biggs has done on each language;

Table 4.5 Known Polynesian cognate vocabulary

	TON	*NIU*	*SAM*	*EFU*	*EUV*	*TUV*	*EAS*	*MQA*	*HAW*	*TAH*	*MAO*
CWA	2043	1333	1939	1842	1459	1313	732	1422	1688	1804	2462
TON	–	1099	1466	1437	1268	952	442	785	813	870	1100
NIU	1099	–	998	947	817	718	375	642	649	691	847
SAM	1466	998	–	1369	1105	961	451	802	826	902	1116
EFU	1437	947	1369	–	1214	951	420	733	750	786	987
EUV	1268	817	1105	1214	–	814	362	599	614	648	814
TUV	952	718	961	951	814	–	369	584	601	626	773
EAS	442	375	451	420	362	369	–	516	516	537	599
MQA	785	642	802	733	599	584	516	–	981	976	1118
HAW	813	649	826	750	614	601	516	981	–	1103	1395
TAH	870	691	902	786	648	626	537	976	1103	–	1405
MAO	1100	847	1116	987	814	773	599	1118	1395	1405	–

Key: CWA: words 'cognate with any' other word from any other Polynesia language in Biggs (1994a); TON: Tongan; NUI: Niuean; SAM: Samoan; EFU: East Futunan; EUV: East Uvean; TUV: Tuvalu; EAS: Rapa Niu (Easter Island); MQA: Marquesan; HAW: Hawai'ian; TAH: Tahitian; MAO: NZ Maori.

* A sixth variable is borrowing, such as borrowing of Tongan by East Uvean. Some of this is undetectable, but those cases that are diagnostic of borrowing are not 'counted as cognate' by Biggs (1992, 1993, 1994a).

5 participation in networks of contact that seem to have resulted in a significant sharing of retentions after periods of sharing of innovations ceased.

In the first instance, there are large dictionaries for some languages whilst nearly nothing is known of others. In the second, languages may be more or less closely related. In the third, some languages seem more conservative than others. In the fourth, Biggs has attempted to account for as much of the vocabulary as possibly for some languages, e.g. NZ Maori (Biggs 1994b), whilst others are rather more neglected, e.g. Rapanui (Easter Island), which has borrowed from Marquesan and more extensively from Tahitian in the historic period and is not the first language of choice for speculating about prehistory. Finally, Western Polynesian languages seem to have retained a great deal of common vocabulary uniquely lost in Eastern Polynesia. Some of this is the result of chance (it only need be lost once in Eastern Polynesia) but a significant portion seems shared for reasons we do not presently understand. It may be that Western Polynesian communities were aware that they were cosmopolitan words and were less inclined to supplant them with innovations. There may be some unrecognized innovations shared through the Tonga to Samoa area that post-date the divergence of Eastern Polynesian, but these have yet to be demonstrated diagnostically. The main impression is that vocabulary agreements of Western Polynesian languages seem to be due to the sharing of retentions.

The Eastern Polynesian and Central Eastern groups stand out in Table 4.5 in the sense that each language from those groups has its highest scores within those groups. But the agreement of individual Eastern Polynesian languages with Samoan is never much higher than with Tongan. As with the exercise that limited Samoic-Outlier to two languages, we cannot suggest that Eastern Polynesian vocabulary has a pronounced affinity to Samoic-Outlier as compared to Tongic on the basis of total modern vocabulary comparisons. Hence the suggestion above that Nuclear Polynesian and Tongic may have simply been different Proto Polynesian dialects at the time that Eastern Polynesian diverged, a point that Pawley (1996) has also recently emphasized.

I will now review some etymological groups from Biggs (1992) that slightly expand the published evidence for the Nuclear Polynesian group. Nuclear Polynesian was first named and argued on the basis of uniquely shared innovations in morphology by Pawley (1966). Pawley noted that Elbert (1953) had previously published a list of sound changes for Polynesia indicating that Samoan grouped with Eastern Polynesian as opposed to Tongan; but Elbert did not name or argue for such a group. Some of the morphological arguments for Nuclear Polynesian put forward by Pawley (1966) have been attacked by Dyen (1981). Harrison (1981) has noted that most of Dyen's arguments against Nuclear Polynesian were lexicostatistical and non-diagnostic, but some of the matters Dyen raised involved alternate morphological interpretations. Neither Harrison nor anyone else has spoken regarding those alternative interpretations (but see Marck 1999: 161). This section presents new data from Biggs (1992, 1994a) that *does* provide diagnostic support for:

1 Nuclear Polynesian and
2 the inclusion of Samoan, Eastern Polynesian and Ellicean Outlier in a subgroup similar to one proposed by Wilson (1985).

Of Biggs' (1992) 420 reconstructions to Proto Nuclear Polynesian, I will discuss only five pairs, as they involve irregular sound changes (which are often diagnostic of subgroups) and clear evidence of the situation in Proto Eastern Polynesian. Only two of those five pairs are entirely consistent with the Nuclear Polynesian hypothesis as it is generally received. The other three 'isoglosses' around the Western Polynesia area indicate that Eastern Polynesian speech may have emerged more out of the Samoa, Tokelau and Tuvalu area than out of 'Samoic-Outlier' in general (as Figure 4.1 implies).

Figure 4.3 gives the forms concerned. The lines through the maps show the limits of an older pronunciation to the south of the line in comparison to the distribution of an innovative pronunciation north of the line that is known from Eastern Polynesian as well. The map plots two distributions as occurring in accordance with the Nuclear Polynesian hypothesis as presently received: **hui* 'bone' and **tafu-raqa* 'whale' are reconstructed for Proto Polynesian, whilst Polynesia languages other than Tongan and Niuean reflect Proto Nuclear Polynesian **iwi* and **tafo-laqa*.

The first contrast in Figure 4.3 is Proto Polynesian **hui* 'bone' versus Proto Nuclear Polynesian **iwi*[3], seen on the line separating Tonga from the other islands. Proto Polynesian **hui* is reconstructed by comparison of Proto Tongic to external evidence. Proto Nuclear Polynesian **iwi* is reconstructed on the basis of all Nuclear Polynesian languages save East Uvean, which has *hui* like Tongan. East Uvean is expected to lose the initial consonant (cf. Biggs 1980: 118–119), although even if it did not share in the irregular Proto Polynesian *-*ui* to Proto Nuclear Polynesian *-*iwi* change, we still expect East Uvean ***ui*. I consider this a Tongan loan, as do Pawley (pers. comm.) and Biggs (1992, 1994a) who does not count it as cognate in his PPN **hui* 'bone' reconstruction.

In addition to the arguments for two distinct patterns of consonant inheritance in East Uvean, one which is seen as the original pattern and the second which is due to borrowings from Tongan (Biggs 1980), there is now further evidence that one of them is specifically borrowed from Tongan rather than a residual of an old dialect chaining with Tongan, as Rensch (1987) suggested. Tongan has vowel changes that are exhibited in certain East Uvean words. Within this group, certain words offer an opportunity to examine whether East Uvean shared the vowel change process with Tongan or if the vowel changes seem limited to words borrowed from Tongan. Where diagnostic differences occur in the consonants:

1 East Uvean always follows the Nuclear Polynesian pattern in the vowels when it follows the Nuclear Polynesian pattern in the consonants and

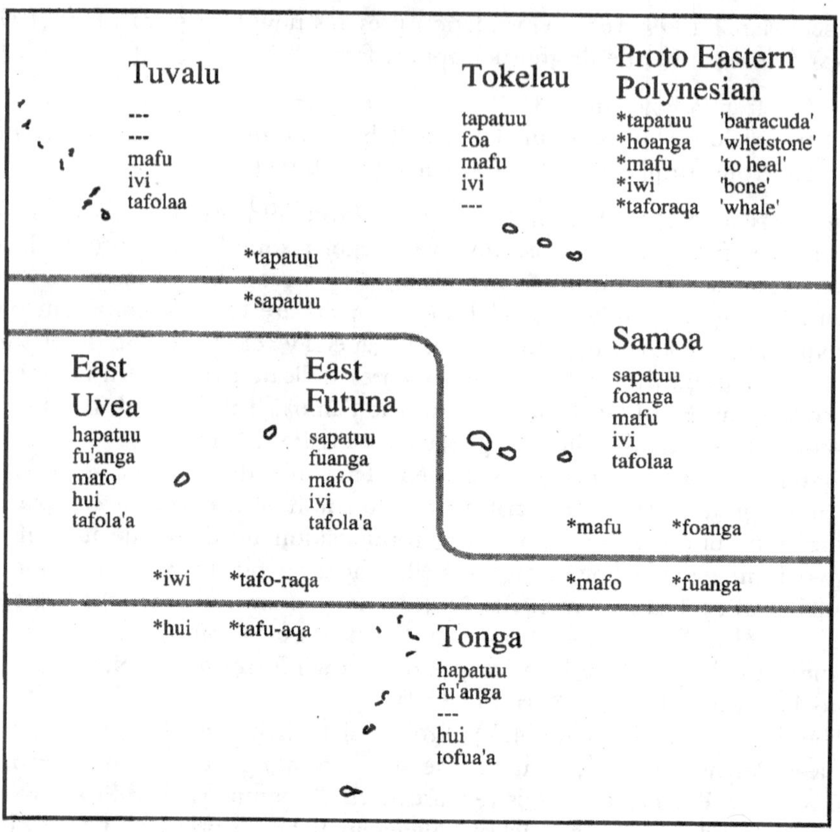

Figure 4.3 Some Proto Eastern isoglosses with Western Polynesian languages
Source: Marck.
Note: Author (Marck 1999: 132–133) now considers *tapatuu/*sapatuu a result of conflation.

2 it always follows the Tongan pattern in the vowels when it follows the Tongan pattern in the consonants (Marck, 1999: 137–139, based on Biggs 1993).

Thus to Biggs' (1980) observation that East Uvean basic vocabulary lacks the Tongan consonant reflexes, we can add the observation that East Uvean vocabulary in general lacks vowel changes of the Tongan type in words that follow the consonant correspondences of its basic vocabulary.

The second word pair on the line in Figure 4.3 is the comparison of Proto Tongic **tafu-aqa* 'whale' with Proto Nuclear Polynesian **tafo-laqa* 'whale', which also involves a demonstrable innovation of Proto Nuclear Polynesian. The change of Proto Polynesian **r* to Proto Nuclear Polynesian **l* and its loss in Tongic are expected. It is the change of the second vowel that is unexpected. Proto Tongic **tafu* agrees with Fijian (Bauan ***tavuto*** 'sperm whale' and Wayan ***tavuto*** 'large whales'). When there is such an agreement between a first order subgroup and external evidence, the agreement is reconstructed for the proto language (PPN).

The second east–west line in Figure 4.3 shows an isogloss where there are innovations in Proto Polynesian **fuanga* 'whetstone, grindstone' and **mafo* 'healed' north of a line that groups Samoan, Tuvalu and Tokelauan with Eastern Polynesian, but excludes East Futunan and East Uvean, which show the older pattern like Tongic. Biggs (1992, 1994a) gives no external evidence for Proto Polynesian **fuanga*, but evidence internal to Polynesia comes from Tongan and many Outliers and is probably the older pattern. In the instance of Proto Polynesian **mafo*, there is abundant external evidence that Tongan and numerous 'Futunic' Outliers retain an older pattern, whilst Eastern Polynesian, Tokelauan, Tuvalu, Samoan and Ellicean Outliers (see Figures 4.4 and 4.5 for lists of members of the outlier groups) innovated with **mafu*.

Finally, Figure 4.3 shows an isogloss that has **tapatuu* 'barracuda', shared by Eastern Polynesian, Tokelauan, Tuvalu and Ellicean Outliers, whilst an apparently older pronunciation is found to the south where Tongan, Samoan, East Uvean and East Futunan share reflexes of **sapatuu* with other Outliers. No external cognate is known, as with **fuanga* above. Biggs (1994a) also gives SIK *saaputu* 'fish sp.' and TIK *saputu* 'sea fish of reddish colour; Maori Snapper (*Lutjanus rivulatus*) or an Emperor (*Lethrinus kallopterus*)' in his PPN **sapatuu* 'barracuda' reconstruction, but there are reasons to question whether either are cognate with the forms from TON, SAM, EFU and EUV. They lack the final long vowel, have a different second vowel and do not mean 'barracuda' (see note to Figure 4.3).

The irregular (sporadic) sound changes that Eastern Polynesian shares with Tokelauan and Tuvalu, and to a lesser extent Samoan, are also shared with certain Outliers. They are the same Outliers (Ellicean) that Wilson (1985) found to share uniquely certain aspects of the Eastern Polynesian pronominal system. Table 4.6 gives the relevant data, and, for the purposes of the table, the proto language common to Samoan, Eastern Polynesian and certain Outliers is called 'Proto Ellicean', as in Wilson (1985: 89).

As can be seen, the fit is not perfect. If Ellicean is truly a subgroup in the way that Wilson (1985) defined it, there may have been mixing of dialects or some of the changes may have taken place more than once. In the instance of the dual reflexes of **fuanga* and **foanga* in Tokelauan and Takuu, there are slight differences of meaning. Such doublet formation normally occurs under conditions of borrowing. In any event, it is only in *certain* Nuclear Polynesian languages that the change is found at all. In the case of the **mafo* and **mafu* pair, there is a certain propensity for languages in general to make this kind of change (**o* > **u* in word final position). Thus the change may have occurred more than once. Another explanation for this distribution is dialect mixing in the establishment of West Uvean, Tikopian and Rennellese.

In the case of the Proto Polynesian and Proto Nuclear Polynesian correspondence of **sapatuu* to Proto Ellicean **tapatuu*, a very unusual kind of sound change is involved. The early Polynesia **s* sound does not otherwise become

Table 4.6 Innovations charted for Western Polynesia in Figure 4.3 as compared to certain Outlier languages

	whale	*bone*	*whetstone*	*heal*	*barracuda*
PPN	*tafu-raqa	*hui	*fuqanga	*mafo	*sapatuu
PTO	*tafu-aqa	*hui	*fuqanga	-	*hapatuu
TON	tafu-a'a	hui	fu'ofu'anga	–	hapatuu
NIU	tafu-aa	hui	–	–	–
PNP	*tafo-laqa	*iwi	*fuqanga	*mafo	*sapatuu
EFU	tafo-la'a	ivi	fuaga	mafo	sapatuu
EUV	tafo-la'a	*hui	fu'aga	mafo	hapatuu
WFU	tafo-ra	ivi	fuaga	hmafo	tapatu
WUV	tafo-laa	ivi	fuanga	mafu	–
TIK	tafo-raa	ivi	fuanga	mafu	–
REN	taho-ga'a	ibi	–	mafu	–
SAM	tafo-laa	ivi	foanga	mafu	sapatuu
PEC	*tafo-raqa	*iwi	*foanga	*mafu	*tapatuu
TOK	–	ivi	foa, fuaga	mafu	tapatuu
TUV	tafo-laa	ivi	–	mafu	tapatuu
TAK	tafo-raa	ivi	foana, fuana	mafu	tapatuu
SIK	taho-laa	ivi	–	–	tapatu
OJA	–	ivi	–	mahu	kapaku
NKR	doho-laa	ivi	hooanga	mahu	dabatuu
KAP	doho-raa	iwi	hooanga	–	dabaduu
PEP	*tafo-raqa	*iwi	*foanga	*mafu	*tapatuu

Key: * Borrowing from Tongan. PPN: Proto Polynesian; PTO: Proto Tongic; TON: Tongan; NIU: Niuean; PNP: Proto Nuclear Polynesian; EFU: East Futunan; EUV: East Uvean; WFU: West Futunan; WUV: West Uvean; TIK: Tikopian; REN: Rennellese; SAM: Samoan; PEC: Proto Ellicean; TOK: Tokelauan; TUV: Tuvala; TAK: Takuu; SIK: Sikaiana; OJA: Ontong-Java; NKR: Nukuoron; KAP: Kapingamarangi; PEP: Proto Eastern Polynesian.

t in any of seven Polynesian languages I examined in detail (Tongan, Samoan, Rapanui, Marquesan, Hawai'ian, Tahitian and NZ Maori). The exception is the spontaneous change of Proto Polynesian **mosokoi* > Tahitian *moto'i* 'ylangylang (*Cananga odorata*)', a change shared by a few Cook Island languages (Marck, 1999 based on Biggs 1992, 1993, 1994a). The change of * *s* > *t* is also noted in numerous instances for Kapingamarangi (Biggs 1994a, note in PEC **si(si)u* reconstruction). The chance of two such independent changes affecting the same word is very small. The modern distribution is best explained by claiming a single change, although this does not allow us to rule out, for instance, Samoan borrowing back the older pronunciation from Tongan, East Uvean or East Futunan. It is also possible that the earlier form of the word was **tapatuu* and that Tongan, East Uvean, East Futunan and Samoan changed the initial consonant to *s* and *h*.

Three words suggest Eastern Polynesian emerged out of Ellicean or a dialect of Nuclear Polynesian that had developed distinctive pronunciations compared to Tongic, East Uvean, East Futunan and possibly even Samoan. These words are the total set of cases found in Biggs' (1992) 420 Proto Nuclear Polynesian reconstructions in which sporadic sound changes of Eastern Polynesian are shared with all or part of what has been called Samoic-Outlier. Some others, such as the reflexes of Proto Polynesian **fanua* 'land, island' and **nguu-feke* 'squid', have sporadic sound changes or distinctive changes in morphology, but it is more difficult to suggest how changes were established. There is also Biggs' (1994a) PPN **kawaiki* 'crab sp.' reconstruction, where the second **k* > *t* in MVA (the only EP for which there is a cognate given) and in all Ellicean Outliers where cognates are known (but not Samoan which retains a regular **k* reflex). The form is not included here due to lack of further EP evidence.

Such changes are always subject to multiple interpretations. The clear, demonstrable, uniquely shared, sporadic sound changes so abundant for Proto Polynesian, Proto Eastern Polynesian, Proto Central Eastern Polynesian, Proto Tahitic and Proto Marquesic are less evident for Proto Nuclear Polynesian. The five that are most secure support for the Nuclear Polynesian hypothesis suggest the classification of Samoan and Eastern Polynesian as Ellicean.

SAMOIC-OUTLIER

Looking at the problem from a different perspective, are there any sporadic sound changes within Samoic-Outlier that shed light on the disintegration of Samoic-Outlier? The study of Nuclear Polynesian suggests that they may not be found, since Eastern Polynesian appears more closely related to some Samoic-Outlier (Ellicean) languages than others. But there could be sharings within Samoic-Outlier *after* the divergence of Eastern Polynesian, or overlapping distributions of innovations *before* the divergence of Eastern Polynesian.

There are commonly more sporadic vowel changes in any given Polynesian language than sporadic consonant changes (Marck 1999). Taking the vowels first, the only clusters of shared changes in Biggs' (1992) reconstructions occur within Pawley's (1967) Central Outlier and Northern Outlier groups (assigned above to Ellicean following Wilson 1985). These changes do not, by themselves, support the notion that the central and northern groups are linked, and they do not link any new languages or groups.

The 109 Proto Samoic-Outlier forms of Biggs (1994a) that have yet to be linked to any language outside the purported group, remain a problem. A Proto Polynesian word need be lost only twice (once by Proto Tongic times and once by Proto Eastern Polynesian times) to be reconstructed to Proto Samoic-Outlier on the basis of modern evidence (in the absence of cognates external to Polynesian). Similarly, a Proto Nuclear Polynesian word need be lost only once (by Proto Eastern Polynesian times) to be reconstructed to Proto Samoic-Outlier on the basis of modern evidence (in the absence of cognates external to Polynesia). Thus we must expect that many Proto Samoic-Outlier reconstructions are simply uniquely shared retentions from Proto Polynesian or Proto Nuclear Polynesian.

Secondly, based on the evidence in the section above and from Wilson (1985), Samoic-Outlier may not be properly conceptualized. When we look closely at Biggs' Proto Samoic-Outlier reconstructions, we find no diagnostic evidence for the notion that such a group exists independently of Proto Nuclear Polynesian itself.

Biggs (1992, 1993, 1994a) organized his data according to the standard theory of Polynesia subgrouping (Figure 4.1), and so his assumptions are clear and open to examination. He does not always find the standard theory acceptable and says so in notes concerning some reconstructions. For instance, in a typical note in the Proto Nuclear Polynesian **maka* 'sling' reconstruction, Biggs (1994a) states:

> Note: The innovation is loss of the meaning '*stone* n.'. As with a number of other apparent innovations it is shared exclusively between Eastern Polynesian and with some outliers. Cf. Bill Wilson's paper on the pronouns which no one has responded to.

Many of Biggs' (1994a) Proto Samoic-Outlier reconstructions involve unusual sound changes much like those in Figure 4.3, but they form no particular geographical pattern as a group. I review them individually below to demonstrate that they do not, individually or as a group, define a closed subset.

Proto Polynesian **fanua* 'land, placenta' is reconstructed on the basis of internal and external evidence. Although Nuclear Polynesian languages generally reflect a change to **fenua*, Samoan does not and neither does Tuvalu or West Futunan. Chants and poetry may have preserved the older pronunciation in Samoan, Tuvalu and West Futunan, just as Tongan *fanua* is preserved in chants where it is *fonua* in common speech (Churchward 1959). Three languages share a further change: Kapingamarangi, Nukuoro and Tongarevan (Penrhyn) have *henuu*, where the final vowel has come to agree with the semi-final vowel. Tongarevan (Penrhyn) is an Eastern Polynesian language in the Northern Cooks. Since there is no other evidence to link Kapingamarangi and Nukuoro with Tongarevan (Penrhyn), it would seem that the change has taken place twice and the observation is of restricted subgrouping value.

As Biggs (1994a) notes, the reflexes of Proto Polynesian **ngata* 'ended (of matters of concern, emotions)' are in complementary distribution with Proto Samoic-Outlier **ngato* 'used up, finished'. Reflexes of **ngata* are found in Tongan, Niuean, Samoan and NZ Maori, and reflexes of **ngato* are found in East Futunan, Mae, Rennellese and Nukuoro. Although there is only one Eastern Polynesian witness, NZ Maori, Biggs (1994a) has made extraordinary efforts to account for the vocabulary of NZ Maori, the sources are very good, and the history of more NZ Maori vocabulary is accounted for better than any other language in POLLEX. So **ngata* would seem to be the older pronunciation, although there are not enough Samoic-Outlier languages with reflexes of either to form a clear picture of distributions.

Proto Samoic-Outlier **maanatu* 'think, remember' is reconstructed in Biggs (1994a) and contrasts with Proto Nuclear Polynesian **manatu* 'think, remember'. The distribution of the Proto Nuclear Polynesian short vowel reflexes consists of forms in NZ Maori, Pukapukan, Samoan, Takuu, Tokelauan and West Futunan. The distribution of the Proto Samoic-Outlier long vowel reflexes are Samoan, Mele-Fila, Nukuoro, Pukapukan, Rennellese, Takuu and Tikopian. As Pukapukan and Samoan have reflexes of both, it may be an old alternation that probably still exists in many of the languages but was not recorded. Such incompletely documented distributions have no value for subgrouping.

Four related words for 'yawn' are reconstructed in Biggs (1994a): Proto Polynesian **mamawa*, Proto Polynesian **mama*, Proto Tongic **mamao* and Proto Samoic-Outlier **maavava*. The Proto Oceanic reconstruction (POC **mawap*) suggests Pre-Polynesian[4] **mawa*, which was reduplicated differently in some languages. Rapa Nui (Easter Island) *haka-mama* and Marquesan, Tahitian *mama* are the only Eastern Polynesian cognates known for any of the reconstructions, so surely Proto Eastern Polynesian was **mama*. This form is similar to Tongan and Niuean which reflect Proto Tongic **mamao* (> Pre Tongic **mamawa*). **Maavava* is reflected only in Samoan, East Futunan, Tokelauan, Tuvalu and Rennellese, whilst evidence for **mamawa* comes from Mae, Rennellese and West Futunan. If we assume a variant with a reduplicated initial syllable by Proto Polynesian times, and so reconstruct Proto Polynesian **mamawa*, it is not hard to imagine **-wa* becoming **-o* in Proto Tongic and Proto Eastern Polynesian shedding the final syllable. We are then left to account for **maavava* reflexes in some Samoic-Outlier languages. With the exception of Rennellese, these reflexes cluster around Western Polynesia: Samoan, East Futunan, Tokelauan and Tuvalu. Possibly it is a development in the area that post-dates the divergence of Eastern Polynesian and was either borrowed by Rennellese or independently developed there.

Proto Polynesian **kapakau* 'wing' is reconstructed on the basis of agreements between Tongan, Niuean, Samoan, East Futunan, East Uvean, Tuvalu and numerous Outliers, whilst other Outliers suggest a metathesized (i.e. one where some sounds have changed order) form **pakakau*. 'Futunic' Outlier languages seem to retain either form; but for West Uvean, the distributions are mutually exclusive: languages with **kapakau* reflexes do not have **pakakau* reflexes. Possibly this metathesis has occurred more than once, or possibly there are more doublets than the sources record. At any rate, the group is difficult to subgroup internally due to overlapping distributions of competing forms.

Proto Polynesian **pilau* 'decayed, stinking' is reconstructed on the basis of Proto Eastern Polynesian **pilau* and regular reflexes from Tongan and some Samoic-Outlier languages. However, Samoan, East Futunan, East Uvean, Tuvalu, Tokelauan and other Outliers do not have reflexes of **pilau*. Instead, they have reflexes of **pilo* 'stench of faeces'. This is fair evidence for a possible modification to the standard subgrouping of Polynesian, although it might simply be an innovation that has occurred more than once. If the change happened only once, there could be several explanations for the modern distributions. It could have been the case that **pilo* 'stench of faeces' developed around the East Futuna/Uvea–Samoa–Tokelau–Tuvalu area after the divergence of Eastern Polynesian or that Eastern Polynesian borrowed **pilau* under Tongan influence, i.e. that Tongans also contributed to the settlement of Eastern Polynesia and to the speech that arose there. This is something I have pondered but never been able to demonstrate beyond a few such forms as this. Langdon (1989) suggests a couple of Tongan loans into Marquesan and Proto Eastern Polynesian, but these rest on the irregular loss of the glottal stop in Marquesan which is a common sporadic occurrence, indicative but not diagnostic. We cannot demonstrate such a loan into Proto Eastern Polynesian diagnostically with the current distribution, but it is at least one possible explanation.

Proto Polynesian **qalito* 'core of a boil' is reconstructed on the basis of Proto Tongic, Samoan, Nukuoro and Rennellese, whilst three Samoic-Outlier languages reflect a metathesized form, Proto Samoic-Outlier **qatilo* 'core of a boil' (Samoan, Nukuoro, Rennellese). This seems an old doublet as both forms are found in Samoan, Rennellese and Nukuoro.

Proto Polynesian **sua* 'turn over, lever up, as soil with stick when weeding' is reconstructed securely on the basis of external agreements to Proto Tongic, Proto Eastern Polynesian and East Uvean, Mae, Mele-Fila, Nukuoro, Pukapukan, Samoan, Tikopian, Tuvalu and West Uvean. Numerous Samoic-Outlier languages reflect **sue* 'uncover', but the reflexes of the two are not in complementary distribution. **Sue* reflexes are found in Anutan, East Futunan, East Uvean, Mae, Mele-Fila, Luangiua (Ontong Java), Rennellese, Sikaiana, Tikopian and Tuvalu. This appears to be an old doublet.

This is the extent of materials from Biggs (1994a) with some irregularity of sound change reflecting the development and breakdown of Proto Samoic-Outlier. They do not support the notion of a closed subgroup (one that has no other members) but suggest that Samoic-Outlier was internally diverse by the time of the divergence of Eastern Polynesian. Samoic-Outlier is thus simply co-ordinate with Nuclear Polynesian and no distinct subgroup exists, i.e. Eastern Polynesian subgroups with a particular Samoic-Outlier group before it groups with the others. No new light has been shed upon the question of which Futunic Outliers came from which part of the Samoa, East Futuna, East Uvea, Tuvalu and Tokelau area.

It is therefore of interest to return to Pawley (1967) and to ask what were the original reasons for postulating Samoic-Outlier and to what extent they hold today (Marck 1999). Pawley's (1966) Nuclear Polynesian group was defined on the basis of shared innovations in morphology (which can be contrasted with shared retentions elsewhere), whilst Pawley's (1967) Samoic-Outlier group was suggested, not defined, on the basis of uniquely shared features (which cannot be contrasted with shared retentions elsewhere). Where different methods were used, the two groups tended to be accepted at once by other researchers (but see Howard 1981: 115).

The evidence for Samoic-Outlier was developed with sparse information for many languages. Pawley (1967) found reasons, as Bayard (1966) had in a lexicostatistical study, for believing that Outliers from Micronesia south towards the middle of the Outlier distribution in Melanesia came from Tuvalu, whilst Outliers from the south of the distribution in Melanesia north towards the middle may have come from East Futuna. Tuvalu was called Ellice at the time, and Ellicean was the name given to the group apparently emanating out of Tuvalu, whilst Futunic refers to the others. But with little data for many languages, it was not clear which Outliers were members of which group or if they were members of either.

The evidence for the integrity of Samoic-Outlier (Pawley 1967: 274-281) is given as sixteen exclusively shared features. I think we must now concede that there are potential problems with all of them in terms of their value for defining a closed group. The summary reasons are:

1. 2. 3. 8. 14. Could simply be shared retentions of Proto Nuclear Polynesian, lost in Eastern Polynesian.
4. 10. 11. 12. Are not shared by the entire group. Other 'SO' languages show retention of older patterns.
5. 6. Were specified as possible uniquely shared retentions in 1967, and this holds true today.
7. 9. 15. 16. The characteristics in question are limited to Ellicean or Ellicean plus Samoan and East Futunan and/or East Uvean.
13. Is now reconstructed by Biggs (1994a) to Proto Polynesian and could simply have been lost in Eastern Polynesian.

As there are no diagnostic arguments for Samoic-Outlier, I feel it must be abandoned in favour of the Samoan Ellicean-Eastern Polynesian group. East Futunan, East Uvean, Outliers other than Ellicean Outliers and Pukupuan therefore become unclassified Nuclear Polynesian languages.

REDEFINING NUCLEAR POLYNESIAN AND ABANDONING SAMOIC-OUTLIER

Uniquely shared, sporadic, irregular sound changes can be taken as one rough measure of a group's period of common development. Sporadic irregular sound changes are known from all Polynesian languages and presumably occurred in all Polynesian languages in the past. Our ability to demonstrate a host of them for Proto Polynesian but a relatively meagre number for Proto Nuclear Polynesian and none for Proto Samoic-Outlier suggests that:

1 the sporadic sound changes that we attribute to Proto Polynesian were still spreading easily through Western Polynesia up to the time at which Eastern Polynesian diverged, or
2 Eastern Polynesian diverged rather quickly after the period of easy sharing through Western Polynesian ended.

Wilson's (1985) pronoun study resulted in the subgrouping for Nuclear Polynesian given in Figure 4.4. We obtain a similar phylogeny with the

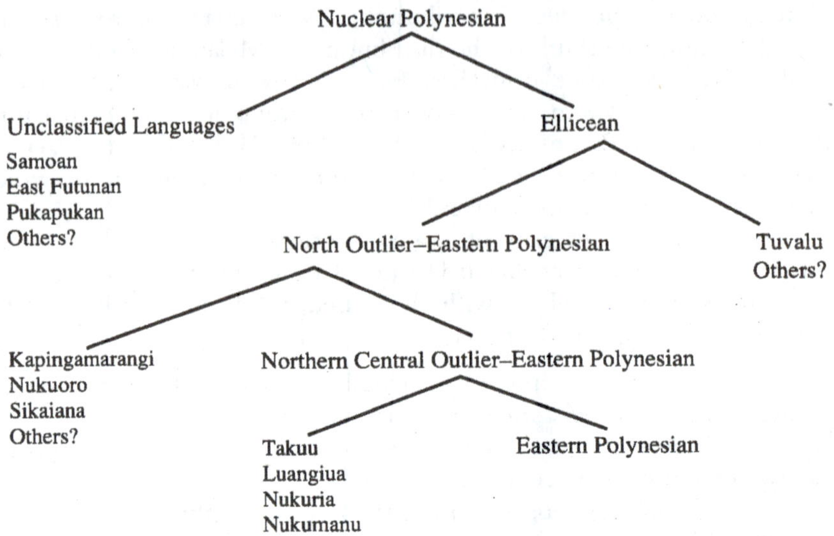

Figure 4.4 Subgrouping of Nuclear Polynesian based upon shared innovations in pronouns

Source: after Wilson (1985: Table 1)

present study of uniquely shared sporadic sound changes. Shared sporadic sound changes and shared innovations in the pronominal system may have distributed somewhat differently through time and space (slightly different phylogenies might have been involved). Shared sporadic sound changes are subtle differences in speech such as commonly exist between English dialects today and were possibly shared only between dialects that were essentially continuous. These scarcely noticeable markers of local dialects may have been contrastive in distribution with the pronominal systems. These may have been drawn into later areal systems after the sharing of sporadic sound changes ceased, or the pronouns may have changed over a wider area than the sporadic sound changes.

Samoic-Outlier is not defined by any uniquely shared sporadic sound changes (pp. 107–111). Eastern Polynesian, Samoan and the Ellicean Outliers are found to share two changes not shared by East Uvean, East Futunan or other non-Ellicean Outliers (pp. 100–107), and an otherwise universal Ellicean Outlier–Eastern Polynesian change is not shared by Samoan (pp. 100–107). The direction of change, in that final instance, is not secure, and it may not be the Ellicean Outliers and Eastern Polynesian that changed. But there is another such change where the direction is clear, and it was Ellicean Outlier and Mangarevan that innovated from a previous pattern. In any event, two and possibly three innovations of Proto Eastern Polynesian as compared to Proto Polynesian can now be shown to be shared with Ellicean Outliers and Samoan, whilst 'Futunic' Outliers, East Uvean and East Futunan show the Proto Polynesian pattern.

The subgrouping implications are that subtle aspects of the earliest surviving Eastern Polynesian speech developed out of Ellicean, as Wilson (1985) and others (Bayard 1966; Pawley 1967; Howard 1981; Biggs 1992, 1994a) have defined that group. A major difference here is that Samoan is assigned to Ellicean. 'Futunic' (non-Ellicean Outliers) languages have yet to be linked as a group by diagnostic methods and may have been the first Nuclear Polynesian languages to diverge from the others. Although called 'Futunic', since Bayard's (1966) work the people of the Outliers concerned generally recall coming from Uvea (East Uvea). Late borrowing by East Uvean from Tongan has probably obscured the nature of the original relationship between East Uvean and the 'Futunic' Outliers, with East Futunan now showing more obvious resemblances due to its retention of more directly inherited vocabulary and directly inherited pronunciations than East Uvean.

By comparing uniquely shared sporadic sound changes from Eastern Polynesian, Ellicean Outliers and Eastern Polynesian are placed in a group separate from Samoa. The classification of Samoan as the first to diverge within the group is very tenuous, so the subgrouping and subgroup names proposed are given in Figure 4.5.

The 'Samoic-Outlier' section (pp. 107–11) argues that there may have been some sharing of innovations between Samoa and East Futuna/Uvea after the divergence of Eastern Polynesian (see especially the discussion of **pilau*). Whether Tongan and Samoan continued to share innovations after the divergence of some Outliers and Eastern Polynesian is more difficult to establish by simple reference to Biggs (1992, 1993, 1994a). Those materials are organized on the assumption that a feature shared between Tongic and Samoan, East Uvean or East Futunan is inherited from Proto Polynesian. To establish

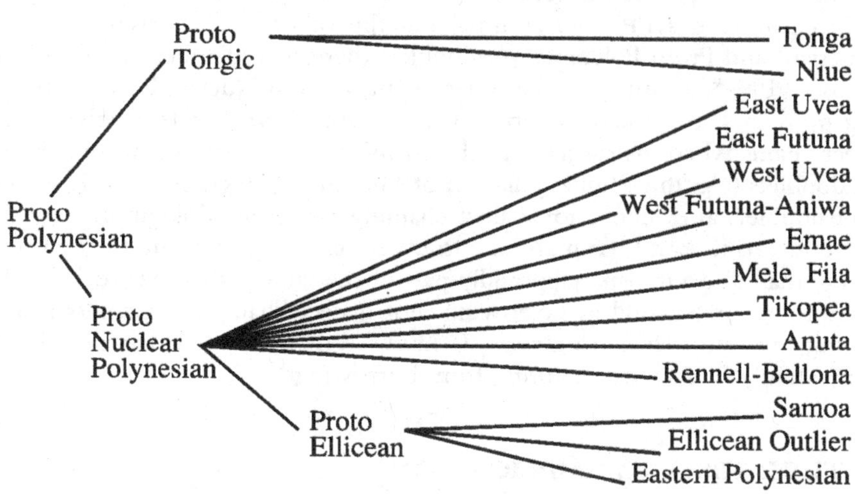

Figure 4.5 The revised subgrouping of Polynesian languages
Source: Marck

that changes may have spread between Tongan and Samoan after the divergence of Eastern Polynesian requires materials in which Proto Eastern Polynesian or numerous Outliers retain a pattern known from external evidence whilst Tongan and Samoan show some common change. A well-known case is the change of POC **sapa* > Proto Polynesian **hafa* > Proto Nuclear Polynesian **afa* < Proto Eastern Polynesian **afa* 'what?', which all involve regular changes and contrast with Tongic, Samoan and numerous Outliers which irregularly lost Proto Polynesian **f*. Pawley (pers. comm.) also cites Proto Polynesian **roo* 'go (pl. subject)' which is *oo* in Tongan and was **loo* in Proto Nuclear Polynesian, as expected, but is *oo* in Samoan and numerous Outliers, which seem to have borrowed the Tongan pronunciation (Samoan/Outlier ***loo* is expected).

This suggests a reconsideration of the possibility of dialect chaining in Western Polynesia, as Dyen (1981) and Rensch (1987) suggest. Dyen's suggestion of Western Polynesian dialect chaining was based upon lexicostatistics showing that Tongan, Samoan, East Uvean and East Futunan are far more closely related to each other than any are to Eastern Polynesian. Lexicostatistics gives equal weight to shared retentions and shared innovations. When we limit our discussion to shared innovations, a more subtle and convincing phylogeny obtains. More and less closely related Western Polynesian languages in proximity have tended to retain common ancient vocabulary for long periods after they cease to share innovations freely. Shared innovations are diagnostic of phylogenetic relations, whilst shared retentions are not, although practitioners of lexicostatistics sometimes imply that they mean more than they do.

Rensch's (1987) arguments for dialect chaining around Western Polynesia after the divergence of Eastern Polynesia are based upon doublets in Tongan and East Uvean. The Proto Polynesian **h* doublets (where one member of a minimal pair retains PPN **h* as usual but the other loses it, as in Nuclear Polynesian) and Proto Polynesian **r* doublets (here one member of a minimal pair loses PPN **r* as usual but the other retains it, as in Nuclear Polynesian) of Tongan are best explained in terms of borrowing from Nuclear Polynesian. Dialect chains where *innovations* are shared more or less broadly do not show such doublets (e.g. the Chuukic dialects of Micronesia). Rensch's (1987) offering of doublets as evidence for dialect chaining was methodologically flawed. If the doublets in East Uvean are due to dialect chaining, why are they rare in East Futunan and Samoan, supposedly part of the same dialect network. Such situations are understood as borrowing, not dialect chaining; to demonstrate chaining one must show isoglosses. It must be demonstrated that they exist independently of doublets resulting from borrowing.

CULTURE HISTORY IMPLICATIONS

Proto Polynesian seems to have developed in a compact area between Tonga, Samoa, East Uvea and East Futuna, with northern and southern dialects (Green

1981) North Pre-Proto Nuclear Polynesian and South Pre-Proto Polynesian respectively (Pawley, forthcoming). Tuvalu shares about the same amount of total vocabulary with Tongan, East Futunan and East Uvean as it does with Samoan (see Table 4.5). Thus we must not assume that the origins of the Tuvalu and Tokelau language in Samoa came at a time when the vocabularies of Proto Tongic and Proto Nuclear Polynesian were highly distinct.

Nuclear Polynesian seems not to have become highly distinct from Tongic before it began to have identifiable dialects of its own. ('Dialects' is used here in the sense of the regionalism that exists in English in the British Isles today: small differences in pronunciation that would have marked people as coming from one island or group of islands rather than another.) There is very little we can attribute to the entire group (cf. Pawley, 1996). Even total vocabulary agreements within the group are only marginally higher than with Tongan (see Table 4.5); but Tongan seems to have a more conservative lexicon than Samoan, so there were probably more distinctions in vocabulary than crude statistical estimates suggest.

The disintegration of Proto Nuclear Polynesian seems to be associated with two events that cannot presently be ordered in relation to each other. One is the internal disintegration of 'Futunic' (East Futunan, East Uvean and the 'Futunic' Outliers (see Table 4.6; Bayard 1966; Pawley 1967)). Our current inability to demonstrate diagnostically a Futunic subgroup or to suggest its internal structure was predicted by Ross (1995), who notes the general phenomenon of 'unclassified' languages in the locality where a proto language developed, Proto Nuclear Polynesian in this instance. The other event is the development of innovations in Ellicean (Samoan, Eastern Polynesian and the traditional Ellicean Outliers), where East Uvean, East Futunan and 'Futunic' Outlier languages show retentions of older patterns. By the measure of uniquely shared sporadic sound changes, Samoan is Ellicean and Eastern Polynesian may have specifically stemmed from an Ellicean language other than Samoan, as there are one and possibly two sporadic sound changes shared between Ellicean Outliers (other than Samoan) and Eastern Polynesian. But one form and its distributions are problematic; the other is not well known from Eastern Polynesian and could be a local development in Mangareva; Samoa may have later borrowed pronunciations from the south, and, therefore, Eastern Polynesian may have emerged directly out of Samoa. The previous ordering is no longer the only one possible.

Biggs (1994a) contains a few Proto Ellicean Outlier forms that appear to be good lexical innovations but could easily be cases of shared loss in Eastern Polynesian and Samoan or ones in which cognate forms exist but are not reported for Eastern Polynesian and Samoan. Like Pawley's (1967) exclusively shared features for 'Samoic-Outlier', cognates outside the group may still be lurking around somewhere or may simply be shared retentions from Proto Nuclear Polynesian.

The subgroupings suggested here and by Wilson (1985) differ from Pawley (1967) in implying that the effective linguistic settlement of Eastern Polynesia took place at a time after or intimately connected with the settlement of the Outliers. Pawley's (1967) subgrouping suggested that 'Samoic-Outlier' languages continued to share innovations after Eastern Polynesian diverged. This now seems unlikely to have been the case, and we seem to be speaking of a narrow time frame in which both the Outliers and Eastern Polynesian diverged: 'Futunic' Outliers from East Uvean or East Futunan, as Bayard (1966, 1976) and their own traditions suggest, and Eastern Polynesian and Ellicean Outliers specifically from Samoa or Tokelau/Tuvalu. Howard (1981: 111–114) has produced an impressive list of apparent lexical innovations shared by the Ellicean Outliers, but these may be shared losses of Eastern Polynesian languages (and Samoan as well). The absence of shared sporadic sound changes in the Ellicean Outlier group as a whole suggests internal divergence before, at the time of or only slightly after the divergence of Eastern Polynesian, and not after, as the unrevised subgrouping implied.

So there is now the question of whether similar ecological, technological and social or political factors may have been involved in the settlement of the Outliers and Eastern Polynesia, where previously there may have been the notion that there were distinct constellations of such variables. Glottochronological dates for the time concerned are about 2000 years BP, but these are not reliable. However, the relative order is now established by diagnostic evidence, and we can be reasonably sure that the *effective* linguistic settlement of Eastern Polynesia (i.e. that which resulted in the modern languages) was after or at about the same time as the Outliers, and not before in the instance of the 'Futunic' Outliers, and not before (or not much before) in the instance of the Ellicean Outliers.

What contacts did the settlers of Eastern Polynesia have with Micronesia at the time they were leaving for their destination? The Micronesia–Polynesia cultural boundary would probably have been established at about this time. Such traditional puzzles as Micronesia sharing certain fishhook types with Eastern Polynesia (but not Western Polynesia) could originate in localized contacts between people from Micronesia and Polynesia as they came to settle the atolls of Kiribati and Tuvalu, with people from the Tuvalu/Tokelau area then taking the technology to Eastern Polynesia.

Irwin's (1992) argument for a continuous settlement is intriguing, and it is indeed mysterious that a pause of 1,000 years or more may have occurred between the settlement of Western and Eastern Polynesia. But such a pause remains apparent in both the linguistic record and dated archaeological dates. If Wilson's (1985) interpretation and the present study are correct, the apparent end of the pause takes on a slightly different character, one that may be linked to the emergence of atolls north of Samoa and in the Cooks as habitable surfaces. This emergence may have occurred by about 2000 BP or earlier (Nunn 1994).

The earliest good dates for an Eastern Polynesian archaeological site stand at AD 300–600 and come from the Marquesas (Spriggs and Anderson 1993). I mentioned in the introduction that the standard linguistic model of

Polynesian settlement had much in common with a *conservative* archaeological model of Polynesian settlement. That model is based simply on the earliest reliable dates for cultural materials in the various groups (as defined by Spriggs and Anderson 1993) and asks why there should be an interdisciplinary model of settlement beginning much earlier when earlier dates have not been encountered. Other archaeologists (e.g. Kirch 1986; Kirch and Green 1987) expect that earlier dates will emerge due to previous sampling error, problematic geological situations, or for other reasons.

Irwin (1992) predicted that an entirely different order of dates going back to the first millennium BC would be recorded. This is based upon the premise that a pause of the magnitude presently suggested by linguistics and the most conservative archaeological assessment is inconsistent with the history of the broader, relentless movement of Austronesians into the island world.

Pawley (1996) replies to Irwin (1992) by presenting the linguistic evidence for this extended pause and noting that replacement of previously established languages around Eastern Polynesia would have been difficult. Polynesian linguists in general probably have little enthusiasm for Irwin's model, since linguistic substratum evidence from around Eastern Polynesia seems so entirely lacking. There is some problem with attributing the long-distance navigation skills of historic Micronesians and Polynesians to the groups emerging out of the Bismarck Archipelago in the mid-second millennium BC, as Irwin implies in his model. Long-distance voyaging to small targets is a uniquely Micronesian and Polynesian problem, and early Oceanic-speaking groups may not have had those skills or they may have developed only locally around the Santa Cruz area, which has the main small targets in Melanesia.

The present linguistic model would be compatible with somewhat earlier archaeological dates for the settlement of central Eastern Polynesia. Most linguists familiar with the relevant issues would probably be comfortable with dates back to about 2000 years BP. The reasons are not so much lexicostatistical, although that is the estimate[5] for the divergence of Eastern Polynesian from other Polynesian. The reasons concern an extended period of common development by Polynesian languages before their internal diversification. If linguists cannot have dates for Polynesia earlier than about 1100 BC, they feel that they need dates running to the end of the first millennium BC to account for Polynesian homogeneity (cf. Pawley, 1996). On the basis of Eastern Polynesian language heterogeneity, most linguists would wonder how those languages could have remained unified past about AD 500. The period from about BC-AD to AD 500 can then be seen as the period in which the innovations of Eastern Polynesian, and especially Central Eastern Polynesian, developed. It is a comfortable scenario for the linguists and one in which there is some room for reinterpretation of exact dates.

New archaeological methods are being applied in Polynesia and are resulting in suggestions of much earlier dates for significant changes in paleoenvironments and their flora (Kirch and Ellison 1994). Paleo-environmental evidence can be well marked and does not rely upon chance discoveries to the extent of those associated with material culture. Thus linguists anxiously await a consensus from the archaeologists on the nature and meaning of the paleoenvironmental evidence for Polynesia as a whole. A rigorous study of Mangaia produced several independent dates of about 2500 BP for the appearance of charcoal in the pollen record for the first time, indicating burning of vegetation by people (Kirch and Ellison 1994). These kinds of studies have the potential for leading to the most reliable profiling of human arrival in the area, though they do not result in the association of this human activity with any particular material culture. Recovery of material culture is problematic for even the 'late-early' and 'early-middle' stages of human settlement. Still, Kirch and Ellison's (1994) interpretation has been questioned, as has the general efficacy of such profiling (Anderson 1995). Perhaps when it has been applied to more localities the archaeologists will move towards greater consensus.

Dates of 2500 BP for Eastern Polynesia are quite disturbing because they challenge two assumptions:

1 that a long period of common development occurred in Western Polynesia because
2 the distance from Western Polynesia to Eastern Polynesia was so great that linguistic unity would have ended upon settlement of Eastern Polynesia.

We can rethink these assumptions by examining the extent to which the apparent mass of 'pause' vocabulary is real or an artefact of the descriptive and comparative situations. A sample of Biggs' (1994a) reconstructions not currently known from outside of Polynesia could be thoroughly searched for external cognates, and the overall uniqueness of Polynesian vocabulary reassessed.

Perhaps Eastern and Western Polynesian languages did not share innovations after the effective linguistic settlement of Eastern Polynesia, i.e. the effective linguistic settlement of Eastern Polynesia ended the Pre-Polynesian and Pre-Nuclear Polynesian periods. (*Pre-* languages are those leading up to a proto language, the language spoken at the time of a linguistic split.) Tahitian and the Society Islands have dominated the linguistics of a vast area (the Tuamotus, Cooks and Australs) for a long period of time. If they ever regularly shared innovations with the Samoan area, the linguistic model would require a dimension as yet unexplored. This is difficult to contemplate, given the distances involved, but we have no substratum hypotheses,[6] none seems likely to develop, and the archaeological evidence for the area may ultimately require the linguists to take one tack or the other, to postulate a shorter period of common development or to claim that there was replacement without demonstrable substrata. The second claim would ultimately be the more difficult to defend, so we may eventually give ground on the first issue or revise the assumption of rapid linguistic divergence once Eastern Polynesia was settled.

Linguistic and archaeological investigations of possible 'atoll corridors' into Eastern Polynesia will be an important subject for future work. On the linguistic side there has been a great deal of useful recent work on Tokelauan (e.g. Simona 1986; Hovdhaugen 1989; Hovdhaugen *et al.* 1989; Vonen 1989, 1991, 1994) and Tuvalu (e.g. Ranby 1980; Jackson 1994), but the Northern Cook languages are not well documented and the archaeology of all those atolls is more or less neglected.

The identification of Eastern Polynesian as Ellicean requires a closer look at the atolls both linguistically and archaeologically. There was linguistically regular transference of Proto Polynesian and Proto Nuclear Polynesian names for high island plants, physiography and such. Atolls may have linked Eastern and Western Polynesia in a way that was not previously suggested. Obviously the atolls themselves were not the source of the high island crops and physical environment vocabulary that found their way to Eastern Polynesia, but more subtle aspects of speech followed the Ellicean pattern. Another possibility is that Eastern Polynesian simply emerged out of Samoa and that Ellicean Outlier did so at about the same time or slightly later. Ordering those events has not yet been accomplished by linguistic methods.

Wilson's (1985) suggestion of an Eastern Polynesian–Ellicean connection has been neglected, but it was not obvious how to support or contest it (Biggs 1992, 1993, 1994a). A search of Biggs (1992) produced sparse supporting evidence from uniquely shared sporadic sound changes, and the result was a somewhat different but supporting phylogeny.

This reassessment of the standard subgrouping does not preclude the previous implication, i.e. that there was little dialect diversity in Nuclear Polynesian at the time of Eastern Polynesian's divergence. There *was* internal diversity and we can add a little to Wilson (1985) in constructing a picture of how much diversity was current. Eastern Polynesian *is* Ellicean, and Samoan is also a member of it. Whether the 'Futunic' Outlier languages had already diverged from East Futuna and/or East Uvea (or even Samoa) when Eastern Polynesian diverged from Ellicean has not yet been ordered by linguistic methods, but it is possible that 'Futunic' Outliers had begun diverging before Eastern Polynesian, where before we had reason to believe that they had not.

Ellicean Outliers and Eastern Polynesian may have split from Samoan before they diverged from each other (also Wilson 1985). This would essentially require settlement of Tuvalu and/or the Tokelaus prior to the settlement of Eastern Polynesia from those atolls (and not Samoa – or East Futuna/Uvea). But the evidence here is a single, uniquely shared sporadic sound change, which may have that particular pattern due to borrowing, and a second, uniquely shared sporadic sound change, for which only a single cognate is known from Eastern Polynesia.

NOTES

1 I am indebted to Andrew Pawley and Matthew Spriggs who commented extensively on earlier drafts of this chapter.

2 As opposed to a 'regular' (across the board) change that occurred in all cases where a certain condition existed in the proto language.

3 Although Pawley has lectured on this distribution since the 1970s, I cannot find it published.
4 The language as it was developing up to Proto Polynesian times, which is the time of internal diversification.
5 The rates of basic vocabulary replacement that lexicostatistics assumes are relatively stable usually turn out to be variable.
6 After three years of intensive work on Biggs' (1992, 1993, 1994a) *POLLEX*, the only substratum hypothesis I can offer for any part of Polynesia is that there may be a non-Central Eastern Polynesian substratum in Mangarevan speech. It is probably an Eastern Polynesian substratum rather than anything more ancient (i.e. the word involved is consistent with the Proto Eastern Polynesian reconstruction but not the Proto Central Polynesian reconstruction). In any event, only one word is involved: Marquesan *'a'ine* 'woman' which is a doublet for the regular Central Eastern *ve'ive* of Mangarevan.

REFERENCES

Anderson, A. (1995) Current approaches in East Polynesian colonisation research. *Journal of the Polynesian Society* 104(2), 110–132.

Anderson, A. and I. Smith (1996). The transient village in southern New Zealand. *World Archaeology* 27(3): 359–371.

Bayard, D.T. 1966. The cultural relationships of the Polynesian Outliers. MA thesis, University of Hawaii.

Biggs, B. 1992, 1993, 1994a. *Comparative Polynesian Lexicon (POLLEX).* Computer file, Department of Maori Studies, University of Auckland. (Copies obtained annually during course of work).

Biggs, B. 1994b. Does Maori have a closest relative? In *The Origins of the First New Zealanders*, D.G. Sutton (ed.), 96–105. Auckland: Auckland University Press.

Churchward, C.M. 1959. *Tongan Dictionary*. Tonga: Government Printing Press.

Dunnell, R.C. 1987. Comment on Kirch and Green 'History, phylogeny, and evolution in Polynesia', *Current Anthropology* 28(4), 444–445.

Dyen, I. 1981. The subgrouping of the Polynesian languages. In *Studies in Pacific Languages and Cultures in Honour of Bruce Biggs*, J. Hollyman and A. Pawley (eds), 83–100. Auckland: Linguistic Society of New Zealand.

Elbert, S.H. 1953. Internal relationships of Polynesian languages and dialects. *Southwestern Journal of Anthropology* 9(2), 147–173.

Emory, K.P. 1938. The Tahitian account of creation by Mare. *Journal of the Polynesian Society* 47, 45–63.

Emory, K.P. 1963. East Polynesian relationships: settlement pattern and time involved as indicated by vocabulary agreements. *Journal of the Polynesian Society* 72(1), 78–100.

Green, R. 1966. Linguistic subgrouping within Polynesia: the implications for prehistoric settlement. *Journal of the Polynesian Society* 75(1), 6–38.

Green, R. 1981. Location of the Polynesian homeland: a continuing problem. In *Studies in Pacific Languages and Cultures in Honour of Bruce Biggs*, J. Hollyman and A.K. Pawley (eds), 133–158. Auckland: Linguistic Society of New Zealand.

Green, R. 1988. Subgrouping of the Rapanui language of Easter Island in Polynesia and its implications for East Polynesian Prehistory. In *First International Congress, Easter Island and East Polynesian, Hanga Roa, Easter Island, 1984, Volume 1: Archaeology*, C.F. Cristino, P.C. Vargas, R.S. Izaurieta and R.P. Budd (eds), 37–57. Universidad de Chile: Facultad de Arquitectura y Urbanismo, Instituto de Estudios Isla de Pascua.

Harrison, S. 1981. Recent directions in Oceanic linguistics: a review of the contributions to *Studies in Pacific Languages and Cultures* (Hollyman and Pawley 1981). *Oceanic Linguistics* 20(2), 151–231.

Hovdhaugen, E. 1989. The morphology of plural verb formation in the Tokelauan dialect of Nukunonu. In *VICAL 1: papers from the Fifth International Conference on Austronesian Languages*, R. Harlow and R. Hooper (eds), 271–282. Auckland: Linguistics Society of New Zealand.

Hovdhaugen, E., I. Hoëm, C.M. Iosefo and A.M. Vonen 1989. *A Handbook of the Tokelau Language*. Oslo: Norwegian University Press.

Howard, I. 1981. Proto-Ellicean. In *Studies in Pacific Languages and Cultures in Honour of Bruce Biggs*, J. Hollyman and A. Pawley (eds), 101–118. Auckland: Linguistic Society of New Zealand.

Irwin, G. 1992. *The Prehistoric Exploration and Colonisation of the Pacific*. Cambridge: Cambridge University Press.

Jackson, G.W. 1994. *A Tuvaluan–English Dictionary*. Suva: Oceania Printers.

Kirch, P.V. 1986. Rethinking East Polynesian prehistory. *Journal of the Polynesian Society* 95(1), 9–40.

Kirch, P.V. and J. Ellison 1994. Paleoenvironmental evidence for human colonization of remote Oceanic islands. *Antiquity* 68(259), 310–321.

Kirch, P.V. and R.C. Green 1987. History, phylogeny, and evolution in Polynesia. *Current Anthropology* 28(4), 431–456.

Langdon, R. 1989. The significance of cultivated plant names in the settlement of Eastern Polynesia. In *VICAL 1: oceanic linguistics, papers from the Fifth International Conference on Austronesian Linguistics, Part I*, R. Harlow and R. Hooper (eds), 305–333. Auckland: Linguistics Society of New Zealand.

Marck, J. 1986. Micronesian dialects and the overnight voyage. *Journal of the Polynesian Society* 95(2), 253–258.

Marck, J. (1996) Eastern Polynesian subgrouping today. In *Oceanic Culture History: Essays in Honour of Roger Green*, J. Davidson, G. Irwin, A. Pawley and D. Brown (eds), pp. 491–511.

Marck, J. (1999) Polynesian language and culture history, Ph.D. thesis, Department of Linguistics, Research School of Pacific and Asian Studies, Australian National University.

Nunn, P. 1994. *Oceanic Islands*. Oxford: Blackwell.

Pawley, A. 1966. Polynesian languages: a subgrouping based on shared innovations in morphology. *Journal of the Polynesian Society* 75(1), 39–64.

Pawley, A. 1967. The relationships of Polynesian Outlier language. *Journal of the Polynesian Society* 76(3), 259–296.

Pawley, A. (1996) On the Polynesian subgroup as a problem for Irwin's continuous settlement hypothesis. In *Oceanic Culture History: Essays in Honour of Roger Green*, J. Davidson, G. Irwin, A. Pawley and D. Brown (eds), pp. 387–410.

Ranby, P. 1980. *A Nanumea Lexicon*. Canberra: Pacific Linguistics C-65.

Rensch, K. 1987. East Uvean, Nuclear Polynesian? Reflections on the methodological adequacy of the tree model in Polynesia. In *A World of Language: papers presented to Professor S.A. Wurm on his 65th birthday*, D.C. Laycock and W. Winter (eds), 565–581. Canberra: Pacific Linguistics C-100.

Ross, M. 1995. Some current issues in Austronesian lingustics. *Comparative Austronesian Dictionary Volume 1*, 45–120. Berlin: Mouton de Gruyer.

Simona, R. 1986. *Tokelau Dictionary*. Apia, Western Samoa: Office of Tokelau Affairs.

Spriggs, M. and A. Anderson 1993. Late colonisation of East Polynesia. *Antiquity* 67(255), 200–217.

Terrell, J. 1987. Comment on Kirch and Green 'History, phylogeny, and evolution in Polynesia'. *Current Anthropology* 28(4), 447–448.

Vonen, A. 1989. Verb phrase versus noun phrase in Tokelauan. In *VICAL 1: papers from the Fifth International Conference on Austronesian Languages*, R. Harlow and R. Hooper (eds), 581–591. Auckland: Linguistics Society of New Zealand.

Vonen, A. 1991. Nouns or verbs in Tokelauan? In *Proceedings of the Twelfth Scandinavian Conference on Linguistics*, H.A. Sigurdsson (ed.), 449–460. Reykjavik: Haskoli Islands.

Vonen, A. 1994. The expression of temporal and aspectual relations in Tokelau narratives. In *Tense, Aspect and Action: empirical and theoretical contributions to language typology*, C. Bache, H. Basboll and C.-E. Lindberg (eds), 371–396. Berlin: Mouton de Gruyter.

Welsch, R.L. 1987. Comment on Kirch and Green 'History, phylogeny, and evolution in Polynesia'. *Current Anthropology* 28(4), 444–445.

Wilson, W.H. 1982. *Proto-Polynesian Possessive Marking.* Canberra: Pacific Linguistics B-85.

Wilson, W.H. 1985. Evidence for an Outlier source for the Proto Eastern Polynesian Pronominal System. *Oceanic Linguistics* 24(1&2), 85–133.

Part II

INTERPRETING CHANGE

5 *Celts and others: maritime contacts and linguistic change*

JOHN WADDELL AND JANE CONROY

INTRODUCTION

There has been a prolonged, if sporadic, debate as to how and when the islands of Ireland and Britain became Celtic speaking. It is a debate that is older than the century, going back at least as far as Sir John Rhys' influential *Celtic Britain* in 1882, and it has waxed and waned as scholars have agreed or disagreed. Until relatively recently, there has at least been some consensus that the presence of a Celtic language in these islands was due to the arrival of one or more groups of Celtic-speaking peoples. This was not an unreasonable supposition, for history, after all, records several instances of the migration of Celtic peoples on the Continent. But history is of little help as far as these islands are concerned. Archaeology, of course, raised other problems, and much has been written in attempts to correlate linguistic developments and archaeological hypotheses. The traditional view has been that the Celts and their language spread over a wide area of Europe in the later part of the last millennium BC (Figure 5.1).

The Celtic family of languages is today divided into Insular Celtic and Continental Celtic, names reflecting the geographical distribution of these groups in Europe and Asia Minor. Insular Celtic is subdivided on linguistic grounds into Goidelic, whose modern descendants comprise Irish, Scots-Gaelic and Manx, and Brittonic, comprising Welsh, Cornish and Breton. The latter is classed as an Insular Celtic language because it derived from Brittonic in the fifth century AD. Continental Celtic includes Gaulish, Lepontic, Hispano-Celtic (or Celtiberian) and Galatian. All were extinct by the seventh century AD. The evidence is very fragmentary indeed and consists mainly of inscriptions, but also coin inscriptions, names, glosses and substratum words. The Gaulish inscriptions date from the third century BC to the first century AD. The Lepontic inscriptions come from a limited area of northern Italy and are written in a variant of the Etruscan alphabet dating to the last centuries BC. Hispano-Celtic is known in the main from stone inscriptions and a few bronze inscriptions from northern Spain; the earliest date to the last three

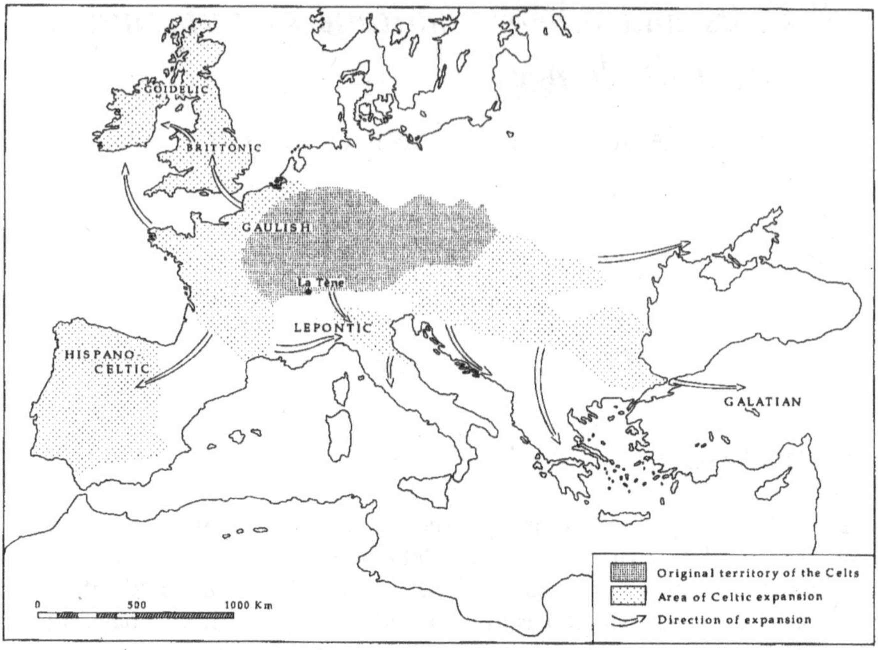

Figure 5.1 The Celtic World
Source: after Duval (1977)

centuries BC. In addition to this sort of evidence, Continental Celtic includes thousands of proper names from a great variety of sources. This whole body of material is the earliest extant primary evidence for the study of the Celtic language family.

The earliest inscriptions in Insular Celtic are to be found in the Ogham inscriptions known mainly from the southern half of Ireland and from Wales; some of these date to the fourth century AD or earlier, some are later. The rich textual evidence, of course, is later still, much of it incorporated in medieval manuscript. Primitive Irish or Goidelic differs from British or Brittonic phonetically in several ways, as well as retaining the archaic Indo-European **qu*. The instances of *q* in Gaulish, as in the river name *Sequana*, or the names of the months *Quimon* or *Equos* in the Coligny Calendar are unexplained but may be archaisms. Archaic elements are a notable feature of Hispano-Celtic where, as in Goidelic, the non-labialization of *q* is found and is but one of a number of primitive features common to both of these languages. A recent authoritative review of the Celtic languages, ancient and modern, is *The Celtic Languages* edited by Ball and Fife (1993). Evans (1995) has also provided a measured assessment of the linguistic debate whilst recognizing 'the labyrinthine and frustrating nature of the subject'.

In the last few centuries BC, the Celtic language family had a remarkable distribution, from the Atlantic to the Carpathian basin, and beyond. Figure 5.1

is a slightly modified version of a distribution map published by P.-M. Duval in his major work *Les Celtes* (1977: fig. 449). The map is entitled 'The Celtic World', and the arrows purport to show the various directions of Celtic expansion. This is, more or less, the traditional picture presented by comparative linguistics and archaeology for over a century: the dispersal of Celtic-speaking peoples from a homeland in western and central Europe.

In Ireland, various writers have equated 'the coming of the Celts' with the introduction of such archaeological phenomena as the knowledge of iron or a La Tène art style. For example, Macalister (1928) stated in his *Archaeology of Ireland* that the Celts came to Ireland at the inception of the Iron Age *c.* 400 BC; they were few in numbers but subdued the pre-Celtic aborigines with their superior iron weapons. Others in the 1920s and 1930s attributed early Celticization to late Bronze Age 'sword-bearers'. In the late 1960s, Myles Dillon, who had studied Indo-European elements in early Irish tradition, reverted to older notions of early Celts and suggested that they might even be identified with the 'Beaker Folk' around 2000 BC (Dillon and Chadwick 1967). He was aware that most scholars dated the first Celtic settlements in Britain as late as 600 BC, but he argued that the great archaism of Irish tradition in language, literature and social organization made such an early date a probability. Today such wide linguistic disagreement seems to be a thing of the past; now, at least in this one area, a measure of consensus is clear: these islands and parts of Continental Europe became Celtic speaking in later prehistoric times. How this came about is still the subject of some disagreement in both archaeological and linguistic circles, and it is probably fair to say that the island of Ireland presents this problem in a most perplexing form because its archaeological record offers no unambiguous evidence for the Celtic settlements proposed by historical linguists.

The long saga of controversy and argument as to how Ireland became Celtic speaking has been summarized in Waddell (1991), a paper that developed from a contribution to a debate on the origins of the Irish initiated by J.P. Mallory at a conference in Belfast in 1984. There was general agreement there that Celtic emerged in the last millennium BC, and some consensus that this could not be explained satisfactorily without some reference to an intrusion of people, though this was not recognizable in the archaeological record of either the later Bronze Age or the Iron Age. The suggestion that a prestige goods economy, reflecting the interaction of regional élites, may have been a primary factor in the emergence of an insular Celtic language was not received with great enthusiasm. However, alternative models of language diffusion were at least being canvassed. Piggott (1979, 1983) had suggested that prestige gift exchange formed a mobile upper-class archaeology and that the transmission of the Celtic languages might also have occurred in this way. Koch (1986) also argued that the *koiné* of the later Bronze Age metal trade may have become a prestige speech.

Renfrew (1987) proposed a convergence model (*sensu* Renfrew 1989) of Celtic linguistic development and adapted Christopher Hawkes' concept of 'cumulative Celticity', which ascribed the Celticization of Britain and Ireland to the continuing accumulation of new, upper-class, Celtic-speaking masters. Renfrew argued that peer polity interaction contributed to the emergence of the Celtic languages from generalized Indo-European 'essentially in those areas where their speech is later attested', there being no one, localized Celtic 'homeland'. The homelands of the Celts would be constituted by the full extent of the area where Celtic languages came to be spoken (always excluding such later offshoots as Galatia).

The later European Bronze Age may have been a period of significant linguistic change and development – a widespread pattern of regional interaction at an élite level and a vertical linguistic continuum in the ranked societies in question being but some of a range of factors that could have produced larger, discrete language groups (Waddell 1991). Koch (1991: 18) has suggested that proto-Celtic could have been consolidated in the prestige economies of the late Bronze Age (*c.* 1300–600 BC) 'in which an Atlantic Zone with centres in Armorica, south-east England, south Wales, Ireland, and later on Iberia, was in a continuous close contact with, and generally followed the cultural lead of, Urnfield/Hallstatt C west-central Europe'.

In fact, the emergence of a Celtic language or languages in later prehistory would be due to an intensification of a complex series of processes operating across parts of Europe since the third millennium BC. These would have included economic intensification, increasing polity size, developing gender and social stratification, and increasingly active trade along coasts and rivers (Robb 1993). Whilst Ireland, as elsewhere, had its own regional idiosyncrasies, there is good evidence that it participated at an élite level in wider contemporary fashions. A range of metal types (and at the moment rather little settlement evidence) indicates that Ireland shared in wider European élite fashions from at least the latter part of the second millennium BC. Some of these items, notably bronze spears and shields, and finely decorated or crafted objects of gold or bronze, of various dates, are justifiably seen as prestigious élite possessions, some probably for ostentatious display. There is some evidence too that social stratification became more marked in the later Bronze Age. There are significant gaps in the archaeological record but the broad picture is a fairly consistent one.

The degree to which the island of Ireland participated in this process is probably one of the difficulties uppermost in the minds of many Irish archaeologists and historical linguists. Most would probably agree with Koch (1991: 17) that 'by any reckoning the Celticisation of the British Isles was one of the great events of Insular prehistory', but would still believe, in some form, in 'the coming of the Celts' or 'Celtic movements' (Schmidt 1992). It is interesting, in this regard, to note that Raftery's comprehensive study of the Irish Iron Age, *Pagan Celtic Ireland*, deals only briefly with this problem and concludes regretfully that 'it seems almost heretical to conclude that a Celtic invasion of Ireland never happened' (Raftery 1994: 228). Elsewhere in Europe, the population movement model is still favoured (Barford 1991).

The waters of the Irish Sea, and indeed those of the English Channel, are more often seen as an obstacle than anything else, but it has been argued that the intensity of Ireland's contacts with Britain has been underestimated (Waddell 1992). Archaeological evidence suggests the distinct possibility that the Irish Sea, a relatively modest body of water, far from being a barrier to communication may have been a focal area for interaction and exchange, displaying evidence for recurring cycles of contact over long periods of time. It is also possible that the role of maritime contacts in precipitating language development is not generally appreciated in some circles.

Ruiz-Galvez (1991) has argued that the Atlantic seaways of the Iberian peninsula may have been important vectors of linguistic change and development. She suggests that Lusitanian in western Iberia developed as a pre-Celtic trade language on the Atlantic coasts of the peninsula and became the dominant language between the Tagus and the Douro. The status of Lusitanian is debated – some believe it to be pre-Celtic as she claims, but others have argued that it is a Celtic dialect related to Hispano-Celtic (see also discussion in McCall and Fleming, Volume III, Ch. 8). In any event, the development of stable settlement, of technological improvements including complex shipbuilding techniques and of long-distance exchange were all significant factors in the process of its development. Changes in social structure, including greater ranking and greater social complexity, were equally important. Ruiz-Galvez emphasizes that, as history demonstrates, intense and continuous trade contacts imply the arrival and establishment of small groups of people who are immensely influential and active in the communities in which they find themselves.

There is abundant sociolinguistic evidence to illustrate the importance both of trade and of sea routes in the spheres of language diffusion and language change. Among the conditions that determine the nature and outcome of language contact, new language needs and consequent patterns of use, 'those of the economic environment play a crucial role' (Coulmas 1992: 154). The importance of maritime routes as agents of language spread, contact and change, in both the Old and New Worlds, is well established for the historical period. If we look at the areas of language spread and language mixing – the latter is of particular interest to us here because, even where historical records of cultural interaction are deficient or non-existent, it offers clearly discernible evidence of language contact and compromise – we can see the importance of seas and rivers as vectors of contact and change. This is something of a truism since, prior to the development of extensive road networks and mechanized transport, travel by water was both faster and potentially further ranging; it follows that contact with unfamiliar cultures frequently came about in coastal, insular or riverine areas. At the more extreme end of contact-induced language change, a glance at Hancock's (1981) listing of pidgin and creole languages indicates how frequently non-genetic language change occurred as a result of both short- and long-haul mariners' activities.

This chapter stresses that not all of these activities involve socio-economic dominance, along the lines of the colonial or master–slave relationship often associated with these non-genetic types of language. In Europe alone, a number of languages arose in this manner. One of the more durable was Sabir, extensively used in Mediterranean ports and the Near East for several centuries at the time of the Crusades and said to survive today in the Balearics. Another maritime trade language that became a stable pidgin is Russenorsk, used in northern Norway (Finnmark and Troms) around the Norway Sea and the Barents Sea by Russian merchants and Scandinavian and Lapp fishermen during the Pomor trade. It appeared *c.* 1785 and died out about the time of World War I, when trading ceased following the Russian Revolution. Yet another European example is an Icelandic Basque pidgin born in the seventeenth century as a consequence of Basque fishing trips in the North Atlantic (omitted by Hancock; see Hualde 1991). Similarly, though further afield, Basque fishermen in their trading forays in northeast America and along the banks of the Saint Lawrence River in the sixteenth and seventeenth centuries created a Basque-based pidgin with their Amerindian and European trading partners (Bakker 1991).

Not all such contact languages are necessarily doomed. Indeed, one theory would have it that proto-Germanic, and so ultimately all Germanic languages, arose from a creolized version of an older Indo-European language, as a consequence of trading in the Baltic area with sea-going peoples from some non-Indo-European linguistic communities (Feist 1932), although this view has been criticized (Polomé 1970). Examples of fairly straightforward language diffusion across bodies of water, along coasts and rivers, are at least as numerous as those where significant mixing occurred; Malay and Swahili are prime cases. A somewhat different and highly interesting case is provided by the language pattern of the Vanuatu archipelago. Darrell Tryon demonstrates that in some instances the correlation between language/dialect chaining and trade networks through sea-borne inter-island trading outweighs terra firma contiguity in chaining, as in the transmission of material culture (Tryon 1976 and Volume III, Ch. 3). Marck (this volume, Ch. 4 especially under 'Nuclear Polynesian') also discusses the importance of maritime linkages in understanding the linguistic patterning of Polynesia.

A further point of relevance when considering a Bronze Age context is that once contact is made, it need not be particularly intensive, either for a language to gain new speakers in a new territory, or for a contact language to develop. Sometimes in trading and fishing situations the difficulty of the voyage and adverse weather conditions have actually contributed to both spread and mixing. The fact that many Russians stayed on in Norway during the winter after the trading season probably caused an expansion of the functions of Russenorsk to non-trading activities; the prolonged stays of French fishermen in Iceland gave rise to a pidgin (Noreen 1911); the spread of Malay, as a lingua franca, on the Malayan peninsula and the Indonesian archipelago was favoured by the fact that the climate, and the vagaries of the monsoon, forced sailors to make lengthy stays ashore on either side of the Malaccan straits. These examples reinforce the need to view the sea as an adequate bridge rather than an obstacle when looking at the areal diffusion of any language or language group.

Many possibilities may present themselves, and these examples and suggestions are presented primarily to illustrate some of the range of patterns that arise and the vibrancy of the mechanisms of language formation and change. Trade and related contacts do not, however, operate in isolation.

An élite dominance model that still allows some intrusion of an influential minority into Ireland would resolve the qualms of some Irish archaeologists, even if this is still undetectable in the archaeological record (see Mallory 1991, 1992; Warner 1991). This is, of course, a possibility, particularly with the recognition of what are claimed to be significant discontinuities in the late prehistoric archaeological record. Whilst Greene (1966), among others, thought that there was very limited evidence for any substrate language, it has been difficult to envisage how a pre-Celtic population could be so completely subsumed by new linguistic masters. In the opinion of Waddell (1991), population intrusion probably only enhanced diachronic linguistic developments already under way.

The vexed question of the substratum in Insular Celtic offers a useful example of how linguistic questions might be approached from a slightly different angle. For archaeologists grappling with the conundrum of the Celtic problem, some hope may be offered by the general approach adopted by John Robb who, in a significant contribution to the whole question of how linguists and archaeologists can establish linkages between prehistoric society and language, chose to 'ignore the question of where modern European languages came from in favour of the question of what happened to them *en route*' (Robb 1993). Among others advocating a similar approach see Sutton (1991).

We know that there were other non-Celtic people in the parts of Europe subsequently affected by the Celtic phenomenon. The presence or absence of substratum influence in Insular Celtic, and the identification of the source of such influence, if it exists, are points on which specialists do not agree. It may be that they never will. If Thomason and Kaufman (1988) are correct in saying that in shift-induced interference, lexical diffusion may be negligible, and if knowledge of morphosyntactic features of proto-Celtic remains as speculative as it currently is, it will be impossible to sustain or refute the arguments for or against a substratum influence.

Let us suppose that there was no substratum influence, even though this position has largely been abandoned. To put it at its mildest, as Koch has done in his convincing account of the transition from Old Celtic to Primitive Irish, whilst one is not obliged to assume the presence of pre-Celtic substrata, 'the data is somewhat more intelligible' if one does (Koch 1995: 48). A historical linguist in search of a genetic affiliation might reasonably feel that the subject is closed, if it is true that all the features of the varieties of Celtic can be explained in terms of inherited Indo-European traits and 'natural' internal innovation. However, since the focus of attention for the archaeologist is not solely the language itself but also the social processes that are implied by such total linguistic deculturation of the autochthonous population, even the absence of substrate transfer would be significant. Put another way, one must ask what types of situation might have given rise to such a complete shift on the part of the earlier inhabitants that there was no perceptible lexical or structural (phonemic or morphosyntactic) transfer from their language(s) to Insular Celtic. The present state of sociolinguistic research offers some insights that go beyond the assumption of migration so often made. For instance, one might here consider Whinnom's 'barriers to hybridization' to see which best fits the picture conveyed by the material evidence. In Whinnom's analysis, if no hybridization occurs – in this instance, if we are to believe that Insular Celtic contains no pre-Celtic elements, whether Indo-European or not – then the grounds might be ecological, ethological, mechanical or conceptual (Whinnom 1971).

The other, and now predominant, position accepts that there are features in the Celtic languages, particularly Insular Celtic, that are not readily ascribed to exclusively internal change, and must reflect another linguistic presence. However, the quest for this shadowy partner has so far proved inconclusive. D. Ellis Evans, whilst granting the possibility, indeed the probability, of non- or pre-Celtic substrata or adstrata, is surely right to stress simultaneously 'the impossibility of identifying or isolating these features or elements with any precision and the danger of attaching to them a linguistic label, whether it be "Larnian", "Eurafrican", "proto-Berber" or even "Old West Indo-European"' (Evans 1983: 954). Until new evidence comes to light, it seems unlikely that the data available will allow identification of the language involved for various reasons: either because it was a doomed genetic isolate; or because all its relatives perished too; or because, despite suggestive similarities between a present or recent language and Insular Celtic, too little is known of the former's ancient form; or because the nature of the presumed intrusive features does not allow us to see from what they developed. In the target language, a transferred structure may be so adapted that its origin is camouflaged – the basic linguistic concept may be from language A, its form may be provided by language B – unless historical evidence provides the clue to track it down. The Hamito-Semitic hypothesis, for instance, remains an object of scepticism, despite almost a century of exploration since Morris-Jones first mooted it, and despite recent research (Jongeling 1991, in an analysis of the basic constituent order of Insular Celtic; Gensler 1993 as cited in Koch 1995).

Some linguists may, understandably, lose interest if, as seems to be the case, no genetic affiliation can be traced. Nonetheless, archaeologists can still find useful indications in linguists' interpretations of the *quantity* and *kind* of these features, in that they may suggest whether shift alone occurred, or shift with borrowing, or whether the process was abrupt or gradual, and the likely respective size of the interacting language communities. The use of the classification system suggested by Thomason and Kaufman (1988) – while the authors are rightly wary of making claims about the predictive potential of many of the principles they propose – and the correlation of data that is increasingly being compiled on the changes taking place in languages currently in contact (though the outcomes of contact may not yet be known) with data from historically attested changes, may offer more solid ground for hypotheses about the prehistoric or fragmentary evidence available for Insular Celtic.

Clearly we are raising here some old questions that have been passed around the world of Celtic scholarship for generations and been examined by very considerable intellects. The point is to suggest that a slightly different focus on these questions, prioritizing processes based on present-day linguistic observations, may yet provide some fresh insights into social matters about which archaeologists could then speculate.

The archaic features found in Celtic suggest an early split of proto-Celtic from the parent Indo-European stem, with preservation of features also to be found in a range of eastern Indo-European languages such as Sanskrit, Indo-Iranian, Greek, Phrygian, Baltic and so on. It is further suggested that the marginal position of Celtic as a colonizer's language was conducive to its preservation of certain archaisms, since colonized areas often maintain forms after they have been abandoned in the epicentre. A good latter-day example is the '*r*-dropping' that occurred after the seventeenth century in British English and in those parts of America (New England and the Southern Atlantic Seaboard) where close ties were maintained with England, but that did not occur elsewhere in America. There is a seeming contradiction in the fact that, as Schmidt pointed out, Celtic languages in terms of traceable age (as evidenced by written sources) occupy 'an intermediate position' within the Indo-European languages, yet are 'characterised by certain archaic features in spite of their contacts with quite a number of different languages, beginning in the Urnfield and Hallstatt periods and ending with the influence of more vital languages [up to the present day]' (Schmidt 1994: 69).

In the type of *Sprachbund* scenario that is increasingly postulated, the possible effects of these contacts with other languages may benefit from some re-examination. Is it possible that the retention of certain archaic features in Celtic may in some instances owe something to other (non-Celtic) Indo-European speakers, during their shift to the superordinate Celtic language? An example of shifting speakers transmitting phonological, syntactic and morphological features of their first language to another (without the tell-tale presence of old loanwords) is to be found in the Finnic subgroup (Uralic) influence on Latvian, Lithuanian and northern Russian dialects, possibly Russian and even Slavic as a whole, if we accept the analysis of Thomason and Kaufman (1988: 238–251). For this to happen in the case of Celtic, one would have to assume that the nature of the contact was of a sort that permits structural transfer (i.e. probably involving considerable numbers of shifting speakers and not in a situation of intense cultural pressure).

One of the puzzling features of Insular Celtic that does not run true to Indo-European type is the increasing tendency towards periphrasis. Periphrasis could perhaps be a 'natural' linguistic innovation possibly reflecting a collective mindset (the dangerous territory of the *génie de la langue*), or a highly ritualistic society, or one with a taste for linguistic experimentation (poetry), or any combination of these. There are some problems with the notion of a language evolving towards periphrasis, as it goes against the trend towards economy of expression, although where it stands in relation to markedness is another question. Hence the tendency to include a psychological or sociological explanation. On the other hand, scholars such as Morris-Jones, Pokorny and Wagner have all argued that it derives from a non-Indo-European substratum. It is worth noting too that periphrasis is a feature not unusual in speakers who have imperfectly mastered a new language. It is to be found in mixed languages, and also in interlects (speech patterns characteristic of an individual's transitional linguistic phase). The most noticeable tendency is to compensate for lacunae in vocabulary by means of highly motivated, transparent circumlocutions, which may subsequently be abandoned or shortened. Gaps in the morphosyntactic system may be solved in a similar way, e.g. by the use of function words in place of a single inflection. This compensatory verbal agility could have some attractions for the native speakers of the target language and be adopted by them. However, it is more likely to take hold if the shifting speakers are present in quite significant numbers. If this view of the periphrasis phenomenon is taken, it would support the idea that small numbers of Celtic speakers interacted in a relatively peaceable way with large numbers of non-Celtic speakers, not necessarily non-Indo-European ones.

In arguing for the role of socio-economic factors such as long-distance trade, élite interaction and maritime contacts, it is easy to forget other significant elements such as religion. In his important contribution to this debate, Koch (1991) has also suggested that a significant shift in religious practices occurred from about 1600 BC; an earlier theocratic religion with a celestial orientation was, in time, supplanted by a religious system in which a hierarchical society, incorporating warriors, craftsmen and a priestly caste, and depositional practices in watery contexts all figure prominently. It is worth recalling that religious activity is one of the functions reserved for the High language in diglossic situations. Even in unilingual situations, ritual/religion is often associated with reserved language forms not in normal use.

In the light of these facts, the reliability of inscriptions found in a religious or ritual context as a basis for generalizations about the language favoured in a particular locality can be questioned. Religion is also an important factor in language spread. It is due to the spread of Islam that Arabic has diffused from the Arabian peninsula to the Atlantic coast and Spain, to Asia Minor, Central and Southeast Asia. Buddhist Pali and Christian Latin are other religiously driven languages. Does the extraordinary diffusion of the rite of cremation in Urnfield Europe incorporate some widespread religious development? Languages acquire new users not only through natural population increase or language shifts but also 'through speakers using a language for either a new function (e.g. literacy, religion, or scholarship) or replacing one language with another for a specific function (e.g. trade and commerce)' (Romaine 1988). An obvious example of this is English replacing French and Latin in law and religion. It is also quite possible that inherent literary and scholarly qualities can be postulated as one of the attractions of the Celtic languages, given their early appearance in written form, compared to other languages (other than Greek and Latin) on the European continent.

Many of the exchanges between archaeologists and linguists on the matter of the Celts, their material culture and their language have been posited on some rather doubtful assumptions regarding correlations between ethnic, cultural and linguistic identity. 'Language groups should never be confused with ethnic groups' (Robinson 1992: 13), and apart from large-scale movements of peoples, languages may spread in a variety of ways. The presence of a broadly common language in a series of geographical locations does not of itself imply shared ethnic identity or even a *sense* of shared identity. The presence of forms of Celtic in a range of locations cannot be taken to mean that the inhabitants of one Celtic-speaking region were a splinter group from another. Indeed, it is worth noting that a number of writers are now seriously questioning the concept of an ethnically unified 'Celtic World' (for example Chapman 1992, Waddell 1995 and Hill 1995 and references).

We would agree with Koch (1991: 19) that Celticization, as an instance of language shift, was not an event but a process. It is now a challenge to historical linguists and archaeologists alike, in Ireland and further afield, to attempt to harmonize their respective interpretative models.

REFERENCES

Bakker P. 1991. 'La lengua de las tribus costeros es medio vasca'. Un pidgin vasco y amerindio utilizado por europeos y nativos americanos in Norteamérica, h. 1540–h. 1640. In *Basque pidgins in Iceland and Canada. Anejos del Annuario del Seminario de Filología Vasca 'Julio de Uriquijo'*, XXIII, 137–165. San Sebastián: Disputación de Gipuzkoa.

Ball, M.J. and J. Fife (eds) 1993. *The Celtic Languages*. London and New York: Routledge.

Barford, P. 1991. Celts in Central Europe and beyond. *Archaeologia Polona* 29, 79–98.

Chapman, M. 1992. *The Celts: the construction of a myth*. Basingstoke: Macmillan.

Coulmas, F. 1992. *Language and Economy*. Oxford: Blackwell.

Dillon, M. and N.K. Chadwick. 1967. *The Celtic Realms*. London: Weidenfeld & Nicolson.

Duval, P.-M. 1977. *Les Celtes*. Paris: Gallimard.

Evans, D. E. 1983. Language Contact in Pre-Roman and Roman Britain. In *Aufstieg und Niedergang der Römischen Welt*. II. H. Temporini and W. Haase (eds), 949–987. Berlin and New York: De Gruyter.

Evans, D.E. 1995. The Early Celts. The evidence of language. In *The Celtic World*, M. Green (ed.), 8–20. London: Routledge.

Feist, S. 1932. The origin of the Germanic languages and the Europeanizing of North Europe. *Language* 8.

Gensler, O. 1993. A typological evaluation of Celtic/Hamito-Semitic syntactic parallels. Ph.D. thesis, University of California: Berkeley.

Greene, D. 1966. The making of Insular Celtic. In *Proceedings of the Second International Congress of Celtic Studies, Cardiff, 1963*, H. Lewis (ed.), 123–136. Cardiff: University of Wales Board of Celtic Studies.

Hancock, I.A. 1981. Répertoire des langues pidgins et créoles. In *Les langues dans le monde ancien et moderne* [no named editor], 631–647. Paris: CNRS.

Hill. J.D. 1995. How should we understand Iron Age societies and hillforts? A contextual study from southern Britain. In *Different Iron Ages: studies on the Iron Age in temperate Europe*, J.D. Hill and C.G. Cumberpatch (eds), 45–66. British Archaeological Reports, International Series S602. Oxford: Oxbow.

Hualde, J.I. 1991. Icelandic Basque pidgin. In *Basque pidgins in Iceland and Canada. Anejos del Annuario del Seminario de Filologia Vasca 'Julio de Uriquijo'*, XXIII, 123–133. San Sebastián: Disputación de Gipuzkoa.

Jongeling, K. 1991. *The Character of the (Non Indo-European?) Substratum of the Insular Celtic Languages*. (Abstract). p. 39, Programme of the IXe Congrès International d'Études Celtiques, Paris.

Koch, J.T. 1986. New Thoughts on *Albion*, *Iemé*, and the Pretanic Isles. *Proceedings of the Harvard Celtic Colloquium* 6, 1–28.

Koch, J.T. 1991. Ériu, Alba and Letha: when was a language ancestral to Gaelic first spoken in Ireland? *Emania* 9, 17–27.

Koch J.T. 1995. The transition and conversion from Primitive to Old Irish *c.* 367–*c.* 637. *Emania* 13, 39–50.

Macalister, R.A.S. 1928. *The Archaeology of Ireland*. London: Methuen.

Mallory, J.P. 1984. The origins of the Irish. *The Journal of Irish Archaeology* 2, 65–69.

Mallory, J.P. 1991. Two perspectives on the origins of the Irish. *Emania* 9, 53–58.

Mallory, J.P. 1992. Migration and language change. *Peregrinatio Gothica III. Universitetets Oldsaksamlings Skrifter, Ny rekke* 14, 145–153.

Noreen, A. 1911. Scandinavian languages. *Encyclopedia Britannica, Vol. XXIV.*

Piggott, S. 1979. South-west England – North-west Europe: contrasts and contacts in prehistory. In *Prehistoric Dartmoor in its Context.* V. Maxwell (ed.), 10–20. Exeter: Devon Archaeological Society.

Piggott, S. 1983. The coming of the Celts: the archaeological argument. In *Proceedings of the Sixth International Congress of Celtic Studies, Galway 1979*, G. MacEoin (ed.), 138–148. Dublin: Institute for Advanced Studies.

Polomé, E. 1970. The position of German among West-lndo-European languages. *ACIL* 10, 49–56.

Raftery, B. 1994. *Pagan Celtic Ireland: the enigma of the Irish Iron Age*. London: Thames & Hudson.

Renfrew, C. 1987. *Archaeology and Language: the puzzle of the Indo-Europeans*. London: Jonathan Cape.

Renfrew, C. 1989. Models of change in language and archaeology. *Transactions of the Philological Society* 87, 103–155.

Robb, J. 1993. A social prehistory of European languages. *Antiquity* 67, 747–760.

Robinson, O. 1992. *Old English and its Closest Relatives: a survey of the earliest Germanic languages*. London: Routledge.

Romaine, S. 1988. *Pidgin and Creole Languages*. London: Longman.

Ruiz-Galvez, M. 1991. Songs of a wayfaring lad: Late Bronze Age Atlantic exchange and the building of the regional identity in the west Iberian peninsula. *Oxford Journal of Archaeology* 10, 277–306.

Schmidt, K.H. 1992. Celtic movements in the first millennium BC. *The Journal of Indo-European Studies* 20, 145–178.

Schmidt, K.H. 1994. Insular Celtic: P and Q Celtic. In *The Celtic Languages*, M. Ball (ed.), 64–98. London: Routledge.

Sutton, M. 1991. Approaches to linguistic prehistory. *North American Anthropologist* 12(4), 303–324.

Thomason, S.G. and T. Kaufman 1988. *Language Contact, Creolization, and Genetic Linguistics*. Berkeley, Los Angeles and London: University of California Press.

Tryon, D. 1976. *New Hebrides Languages: an internal classification*. Pacific Linguistics, Series C, No. 50. Canberra: Australian National University.

Waddell, J. 1991. The Celticization of the West: an Irish perspective. In *L'Âge du Bronze Atlantique. Actes du 1er Colloque de Beynac*, C. Chevillot and A. Coffyn (eds), 349–366. Beynac: L'Association des Musées du Sarladais.

Waddell, J. 1992. The Irish Sea in prehistory. *The Journal of Irish Archaeology* 6, 29–40.

Waddell, J. 1995. Celts, Celticisation and the Irish Bronze Age. In *Ireland in the Bronze Age. Proceedings of the Dublin Conference, April 1995*, J. Waddell and E.S. Twohig (eds), 158–169. Dublin: Government of Ireland Stationery Office.

Warner. R.B. 1991. Cultural intrusions in the Early Iron Age: some notes. *Emania* 9, 44–52.

Whinnom, K. 1971. Linguistic hybridization and the 'special case' of pidgins and creoles. In *Pidginization and Creolization of Languages*, D.H. Hymes (ed.), 91–115. Cambridge: Cambridge University Press.

6 *Archaeological–linguistic correlations in the formation of retroflex typologies and correlating areal features in South Asia*

Bertil Tikkanen

ABBREVIATIONS

MIA Middle Indo-Aryan
OIA Old Indo Aryan
PIA Proto-Indo-Aryan
PIE Proto-Indo-European

INTRODUCTION

According to a hypothesis set forth by Parpola (Volume III) on the basis of recent archaeological findings, the ancestors of the Proto-Indo-Aryans (or, more generally, Proto-East-Aryans) can be equated with the Early Andronovo culture of southern Urals and northern and central Kazakhstan (Petrovka, *c.*2000–1800 BC). From this region, the Proto-East-Aryans would have spread in two successive migrations south(east)wards, entering the Indian subcontinent by way of both Bactria–Balochistan and Tajikistan–Hindu Kush. The Ṛgvedic Indo-Aryans (Proto-Sauma-Aryans) are probably to be equated with the Early Gandhara Grave Culture (Swat, Ghaleghay IV, *c.*1700–1400 BC).

Linguistically, the most salient feature of nearly all modern and ancient languages of South Asia (the Indian subcontinent) is the presence of a phonological opposition of retroflex vs. dental obstruents and sonorants (e.g. Sanskrit áṇu 'atom(ic)' vs. ánu 'afterwards'). The retroflex phonemes are in many cases the outcome of ancient loans and combinatory phonological processes, some of which reflect complex (prehistoric) patterns of convergence.

In 1969, Ramanujan and Masica published a pioneering if somewhat tentative article called 'Toward a phonological typology of the Indian linguistic area'. Džoj Èdel'man preceded them with a monograph in Russian on the linguistic geography of South and Southwest Asia on the basis of the Indo-Iranian languages (Èdel'man 1968). These studies are part of the century-old research on the synchronic and diachronic aspects of the linguistic features characteristic of South Asia (for some critical surveys and discussions, see, e.g. Hock 1975, 1984; Heston 1980, 1981; Dasgupta 1984).

Linguistic convergence has also been used whilst delineating the linguistic prehistory and stratigraphy of South Asia (cf. Southworth 1974). This chapter[1] is an attempt to pursue further the types, origins and patterns of convergence of retroflexion in the various languages of South Asia. The purpose is to find correlations between retroflex typologies and other features of convergence and to relate these correlations to recent archaeological findings concerning the migrations of ethnic groups in South Asia. A preliminary version of this study was published by the author in Parpola (1994: 166–167).

RETROFLEX TYPOLOGIES AND THEIR GENESIS IN SOUTH ASIA

Of all the dozens of areal linguistic features that include or are found in South Asia, only one demarcates almost all of it as a unit distinct from the adjacent areas. This is the presence of a phonemic contrast between retroflex or (post)alveolar versus dental stops (ṭ/t and/or ḍ/d etc.), for example Burushaski *ṭoq* 'pasture' vs. *toq* 'slime, muddy water'. This contrast is lacking only in the most peripheral northeastern and western zones. It is also lacking or weakly established in some isolated parts of the central tribal district (Korku, Sora) and in most of the (partly endangered) aboriginal languages of the now polyglot Andaman and Nicobar Islands.

Other retroflex consonant phonemes, such as retroflex flaps (ṛ/r), sonorants (ṇ/n, ḷ/l, etc.), sibilants (ṣ/s, ẓ/z) and affricates (ç/c, j̣/j) have either a narrower or broader distribution. Hence we can differentiate between two basic retroflex typologies (A and B) and several subtypes within both of them (see Figure 6.1):

Type A: ṭ/t and/or ḍ/d ± ṛ/r (ṛ is often an allophone of -ḍ(-))	
Subtype A + ṇ/n	= A1
Subtype A + ḷ/l	= A2
Subtype A + ṇ/n + ḷ/l	= A3
Subtype A + ṇ/n + ḷ/l/ḻ (ḻ or r is a lateralized retroflex fricative distinct from ḷ and ẓ)	= A4
Type B: A + /ṣ/š/s ± ẓ/ž/z	
Subtype B + ṇ/n	= B1
Subtype B +ḷ/l	= B2
Subtype B + ṇ/n + ḷ/l	= B3
Subtype B + ç/ċ/č ± j̣/ż/j (ċ is a dental, č a palatal affricate)	= B4
Subtype B + ç/ċ/č + ṇ/n ± j̣/ż/j	= B5
Subtype B + ç/ċ/č + ḷ/l ± j̣/ż/j	= B6
Subtype B + ç/ċ/č + ṇ/n + ḷ/l ± j̣/ż/j	= B7

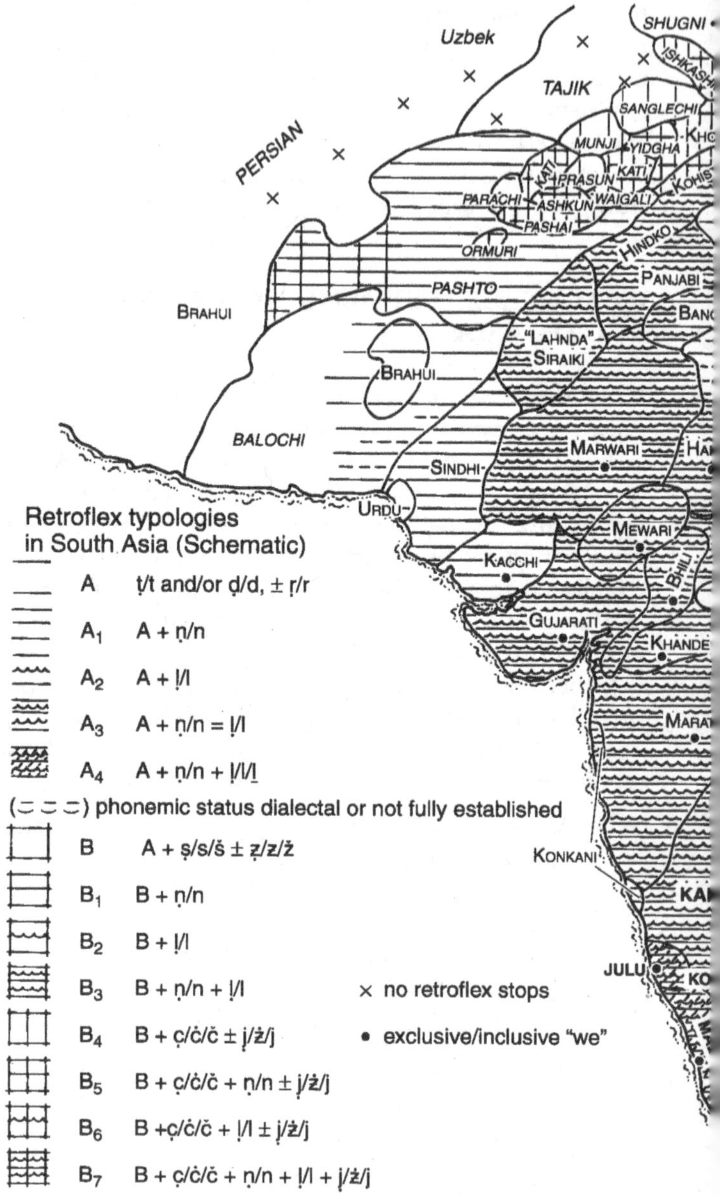

Figure 6.1 Language families of South Asia

Source: Tikkanen and Hameen-Anttila (1995)

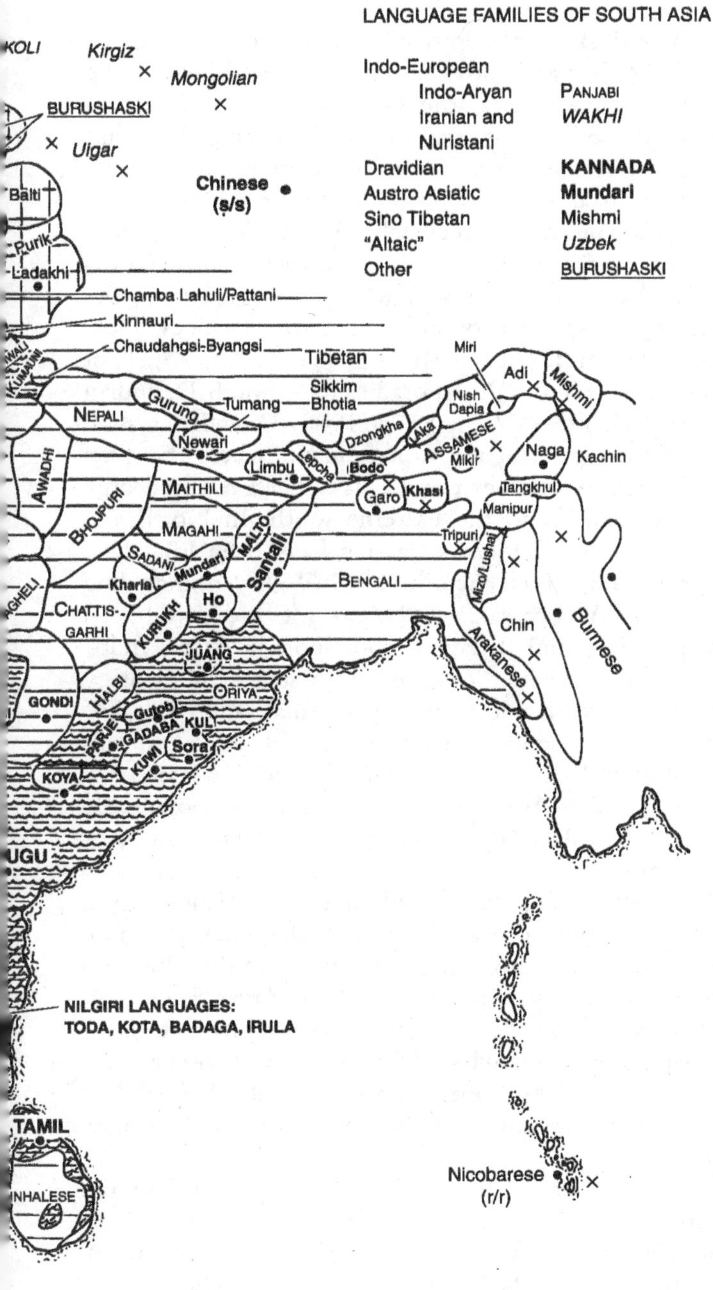
LANGUAGE FAMILIES OF SOUTH ASIA
Indo-European
Indo-Aryan PANJABI
Iranian and Nuristani WAKHI
Dravidian KANNADA
Austro Asiatic Mundari
Sino Tibetan Mishmi
"Altaic" Uzbek
Other BURUSHASKI
KOLI
Kirgiz
Mongolian
BURUSHASKI
Uigar
Chinese (s/s)
Balti
Purik
Ladakhi
Chamba Lahuli/Pattani
Kinnauri
Chaudahgsi-Byangsi
Tibetan
KUMAUNI
Gurung
Tumang
Sikkim Bhotia
NEPALI
Newari
Miri
Adi
Mishmi
Nish Dapla
Dzongkha
Aka
ASSAMESE
Mikir
Naga
Kachin
AWADHI
Limbu
Lepcha
Bodo
MAITHILI
Garo
Khasi
Tangkhul
Manipur
BHOJPURI
MAGAHI
MALTO
Tripuri
SADANI
Mundari
Santali
Mizo/Lushai
Kharia
BENGALI
Ho
CHATTIS-GARHI
KURUKH
Chin
Burmese
JUANG
Arakanese
HALBI
ORIYA
GONDI
Gutob
GADABA
KUL
PARJI
Sora
KUWI
KOYA
UGU
NILGIRI LANGUAGES:
TODA, KOTA, BADAGA, IRULA
TAMIL
Nicobarese (r/r)
NHALESE

For the sake of simplicity, aspiration as an additional distinctive feature has been left out of consideration in the above phonemic subsystems (e.g. ṭ/t//ṭh/th, etc.).

The two basic types, A and B, imply somewhat different processes or evolutions of retroflexion. The former centres around stops and liquids, the latter on sibilants and affricates. Generally, the B-complex is restricted to the mountainous regions of the northwest, the A-complex being dominant elsewhere (with a dividing line at the centre between the more complex A-types toward the west and south and the simpler ones toward the east).

Historically, the complex A-type, more specifically A + ṭ/ṯ/t + ṇ/ṉ?/n + ḷ/ḻ/l (≈ A4), is that of Proto-Dravidian (though not necessarily Proto-North-Dravidian). The Proto-Dravidian retroflex system was, to a great extent, the outcome of pre- or proto-stage combinatory and assimilatory changes across morpheme boundaries, e.g. **ḻ or ḷ + N > *ṇ; *ṇ + t > *ṇṭ; **ḷ + nt > *ṇṭ; *ḷ or ḻ + t > *ṭ (ṭ); ** l + t > *ṯ(ṯ), sporadically > South Dravidian *ṭṭ, etc. (Zvelebil 1970: 102–104, 173–177, etc.). The Proto-Dravidian system has survived best in the southern parts of the subcontinent, where the alien influence has been weakest. The phonemes ṯ and ḻ have been retained only in a few languages or (caste) dialects in the extreme south. In North and parts of Central Dravidian also, the phonemes ḷ and ṇ have been lost.

The B-types, on the other hand, are characteristic of NW Aryan, including Proto- and Old Indo-Aryan (PIA B + ṣ/ś/s + *ẓ/z + ṭ/t + ḍ/d [?-ḷ-/-l-] + ṇ/n, possibly with allophonic ç and j). Apart from sporadic loans, Indo-Aryan retroflexion is mainly the product of (pre-)proto-stage combinatory changes of palatals and dentals, with subsequent phonematization of allophones, e.g. PIE[2] * r, u, $q^{(w)}$, i + dental sibilant > * . . . + palatal sibilant > PIA * . . . + retroflex sibilant; PIA * r, r̥, ṣ ± vowel / velar / labial + n > * . . . ṇ; PIA *ś, ź, ṣ, ẓ + t, d > *ṣṭ, *(*ẓ)ḍ; PIA *-n(d)- > sporadically *-ṇ(ḍ)-; PIA * lt, ln, ls > OIA dialectically ṭ/ṇ/ṣ (cf. Proto-Dravidian *ḷ or l + t > * ṭ(ṭ); ** l + t > *ṯ(ṯ)).

The most complex B-types (B5–B7) are found in eastern Hindu Kush, Kohistan and southern Pamir (= south of the Shughni-Roshani-Sariqoli line), and earlier probably also in northern Pamir. In East Iranian and the Nuristani branch of Iranian, as well as in many neighbouring Indo-Aryan languages, the development of retroflex phonemes is mainly due to proto-stage or secondary combinatory changes of a somewhat different nature than in Proto-Indo-Aryan: * r + stop or sibilant > *retroflex stop or sibilant; *k + sibilant or affricate > *retroflex affricate; *affricate or sibilant or stop + r > *retroflex affricate or sibilant or stop, etc.

The somewhat less complex subtype B + ç/ċ/č + j/j is found in the isolate Burushaski, apparently the most ancient extant language of the northern part of the subcontinent. Before the intervention of the Aryan languages, Burushaski was in close contact with (West) Tibetan (B, B4), where retroflexion is mainly due to combinatory changes involving, e.g. clusters with following or preceding -r-.

There are lexical indications of early contacts also between Indo-Aryan and Burushaski. Retroflexion has sometimes occurred in Burushaski in combination with a preceding r. Yet the Proto-Burushaski system (including phonotactics) is too different from its Proto-Indo-Aryan counterparts to allow us to speculate about Burushaski ad- or substratal influence in the genesis or formation of the latter.

The case for a Burushaski (-type) substratum underlying the genesis or evolution of retroflexion (and certain other peculiarities) in the NW Iranian and Indo-Aryan languages is much better (cf. Èdel'man 1963, 1976, 1980, 1984), but Burushaski need not have been the only language (family) spoken in the northwest mountainous regions in prehistoric and early historical times. There are no Dravidian loanwords in Burushaski apart from those that have been passed on by Old (rather than Middle or New) Indo-Aryan. Though harder to prove, there do not seem to be any Burushaski loanwords in Dravidian either. It is therefore plausible that Burushaski and Dravidian were never in direct contact with each other. If this is so, there are likely to have been other languages in this strategic region between the mainly Dravidian-speaking Indus civilization (cf. Parpola 1994) and the partly Burushaski-speaking Karakorum/eastern Hindu Kush region. Even the oldest Old Indo-Aryan documents contain a number of words that are hard to explain on the basis of the extant south Asian languages.

ORIGIN AND DEVELOPMENT OF INDO-ARYAN RETROFLEXION

The various stages of Indo-Aryan retroflexion and its areal implications can be summarized as follows: PIA: B + ⋆ẓ/z + ṇ/n ± ḷ/l > B + ṇ/n ± ḷ/l > OIA: B + ṇ/n ± ḷ/l > non-NW MIA: A + ṇ/n ± ḷ/l. As far as we know the retroflex systems of the Old Indo-Aryan dialects, they are remarkably alike (the main difference is the absence of ḷ/l in non-Ṛgvedic OIA). Hence it would seem plausible that retroflexion originated in a fairly early phase of 'Proto-Indo-Aryan'. In view of the strictly areal implications of retroflexion and the occurrence of retroflexes in many early loanwords, it is hardly likely that Indo-Aryan retroflexion arose in a region that did not have a substratum with retroflexes.

There are no indications of retroflexion in the Syrian (Mitanni) branch of Indo-Aryan, but the alien orthography may obscure the situation. Mitanni mĭšta(nnu) < Pre-Proto-Indo-Aryan ⋆miždha 'reward' (> ⋆miẓdha > ⋆miẓḍha- > OIA mīḷha/mīḍha-) proves only that the elision of the voiced sibilants had not yet taken place and that Mitanni Indo-Aryan branched off before the development of some fundamental Proto-Indo-Aryan innovations. It is, nevertheless, likely that Indo-Aryan retroflexion did not arise before the western migration of the Mitanni Indo-Aryans from Bactria-Margiana had taken place. Another reason is that retroflexion cannot have originated in Bactria-Margiana, because the early East Iranians who came here around 1500 BC and mingled with the Indo-Aryans (acculturated with the Dāsas) that had remained there did not develop retroflexes (cf. the language of the Avesta).

The Proto-Indo-Aryan retroflex system has remained more or less intact (excepting the Late Proto-Indo-Aryan loss of *ẓ and *z) only in the north-western mountainous regions. Elsewhere, the retroflex series have been reduced to the A-type(s), characteristic of, for example, Dravidian. This tendency to reduced retroflex types increases in the vicinity of Munda and Tibeto-Burman, where the retroflex phonemes are comparatively recent and partly due to Indo-Aryan or Dravidian areal influence.

There is lexical evidence for very early contacts between Indo-Aryan and Austroasiatic (and some anonymous ancient Gangetic language(s)), but the lexical and structural influence of the latter on Indo-Aryan has nowhere been as strong (and early) as that of Dravidian. Both Proto-Dravidian and Proto-Austroasiatic had exclusive vs. inclusive 'we', but in Indo-Aryan this feature is found only in contact with Dravidian.

CORRELATION OF RETROFLEX TYPOLOGIES WITH OTHER AREAL FEATURES IN SOUTH ASIA

Many of the areal features listed below have been treated in detail in the vast literature on South Asia as a linguistic area (Masica 1974, 1976; Emeneau 1980; references in the introduction).

Type A

The A-types especially the subtypes A3–A4, characteristic of Dravidian and (South)western/lower Northwestern Indo-Aryan, correlate with the following areal features, some of which exceed it considerably:

1 *The totalizing use of an indefinite particle or quantifier meaning 'too, even, -ever'* (e.g. '3 too' = 'all 3'): South, Central and Northeast Dravidian (< ProtoDravidian), Marathi. Though not evident in many Indo-Aryan languages, this feature is clearly an inherited Dravidian calque, since it appears also in Classical Sanskrit and Middle Indic.

2 *The past or perfective copulative-adverbial gerund or converb* (mainly from oblique verbal nouns in Indo-Aryan, Tibeto-Burman, North and Central Munda, perfective/preterital/aorist verb stems in Iranian, Burushaski, Dravidian and South Munda), as part of a larger area encompassing especially the non-Iranian part of the B-complex and continuing north (Turkic, Mongolic, Uralic, etc.), east (Tibeto-Burman) and west (Iranian) of the A-complex. This feature has diverse origins in South Asia, but there is syntactico-semantic convergence of the Indo-Aryan converb with the Dravidian converb in the post-Ṛgvedic period (cf. Tikkanen 1987).

3 *The postposed quotative particle* (an anaphoric pronoun 'thus, so' mainly in northwestern, western and central South Asia, elsewhere *verbum*

dicendi: 'having said, saying'), as part of a larger patchy area similar to 2, but excluding Iranian (but not Burushaski) and including the (now extinct) Elamite, Akkadian and Sumerian. Cf. (2).

4 *The lack or scarcity of prepositions* (vs. postpositions or morphological cases), as part of a larger area with the subject-object-verb word order, postpositions, and Standard-Marker-Adjective comparison, encompassing the non-Iranian part of the B-complex and continuing north into central Asia (Tibeto-Burman, Turkic, Uralic, Mongolic, etc.) and east (Tibeto-Burman but not Thai and Mon-Khmer) of the A-area. This feature may be due to Dravidian sub- or adstratum influence in Middle Indo-Aryan.

5 *Morphological second causatives*, as part of a larger, but disrupted area reappearing in central Asia (Tajik?, Turkic, Mongolic, Uralic, etc.). Cf. 4.

6 *Explicative-aspectual auxiliaries with the gerund or converb*, with such literal meanings as 'go, come, give, take, put (down), sit down, throw, leave, rise, fall, finish, see', etc., as part of a larger area encompassing the Indo-Aryan part of the B-complex and reappearing north (Tajik, Turkic, Mongolic, etc.). This feature seems to have spread from Dravidian into Middle and New Indo-Aryan.

Western and southern features

The following features correlate especially with the lower (i.e. mainly Dravidian) part of the A3-area:

1 *Distinction between two forms of the first person plural pronoun 'we' excluding or including the adressee(s)*: Kacchi (dialect of Sindhi), Gujarati, Marathi, Rajasthani (Marwari, etc.); most of the Dravidian languages (excepting modern Kannada, Kota, Gondi, Gadaba, Konda and Brahui, in which languages the loss is relatively recent), but also part of the A-area (Munda and sporadically Tibeto-Burman). Since this feature does not appear in Indo-Aryan in the vicinity of those Austroasiatic and Tibeto-Burman languages that have it (see map), it must be due only to Dravidian influence in Indo-Aryan.

2 *A high proportion of Dravidian-derived place-names.*

3 *Dravidian-type kinship terminology*, continuing in the Munda-speaking A-area.

All these features point to a fairly recent Dravidian substratum in the lower western-central part of the subcontinent (Southworth 1974). Early Dravidian loanwords in Vedic and Classical Sanskrit, together with the areal features listed for Type A, suggest that in the Old and early Middle Indo-Aryan period this substratum extended at least as far north as to the Indo-Gangetic plain.

The following features correlate especially with the eastern part of the A-area:

1 *Numeral classifiers*, as part of a larger patchy area continuing to the north and east of the A-area (Sino-Tibetan, Austroasiatic), but also occurring widely in West and East Iranian. In the eastern and eastern-central parts

of the subcontinent this feature seems to be connected mainly with Sino-Tibetan influence (cf. Barz and Diller 1985).

2 *Noun + attribute or demonstrative as an alternative or earlier word order*, as part of an area excluding Indo-Aryan and extending to the east (Tibeto-Burman, Thai, Austroasiatic) and north (Tibeto-Burman) of the A(+) area.
3 *Lack or semantic (re)interpretation of grammatical gender.*
4 *A high proportion of Austroasiatic toponyms.*

The following areal features correlate approximately with the B-complex:

1 *The presence of a phonemic contrast between dental and palatal affricates*, as extending southeast and slightly north and south of the B-complex (from northern Pamir to Kashmir, with part of western Himalaya; but as an independent area also in the central Deccan). This feature is original in Burushaski and some branches of Sino-Tibetan.
2 *The tendency to secondary deaspiration or absence of (voiced) aspirates* (except in Kalasha, Phalura and Indus Kohistani), with occasional compensatory tone distinction, as part of a larger Western, Central and North Asian area lacking voiced and (starting with Nuristani and Iranian) voiceless aspirates.
3 *The occurrence of initial retroflexes and non-homorganic consonant clusters*, as a feature continuing in northwestern and western South Asia (cf. Lahnda and Sindhi).
4 *Vigesimal basis in higher numerals* (excluding Kashmiri and Iranian [except Wakhi and the Nuristani branch]), as part of a patchy area continuing or reappearing sporadically in northern, central and eastern South Asia. This feature is original in Burushaski.
5 *The lack of case inflection for adjectives* (excepting Kashmiri), as part of a common Central Asian pattern and reappearing in Dravidian and Munda.

CONCLUSIONS AND DISCUSSION

It can be concluded from the above correlations and reconstructions that Aryan retroflexion originated (starting with Proto-Indo-Aryan) only on or at the border of the Indian subcontinent, perhaps independently in both the western and northwestern parts of it, then partly converging. Retroflexion in Iranian is clearly secondary to that of Indo-Aryan.

The problem is that the known substrata of these regions appear to have been rather different with regard to their retroflex systems. Dravidian seems to have been the main substratum in the Indus valley and Panjab, whilst Burushaski was one of the substrata of the high mountain valleys of the north (western Karakorum, eastern Hindu Kush, and possibly Kohistan). Sino-Tibetan may be as ancient as Burushaski in Karakorum and Himalaya, but retroflexion in Sino-Tibetan is secondary.

Proto-Burushaski, as we can reconstruct it, could explain some areally limited phenomena, such as the retention or development of the retroflex sibilants and affricates in the Aryan languages of the northwest (the loss of retro-flex sibilants in Eastern Pashto is recent). However, Burushaski lacks the retroflex nasal, which disappeared only after the Old Indo-Aryan phase in some of the Central and Eastern Indo-Aryan dialects.

Proto-(North-)Dravidian could explain the early Old Indo-Aryan dialectal development of retroflexes from clusters with a lateral (Fortunatov's law) and the loss of the retroflex sibilant in the post-Vedic (late Old Indo-Aryan or early Middle Indo-Aryan) period outside the northwestern region. Possibly this loss had occurred first in those Old Indo-Aryan dialects that had entered the subcontinent by way of Baluchistan, where there was a Dravidian substratum. Then this would be a case of convergence of the post-Ṛgvedic dialects with the non-Ṛgvedic Indo-Aryan dialects of the Indo-Gangetic plains.

We really know too little about these early East Aryan (Dāsa) dialects of the Indus valley (and perhaps also Hindu Kush), since they were in general heavily influenced by the Ṛgvedic Indo-Aryan dialect(s) in the post-Ṛgvedic period. The early Ṛgvedic dialect(s) developed during several centuries in the eastern Hindu Kush and Kohistan region, mainly in the Swat valley.

There is archaeological evidence of ancient cultural contacts between the pre-Aryan culture of Swat and the mainly Dravidian-speaking Indus civilization (whose settlements extended as far northwest as to Shortughai in Bactria). But in the period preceding the coming of the Ṛgvedic Indo-Aryans (Ghaleghay III, *c.*2000–1800 BC), we find cultural influence from the Taxila and Kashmir Neolithic. The Kashmir Neolithic (Burzahom III, *c.*2500–1500 BC) in its turn shows close cultural ties with China (Nakamura 1993) or the East Tibetan Plateau (Xu 1991). The linguistic identity of these cultures is unknown and their dating has been questioned by Nakamura (1993).

Judging from the areal distribution and patterns of convergence of retroflexion in eastern Hindu Kush and Kohistan (involving Proto-Indo-Aryan), it can reasonably be assumed that this language or these languages were of the B-type, and more specifically probably B1 or B3 rather than B4–B7. In other words, the pre-Aryan language(s) of Swat, Kashmir and the adjacent area at the time of the advent of the Indo-Aryans can *hardly* have been either Dravidian, Burushaski or even Sino-Tibetan.

NOTES

1 In the process of writing this chapter I have had numerous fruitful discussions with Asko Parpola and Virpi Hämeen-Anttila, who has drawn the map. I want to thank them both for their invaluable help. Needless to say, they do not necessarily share all the views presented in this chapter.

REFERENCES

Barz, R.K. and A.V.N. Diller 1985. Classifiers and standardization: some South and South-East Asian comparisons. In *Papers in South-East Asian Linguistics*, No. 9: *Language policy, language planning and sociolinguistics in South-East Asia* D. Bradley (ed.), 155–184. Pacific Linguistics, A-67. Canberra: Australian National University.

Dasgupta, P. 1984. On conceiving of South Asia as a linguistic area. *Indian Linguistics* 43: 37–48.

Èdel'man, D.I. 1963. Problema cerebralnyx v vostocšnoiranskix jazykax. *Voprosy Jazykoznanija* V, 67–81.

Èdel'man, D.I. 1968. *Osnovnye voprosy lingvisticšeskoj geografii na materiale indoiranskix jazykov*. Moscow: Nauka.

Èdel'man, D.I. 1976. Strukturnye 'anomalii' vostocšnoiranskix jazykov i tipologija substrata. In *Studien zur allgemeinen und vergleichenden Sprachwissenschaft Karl Ammer zum Gedanken*, 79–93. Jena.

Èdel'man, D.I. 1980. K substratnomu nasledĳu central'noaziatskogo jazykogo sojuza. *Voprosy Jazykoznanija* V, 21–32.

Èdel'man, D.I. 1984. K substratnym javleniam v sintaksise iranskix jazykov. In *Voprosy dialektologii i istorii jazyka*, R.G. Gaffarov and V.B. Silina (eds), 34–41. Dušanbe.

Emeneau, M.B. 1980. *Language and Linguistic Area*. Essays selected and introduced by Anwar S. Dil. Stanford: Stanford University Press.

Heston, W.L. 1980. Some areal features: Indian or Irano-Indian? *International Journal of Dravidian Linguistics* 9, 141–157.

Heston, W.L. 1981. Review of Masica 1976. *Indian Linguistics* 10, 180–187.

Hock, H,H. 1975. Substratum influence on (Rig-Vedic) Sanskrit? *Studies in the Linguistic Sciences* 5, 76–125. [University of Illinois, Urbana-Champaign].

Hock, H.H. 1984. (Pre-)Rig-Vedic convergence of Indo-Aryan with Dravidian? Another look at the evidence. *Studies in the Linguistic Sciences* 14, 89–108. [University of Illinois, Urbana-Champaign].

Masica, C. 1974. The basic order typology as a definer of an Indian linguistic area. *International Journal of Dravidian Linguistics* 3, 154–180.

Masica, C. 1976. *Defining a Linguistic Area: South Asia*. Chicago: University of Chicago Press.

Nakamura, G. 1993. The origins of rice cultivation: problems associated with finds of rice and *Ishibocho* in the northern Indian subcontinent. *Bulletin of the Ancient Orient Museum, Tokyo* XIV, 53–117.

Parpola, A. 1994. *Deciphering the Indus Script*. Cambridge: Cambridge University Press.

Ramanujan, A.K. and C. Masica 1969. Toward a phonological typology of the Indian linguistic area. *Current Trends in Linguistics* 5, 543–577.

Southworth, F.C. 1974. Linguistic stratigraphy of northern India. *International Journal of Dravidian Linguistics* 3: 201–223.

Tikkanen, B. 1987. The Sanskrit gerund: a synchronic, diachronic and typological analysis. Ph.D. thesis, University of Helsinki. *Studia Orientalia* 62. Helsinki.

Xu, C. 1991. The cultural links over the Himalaya ranges in the prehistoric period. *Orient* 27: 12–35.

Zvelebil, K. 1970. *Comparative Dravidian Phonology*. Janua Linguarum, Series Practica 80. The Hague: Mouton.

7 Language change in Southern Melanesia: linguistic aberrancy and genetic distance

JOHN LYNCH

INTRODUCTION

Degrees of superficial similarity or difference, whether in archaeology or linguistics, do not necessarily correspond with degrees of relationship between artefact types or between languages:

> [T]he archaeologist proceeds from descriptive data to contextual inference by demonstrating the existence and validity of various degrees of relation of likeness. This similarity or parallelism of relations is called analogy. Archaeological inference is impossible without recourse to analogy. It is of considerable theoretical importance to recognize that archaeological analogy is based on *a comparison of abstractions rather than a resemblance between individual artifacts.*
>
> (Thompson 1970: 359; emphasis added)

In a number of publications, George Grace (1981, 1985, 1990) has written about a problem in the linguistic prehistory of the Oceanic subgroup of the Austronesian family (Figure 7.1) which concerns the connection – or lack of it – between superficial linguistic similarity or difference and the degrees of relationship between the languages compared. In so doing, he has cast considerable light on the nature of linguistic change in some parts of Melanesia. For example:

> It has long been recognized in comparative Austronesian (AN) linguistics that the comparative method can be applied to some of the languages of the family (call these the 'exemplary' AN languages) with much less difficulty and with much more convincing results than to others (call these the 'aberrant' AN languages). Actually, to suggest that every AN language falls into one of two types is certainly an oversimplification. It would be somewhat more accurate to speak of aberrancy as a matter of

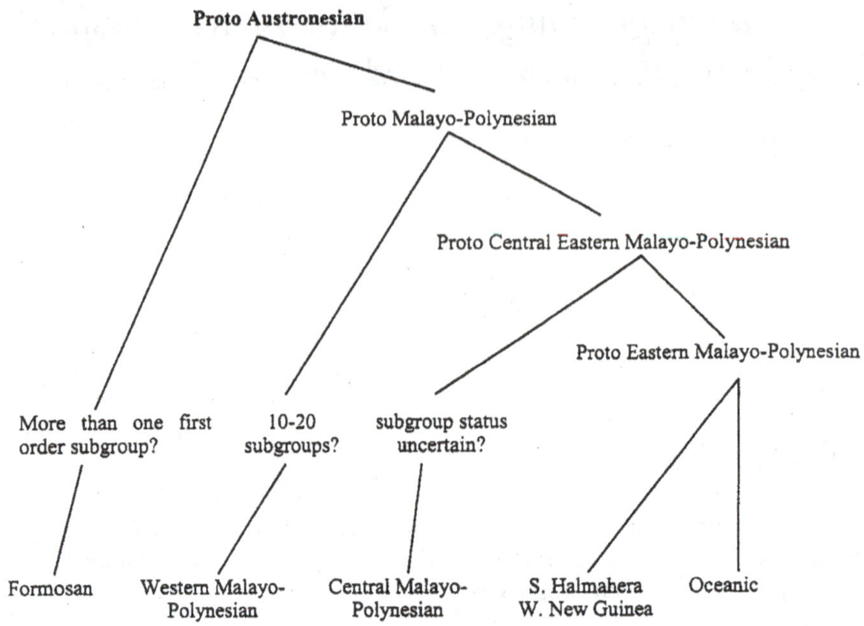

Figure 7.1 Higher-level branches of Austronesian
Source: Lynch

> degree and to imagine a scale such that every AN language could in principle have its degree of aberrancy plotted on the scale.
> (Grace 1981: 255–256)

In this chapter, I will explore Grace's notion of aberrancy and show that aberrancy does not necessarily correlate with genetic distance (and, conversely, that superficial similarity does not necessarily correlate with genetic closeness). I will focus particularly on Southern Melanesia – the area comprising New Caledonia, including the Loyalty Islands, along with the Tafea (southern) district of Vanuatu (see Figure 7.2).

THE NATURE OF 'ABERRANCY'

The languages of Southern Melanesia belong to three separate subgroups of Oceanic (see Figure 7.3). There is a strong possibility that the latter two groups in fact form a single subgroup of Oceanic, which I will call Southern Melanesian (see Lynch 1995), though the validity of its existence is not material to the essence of this chapter.

- The language spoken on Futuna and Aniwa is a Polynesian Outlier language which, along with other (outlier and nuclear) Polynesian

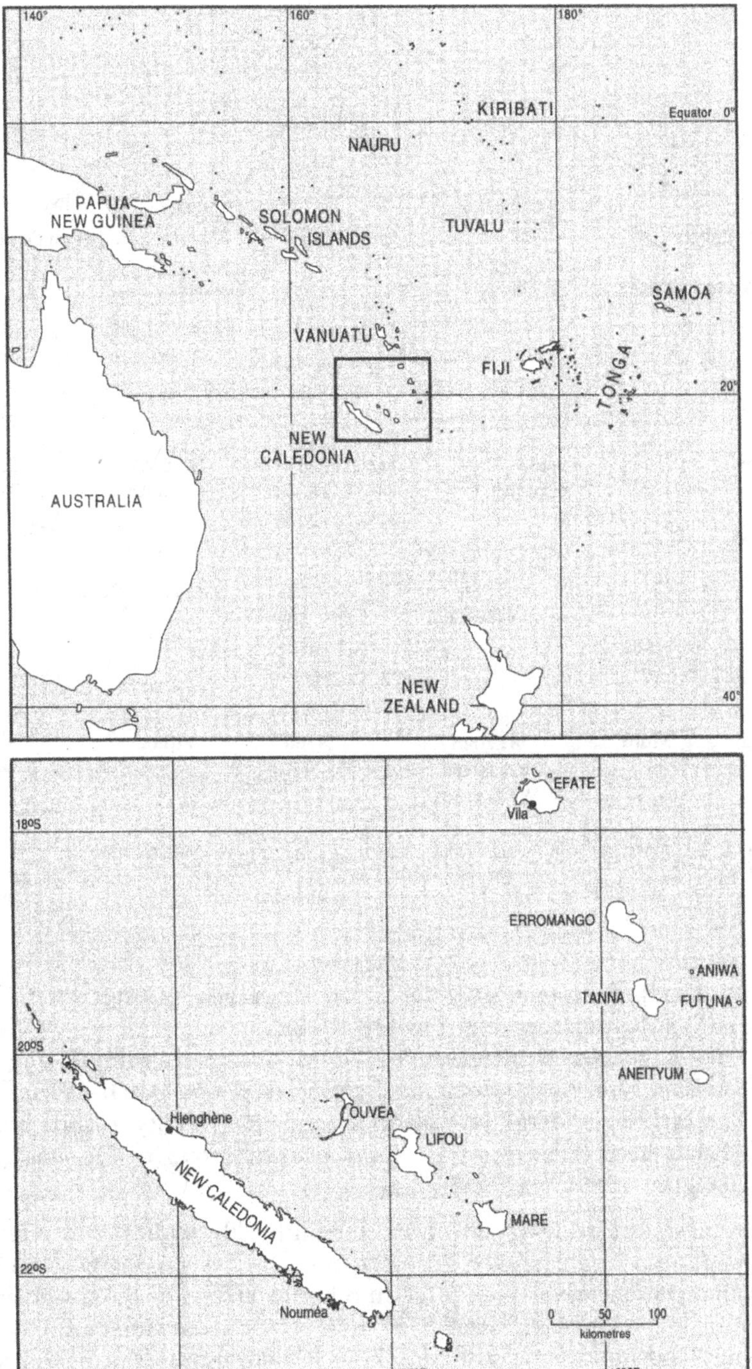

Figure 7.2 Map showing Southern Melanesia

Source: Lynch

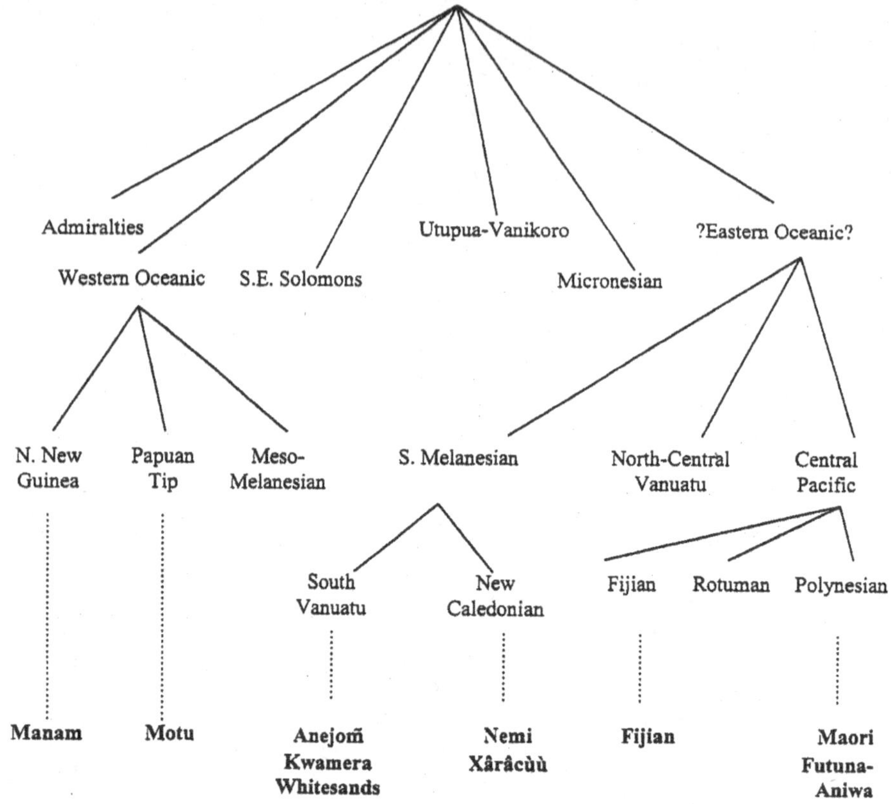

Figure 7.3 Possible Oceanic subgrouping. Languages mentioned in the text are printed in bold

Source: Lynch

languages, Rotuman and the Fijian languages, belongs to the *Central Pacific* subgroup (see e.g. Pawley 1972).

- The languages of Erromango, Tanna and Aneityum belong to the *Southern Vanuatu* subgroup of Oceanic (see e.g. Lynch 1978).
- The languages of New Caledonia and the Loyalty Islands belong to what is here called the *New Caledonian* subgroup of Oceanic (see e.g. Geraghty 1989).

'Aberrancy' can be taken to mean something like radical and often irregular change in a language or a group of languages, as compared with the less radical changes that have taken place in other languages or groups of languages at a comparable genetic 'level'. Its most obvious manifestation comes in phonology and morpheme structure: cognates are more difficult to recognize in 'aberrant' languages than in 'exemplary' languages, and sound correspondences – between these languages themselves, and between them and Proto

Oceanic – are apparently much less regular, or less obviously regular. In discussing his work in New Caledonia, for example, Grace says:

> The Canala and Grand Couli languages have undergone such extensive phonological changes since the dissolution of Proto-Oceanic (their most recent common ancestor for which any extensive body of reconstructions is available) that it is difficult to identify reflexes of Proto-Oceanic etyma in either of them.
>
> (Grace 1981: 259)

The following examples illustrate this point. The set of words in Table 7.1 are from (relatively) 'exemplary' languages, and continue (with transparent sound changes) quite obviously the reconstructed Proto Oceanic (POc) forms (mainly from Ross 1988). Manam (Lichtenberk 1983) and Motu (Lister-Turner and Clark n.d.) are spoken in Papua New Guinea, and belong to the Western Oceanic subgroup, whilst Fijian (Capell 1968) and New Zealand Māori (Williams 1975) belong to the Central Pacific sub-subgroup of the Eastern Oceanic subgroup (see Figure 7.3). Empty cells mean either that no cognate has been located or that there is no word in the language with that meaning (e.g. 'sugarcane' in Māori); ' represents the glottal stop, and slashes separate off accreted non-cognate material.

The set of words in Table 7.2 are from 'aberrant' Southern Melanesian languages, which continue the same POc forms as in Table 7.1, though much less transparently. Kwamera (Lindstrom 1986) and Anejom̃) (Lynch and Tepahae, in prep) are Southern Vanuatu languages, whilst Nemi (Haudricourt and Ozanne-Rivierre 1982) and Xârâcùù (Grace 1975, 1986) are New Caledonian languages.

The nature of the sound changes involved in the development of forms in exemplary and aberrant languages are illustrated in Tables 7.3 and 7.4 by a comparison of two sets of cognates: Motu *ima-mu* and Anejom̃) *nijma-mw* derive from POc **na lima-mu* (ARTICLE hand-2SG:POSSESSIVE) 'your (sg.) hand', whilst Motu *ani* and Anejom̃ *γiñ* derive from POc **kani* 'eat'. (Stress has been marked in forms in the Anejom̃ rules to illustrate the application

Table 7.1 Some POc reflexes in 'exemplary' languages

POc	*Gloss*	*Manam*	*Motu*	*Fijian*	*Māori*
**boŋi*	'night'	*boŋ*_day	*hanua/boi*	^{m}*boŋi*	*pō*
**kani*	'eat'	*'an*	*ani*	*kani/a*	*kai*
**kayu*	'tree'	*'ai*	*au*	*kau*	*kai-*
**kutu*	'louse'		*utu*	*kutu*	*kutu*
**mata*	'eye'	*mata*	*mata*	*mata*	*mata*
**mate*	'die'	*mate*	*mase*	*mate*	*mate*
**pican*	'how many?'	*ira*	*hida*	*viða*	*ϕia*
**rua*	'two'	*rua*	*rua*	*rua*	*rua*
**tolu*	'three'	*toli*	*toi*	*tolu*	*toru*
**topu*	'sugarcane'	*tou*	*tohu*	^{n}*dovu*	

Table 7.2 Some POc reflexes in 'aberrant' Southern Melanesian languages

POc	*Gloss*	*Kwamera*	*Anejom̃*	*Nemi*	*Xârâcùù*
**boŋi*	'night'	*nɨ/pɨn*	*ne/peñ*	*ᵑgen*	*mõ*
**kani*	'eat'	*ani*	*yiñ*	*cani*	*kɛ̃*
**kayu*	'tree'	*n/ei*	*in/γai*	*ceek*	*kʷãã*
**kutu*	'louse'	*ur*	*ne/γet*	*ciik*	*kɨtɨ*
**mata*	'eye'	*nɨ/mrhi-*	*ne/mta-*	*tna/maa*	*kãra/mɛ*
**mate*	'die'	*e/mha*	*mas*	*mac*	*mɛ*
**pican*	'how many?'	*ke/va*	*e/heθ*	*ni/vit*	
**rua*	'two'	*kɨ/ru*	*e/rou*	*he/lu/k*	*baa/ru*
**tolu*	'three'	*ka/har*	*e/sej*	*he/yen*	*ba/ʃee*
**topu*	'sugarcane'	*nɨ/ruk*	*ne/to*		*de*

Table 7.3 Motu rules

Rules	**na lima-mu*	**kani*
Loss of *l before high vowel	*na ima-mu*	
Loss of articles	*ima-mu*	
Loss of *k		*ani*

Table 7.4 Anejom̃ rules

Rules	**na limá-mu*	**káni*
Palatalization of *l, *n before *i	*na jimá-mu*	*káñi*
Rounding of *m before *u	*na jimá-mwu*	
Vowel harmony	*ni jimá-mwu*	*kíñi*
Article accretion	*ni-jimá-mwu*	
Lenition of *k		*yíñi*
Loss of pretonic unstressed vowel	*ni-jmá-mwu*	
Loss of final vowel	*ní-jma-mw*	*yíñ*

of the rule dropping the pretonic vowel; vacuous application of a particular rule is marked by an empty cell.) It will be seen that few rules have applied in Motu, and the forms there are recognizably clear continuations of the POc etyma. On the other hand, there has been considerable alteration of the same POc forms in Anejom̃, making cognation more superficially obscure.

In addition to phonological aberrancy, there is also lexical aberrancy. This refers to the fact that Proto Oceanic lexical reconstructions, which are widespread throughout the family, are often 'inexplicably' missing from the aberrant languages; for example see Table 7.5.

Table 7.5 Languages with 'exemplary' and 'aberrant' reflexes

		'Exemplary'		*'Aberrant'*	
POc	*Gloss*	*Manam*	*Fijian*	*Kwamera*	*Nemi*
**ika*	'fish'	*i'a*	*ika*	*nɨmu*	*nuk*
**niu*	'coconut'	*niu*	*niu*	*napuei*	*t^hep*
**pose*	'paddle'	*ore*	*voðe*	*niveia*	*^{m}ba/haat*
**pudi*	'banana'	*udi*	*vu^{n}di*	*taik*	*pi^{n}jiŋ*
**taŋis*	'weep'	*ta_*	*taŋi*	*asɨk*	*hye*

ABERRANCY AND THE RECONSTRUCTION OF LINGUISTIC PREHISTORY

At the very least, aberrancy of the kind I have been discussing obscures the nature, or even the very existence, of the historical relationship between languages. To illustrate this point in my comparative linguistics course, I give my students a set of data like those in Table 7.6 from Whitesands (Southern Vanuatu) and Fijian, and ask them to decide how many pairs of words are cognate. The answer I usually get is 'none', though the correct answer is 'all'. These words, in both Whitesands and Fijian, derive regularly (through, obviously, a different set of phonological rules) from the following Proto Oceanic reconstructed forms:

**kayu*	'tree'
**qupi*	'yam'
**rua*	'two'
**qacan-gu*	'my name'
**kaRat-i*	'bite it'

A comparison of the data presented in Tables 7.1, 7.2 and 7.5 with Table 7.2 illustrates the point. On the basis of certain shared innovations in phonology and morpho-syntax, the Southern Vanuatu and New Caledonian languages are closer genetically to Fijian and Māori than any of these languages are to Manam or Motu. Yet an examination of the data presented above shows that Fijian and Māori are *superficially* much more similar to Manam and Motu than they are to the Southern Vanuatu or New Caledonian languages. That is to say,

Table 7.6 Whitesands and Fijian

Whitesands	*Fijian*	*Gloss*
nɨɳi	*kau*	'tree'
nu	*uvi*	'yam'
kɨiu	*rua*	'two'
nahŋok	*yaða^{n}gu*	'my name'
us	*katia*	'bite (it)'

relatively superficial comparison of aberrant languages, either amongst themselves, or with Proto Oceanic or exemplary languages, may lead to quite incorrect conclusions, as witness the following:

> I am operating under the assumption that the [Southern Vanuatu] languages form a closed subgroup of Oceanic, *and that they may very well prove to be a first-order subgroup of Oceanic.*
>
> (Lynch 1978: 762; emphasis added)

> These languages [of New Caledonia and the Loyalties] . . . differ, however, so much from common OC [= Oceanic] forms that *it has even been questioned whether they are more than* [sic] *non-Austronesian (NAN) languages with Austronesian (AN) borrowings.*
>
> (Capell 1976: 253; emphasis added)

Even more thorough and detailed comparison of aberrant languages does not always yield promising results:

> The languages of New Caledonia . . . are among the most notoriously aberrant in Melanesia. In the early 1970's I collected data in the field on two of these languages, [Canala (Xârâcùù) and Grand Couli]. . . . [E]ven though the Canala and Grand Couli languages are very similar in structure and give other evidence of being close relatives, *I was not able to give a satisfactory account of the sound correspondences between the two.*
>
> (Grace 1981: 256–257; emphasis added)

Now there are hundreds of non-Austronesian (or Papuan) languages in Western Melanesia, and contact with these has often been invoked as the root cause of aberrancy in Melanesian Austronesian languages. In some parts of the region this is a reasonable, if partial, explanation for divergence from the Proto Oceanic 'norm' (see e.g. Lynch 1981). However, Grace (1981: 256) points out that 'there are aberrant AN languages which are not spoken anywhere near an area in which there is any reason to believe that non-AN languages were ever present', with Southern Melanesia being a prime example. The languages of Santa Cruz and the Reef Islands in Solomon Islands are the modern non-Austronesian languages geographically closest to those of Southern Melanesia; but they are over 1000 km away, and in between Reefs-Santa Cruz and Southern Melanesia almost 100 Austronesian languages are spoken.

Grace goes on to suggest that part of the explanation for the aberrancy of the New Caledonian languages may lie in borrowing between *related* languages. This, he feels, may be responsible for:

> 1 the unexpectedly low cognate percentages,
> 2 the complicated sound correspondences, and
> 3 the large phoneme inventories of New Caledonian languages.
>
> (Grace 1981: 264)

This is, however, by no means the full story. Lichtenberk (1994) has shown that 'complicated sound correspondences' (to use Grace's term) can occur in otherwise 'exemplary' Oceanic subgroups. Intra-subgroup borrowing, or incomplete diffusion of a sound change through a group of closely related languages, can complicate the phonological history of a group of languages *without* bringing about the much more radical changes that have occurred in Southern Melanesia.

The aberrancy of the languages of Southern Melanesia may, therefore, be due to a combination of factors: borrowing from neighbouring closely related languages; borrowing from other, more distantly, related languages (e.g. the Southern Vanuatu languages have borrowed quite heavily from the Polynesian Outlier Futuna-Aniwa); and internal phonological changes and lexical replacements that conspire to make the languages superficially quite different from reasonably close relatives.

ASSESSING GENETIC DISTANCE

In the discussion above (and see also Figure 7.3), I mentioned that the Southern Vanuatu and New Caledonian languages may form a single subgroup of Oceanic, and implied that they are more closely related to the Central Pacific languages than any of them are to Oceanic languages further west, despite the apparent high degree of similarity between words in these western languages and those of the Central Pacific subgroup. We must infer from this, therefore, that the phonological and morpheme structure changes that have taken place in these languages, radical as they are, are really quite superficial when it comes to assessing the linguistic position of these languages – just a 'facelift', in other words, though in this case a fairly drastic remodelling.

This important point is occasionally overlooked by linguists themselves, and is certainly one not appreciated by the majority of non-linguist prehistorians. As with biological or archaeological phenomena, superficial similarities or differences between languages are less important, as far as classification is concerned, than underlying structural-typological similarities or differences.

The languages of Southern Melanesia are a case in point. The closer degree of genetic relationship between the subgroups of Southern Melanesian (Southern Vanuatu and New Caledonian) on the one hand, and between these languages and those of the Central Pacific subgroup on the other, is measurable – though not quantifiable – by shared innovations in the treatment of (certain) Proto Oceanic phonemes, and by shared innovations in grammatical morphemes or in the irregular development of certain morphosyntactic categories or lexical items (see Lynch and Tryon 1985 and Lynch 1995 for some examples of this). Features of this kind, rather than the actual shape and pronunciation of individual lexical items, are more reliable indicators of the degree of genetic distance obtaining between any two languages or groups of languages.

Research on the languages of Southern Melanesia is continuing; unfortunately, as in much of Melanesia, there are too many languages and too few linguists working on them. It is hoped, however, that this research will further elucidate the relationships between the languages of Southern Melanesia and the processes that have led to their superficial aberrancy.

REFERENCES

Capell, A. 1968. *A New Fijian Dictionary* (third. edn). Suva: Government Printer.

Capell, A. 1976. Features of Austronesian languages in the New Guinea area in general in contrast with other Austronesian languages of Melanesia. In *New Guinea Area Languages and Language Study, Vol. 2: Austronesian languages*, S.A. Wurm (ed.), 235–282. Pacific Linguistics C-39. Canberra: Australian National University.

Crowley, T. 1990. *Beach-la-mar to Bislama: the emergence of a national language in Vanuatu.* Oxford: Clarendon Press.

Geraghty, P. 1989. The reconstruction of Proto-Southern Oceanic. In *VICAL 1 – Oceanic Languages: Papers from the Fifth International Conference on Austronesian Linguistics*, R. Harlow and R. Hooper (eds), 141–156. Auckland: Linguistic Society of New Zealand.

Grace, G.W. 1975. *Canala Dictionary (New Caledonia).* Pacific Linguistics C-2. Canberra: Australian National University.

Grace, G.W. 1981. Indirect inheritance and the aberrant Melanesian languages. In *Studies in Pacific Languages and Cultures in Honour of Bruce Biggs*, J. Hollyman and A. Pawley (eds), 255–268. Auckland: Linguistic Society of New Zealand.

Grace, G.W. 1985. Oceanic subgrouping: retrospect and prospect. In *Austronesian Linguistics at the 15th Pacific Science Congress*, A. Pawley and L. Carrington (eds), 1–18. Pacific Linguistics C-88. Canberra: Australian National University.

Grace, G.W. 1986. Hypotheses about the phonological history of the language of Canala, New Caledonia. In *Le coq et le cagou: essays on French and Pacific languages in honour of Jim Hollyman*, C. Corne and A. Pawley (eds), 55–67. [*Te Reo* 29] Auckland: Linguistic Society of New Zealand.

Grace, G.W. 1990. The 'aberrant' (vs. 'exemplary') Melanesian languages. In *Linguistic Change and Reconstruction Methodology*, P. Baldi (ed.), 155–173. Berlin: Mouton de Gruyter.

Haudricourt, A.-G. and F. Ozanne-Rivierre 1982. *Dictionnaire thématique des langues de la région de Hienghène (Nouvelle-Calédonie).* Paris: SELAF.

Lichtenberk, F. 1983. *A Grammar of Manam.* [*Oceanic Linguistics* Special Publication No. 18], Honolulu: University of Hawaii Press.

Lichtenberk, F. 1994. Reconstructing heterogeneity. *Oceanic Linguistics* 33(1), 1–36.

Lindstrom, L. 1986. *Kwamera Dictionary – Nɨkukua sai Nagkiariien Nɨninɨfe.* Pacific Linguistics C-95. Canberra: Australian National University.

Lister-Turner, Revd. R. and Revd. J.B. Clark n.d. *A Dictionary of the Motu Language of Papua* (2nd edn by Percy Chatterton). Sydney: Government Printer.

Lynch, J. 1978. Proto-South Hebridean and Proto-Oceanic. In *Second International Conference on Austronesian Linguistics: Proceedings*, S.A. Wurm and L. Carrington (eds), 717–779. Pacific Linguistics C-61. Canberra: Australian National University.

Lynch, J. 1981. Melanesian diversity and Polynesian homogeneity: the other side of the coin. *Oceanic Linguistics* 20(2), 95–129.

Lynch, J. 1995. *Linguistic subgrouping in Vanuatu and New Caledonia: some preliminary hypotheses.* Paper presented to the Second International Conference on Oceanic Linguistics, Suva, Fiji, July 1995.

Lynch, J. and P. Tepahae In prep. *Anejom̃ dictionary/Diksinari blong Anejom̃/Nitasvitai anijitas antas Anejom̃.*

Lynch, J. and D.T. Tryon 1985. Central-Eastern Oceanic: a subgrouping hypothesis. In *Austronesian Linguistics at the 15th Pacific Science Congress*, A. Pawley and L. Carrington (eds), 31–52. Pacific Linguistics C-88. Canberra: Australian National University.

Pawley, A. 1972. On the internal relationships of Eastern Oceanic languages. In *Studies in Oceanic Culture History, Vol. 3*, R.C. Green and M. Kelly (eds), 1–142. Honolulu: Bernice P. Bishop Museum.

Ross, M.D. 1988. *Proto Oceanic and the Austronesian Languages of Western Melanesia.* Pacific Linguistics C-98. Canberra: Australian National University.

Thompson, R.H. 1970. The subjective element in archaeological inference. In *Introductory Readings in Archaeology*, B.M. Fagan (ed.), 356–361. Boston: Little, Brown and Co.

Williams, H.W. 1975. *A Dictionary of the Maori Language.* (7th edn) Wellington: Government Printer.

8 Linguistic and philological data towards a chronology of Austronesian activity in India and Sri Lanka

WARUNO MAHDI

In a dim distant unrecorded age we had met thou and I, When my speech became tangled in thine and my life in thy life.

Rabindranath Tagore, *Śrī-Vijaya-lakṣmi*

INTRODUCTION

Background

Apart from the Near East, the Asian continent has seen two important centres of civilization: China and India. They have one feature in common, in that they both arose in a multi-ethnic environment, and then experienced repeated, belligerent as well as more peaceable foreign incursions or infiltrations through the course of their history. Having emerged on the basis of peoples of the Yangshao and Longshan cultures, China expanded to include and assimilate neighbouring Miao-Yao, Daic, Tibeto-Burmic, Altaic and perhaps also Austronesian peoples. It experienced invasions of Huns, Churchens, Mongols and Manchus, and accommodated Turkic merchants and craftsmen. Similarly, owing the beginning of its civilization to the meeting of Indo-Aryan, Dravidian and Austro-Asiatic peoples, India experienced incursions by Scyths, Cushans, Greeks, Mongols, Portuguese, Dutch, British and French, and infiltrations by Tibeto-Burmic, Daic and Austronesian peoples. This chapter is dedicated to the neglected Austronesian contribution to Indian culture.[1]

Early contacts between the South Asian subcontinent and Southeast Asia in general, or between India and Malayo-Indonesia in particular, have long been the subject of scientific interest, but at first almost exclusively focusing on the substantial material and spiritual culture influence from India. For reason of space, I shall not review the extensive literature on this subject. With regard to its linguistic aspects, there are the series of publications by van Ronkel (1902, 1903a, 1903b) dedicated to Dravidian loanwords in Malay (see also van Ronkel 1918 on such loans in Batak languages), and a comprehensive survey of Sanskrit borrowings by Gonda (1952). Investigations on reciprocal influence from

Southeast Asia to India have hardly found any echo. It is to archaeology of the last decades that we owe the renewed currency of the problem of cultural influence emanating from Southeast Asia, a process in which Austronesians, with their maritime mobility, played an important role.

Transcriptions

Sanskrit, Arabic, Hebrew and Greek glosses are given in conventional latinizations for these languages, as it is not always possible or desirable to replace them by explicit IPA transcriptions. Thus, for instance, Arabic ğ was pronounced /g/ in some dialects, /j/ in others. It is not always apparent which dialect was involved, and what the pronunciation was in that dialect at that time, or even whether the time can be narrowed down closely enough to be that specific about its contemporaneous phonology. For Chinese, all glosses are given in the standard Pinyin latinization with indication of the tones.

Modified IPA transcription[2] is used wherever possible for comparative lexical data from Austronesian languages. For Malagasy, however, the standard orthographic rendering is used instead,[3] with additional indication of the stressed syllable, and for Paiwan I use the spelling of Ferrell (1982).[4] For Austro-Asiatic (except Vietnamese[5]), Dravidian and Indo-Aryan languages, I have relied on the transcription used in the sources. For reconstructed Austronesian protoforms (marked like all reconstructs by an asterisk before the item), the transcription of Dyen (1971),[6] simplified in that the indices marking hypothetical distinctions have been deleted, is used with the following modifications:

1 *y and *w have been replaced by *i and *u respectively in agreement with Dahl (1977), except in relatively recent protoforms that appeared after desyllabification of the original high vowels in the corresponding positions (see also Mahdi 1988: 90–101, 1994a: 208–209 n. 41);
2 *b is split into *b and *B following Prentice (1974) and Nothofer (1975);
3 the laryngeals *q, *S, *ʔ and *H, the latter including also *h which I do not distinguish as separate proto-phoneme, are reconstructed in accordance with Zorc (1982);
4 the opposition of *d and *D is redefined as in Mahdi (1988: 72–78, 1994a: 170, 1996).

Furthermore, in agreement with Wolff (1974 and 1982), I do not include *c, *r, *T and *z among the original phonemes of Proto-Austronesian, but regard them either as features of various Austronesian meso-languages (lower-order, intermediate proto-languages), or as phantom proto-phonemes reconstructed from sound correspondences from cognate comparisons involving loanwords.

Optional parts of a protoform, e.g. when an affix is not represented in all the reflexes, will be enclosed in round parentheses, so that *(a)bc means *bc and/or *abc. Uncertain and alternative potential items in the reconstruction will be placed in square brackets, whereby *ab[c] means probably *abc, but the *c is uncertain, whereas *ab[cd] means either *abc or *abd. Furthermore, *a[]b means *a. . .b with perhaps something not yet specifiable in the place of the dots. The part of a reflex or cognate that is relevant for the reconstruction or comparison will be separated from irrelevant parts (affixes, other components of compounds) by a hyphen, as e.g. in English *hund-red*, Latin *cent-um*, or in English *yard*, German *Gart-en*, Russian *o-gorod*, Latin *hort-us*. I use ø to denote the zero or 'non-phoneme'.

Protoforms do not necessarily reflect the highest-order proto-language of a language family, since innovations can emerge at all nodes of a language tree. The distribution of reflexes of a protoform is principally determined by two processes: inheritance and borrowing. Inheritance leads to regular reflexion, and reflexes usually exhibit regular sound correspondences (occasional irregularities occur through assimilation, dissimilation, analogy, folk etymology), and are restricted to languages that are direct daughter-languages of the proto-language in which the innovation took place. Borrowing, on the other hand, can often (but not always) be detected by irregularity of sound correspondences. Frequent recurrence of an irregular correspondence may give a misleading reconstruction of a 'phantom' proto-phoneme, with the consequence that all instances exhibiting the irregularity appear to be regular. Borrowed reflexes may be distributed haphazardly with regard to boundaries of language subgroups, and thus suggest a much higher-order proto-language as origin of the innovation, corresponding chronologically to a much earlier time. Thus Latin *catus*, German *Kat-ze* (irregular k, expected h) and English *cat* (irregular c and t, expected h and d respectively), refer to a denotate introduced into Europe much later than the time of separation of the Germanic and Latinic families within Indo-European, whereas English *head*, German *Haupt* and Latin *caput* represent something Indo-Europeans have always had. Failure to take account of the irregular sound correspondences in the forms for 'cat' would have suggested knowledge of the cat since before the Germanic-Latinic split.

EARLY REFERENCES TO SOUTHEAST ASIAN CULTURE CONTRIBUTIONS

Southeast Asia and Southeast China were already recognized as the centre of a neolithic horizon featuring stone adzes with rectangular cross-sections (with or without 'shoulders'), of which the southwestern periphery encompassed much of the Indian subcontinent (Heine-Geldern 1932: 561, 569). Indochina had been identified as the homeland of the Dongson culture with its sophisticated bronze-smelting technology. Related artefacts occur throughout mainland and insular Southeast Asia and in South China (e.g. Goloubew 1929). North Indochina and South China were also seen as the place of origin of an older metallurgical technology producing bronze celts, which spread into insular Southeast Asia. In the Philippines, Otley Beyer estimated a time-depth of 1000–500 BC (Sullivan 1956: 72). Newer and more precise data have led to a reappraisal: Fox (1967: 13) placed the first metal in the Philippines about 500–400 BC, and first iron around 200 BC, whilst Bellwood (1980: 152) considers the Philippine early metal period as 500 BC to 1000 AD. Southeast Asia and the near Pacific have long been identified as one of the world centres of origin of cultivated plants by Vavilov (1927: 417, augmenting an earlier 1926 publication). The transmission of cultigens of Southeast Asian origin to India and Sri Lanka is discussed elsewhere (Mahdi, Volume II).

One notable exception to the prevailing sceptical view on westward Southeast Asian culture influence was in regard to Austronesian shipping. Based on the striking agreement between boat forms in India/Sri Lanka and Oceania, James Hornell suggested an Austronesian (the author used the term 'Polynesian') origin of South Indian outrigger boats, and supported this with data on fishing methods, coconut cultivation, toddy tapping and, with some reserve, betel chewing (Hornell 1920: 140, 221–222, 225–246). The prehistoric and protohistoric Austronesian shipping along the northern perimeter of the Indian Ocean and its implications for the chronology of Austronesian activity in India and Sri Lanka is discussed elsewhere (Mahdi, Volume III, Ch. 5).

INSIGHTS FROM RECENT ARCHAEOLOGY

The 1960s and 1970s witnessed some spectacular archaeological discoveries, profoundly changing the picture of the indigenous Southeast Asian and Western Melanesian contribution to culture development. New finds suggested unexpectedly early beginnings of horticulture in Thailand (Gorman 1969; Yen 1977) and Taiwan (Tsukada 1966: 546; Chang 1970: 181), as also in New Guinea (Golson 1977; Golson and Hughes 1980; Pawley and Green 1973: 5–6). Bronze finds characterized mainland Southeast Asia as one of the earliest hearths of metallurgy, delivering radiocarbon and thermoluminescence dates of 2290 ± 90 BC, 2325 ± 200 BC (Solheim 1968: 61), and prior to 2300 BC at Non Nok Tha, Thailand (Bayard 1972), and 1910 ± 240 BC at Non Chai, Thailand (Charoenwongsa and Bayard 1983: 522). These developments led to new views on the place of Southeast Asia in world culture history (cf. Solheim 1967, 1969, 1970, 1972). Understandably, this also modified the picture with regard to interregional communication, exchange and influence. Solheim (1980: 334) suggested that intensified trade activity spanned the whole Southeast Asian island world between 200 BC and 200 AD, leading to the spread of iron, the use of gold, metal-age megalithism[7] and other culture elements. The author also suggested that Malay or Malay-speaking traders began moving along the coast of the Indian Ocean during the first millennium BC, meeting at the eastern end of the trade between India and the Mediterranean.

Somewhat more sober reappraisals of the data in the 1980s (e.g. Hutterer 1983; Bellwood 1985; Higham 1989; Spriggs 1989) have meanwhile eroded the dramatic edge from this general picture. Nevertheless, the expansion of the

South China–Southeast Asian Neolithic from the mainland must have reached Taiwan at least by 3500, perhaps even 4300 BC, Luzon and Sulawesi by around 3000 BC, and the Bismarck Archipelago perhaps by around 2500 BC, but certainly by 2100 BC (Spriggs 1989: 605). This is in agreement with views among linguists for the first split of Proto-Austronesian at around 4500 BC (Blust 1984–85: 54). In a later publication, the spread of the Neolithic from Taiwan to Timor is placed between 3000 and 2000 BC, arriving in the Bismarcks at around 1800–1600 BC (Spriggs 1991: 308–309). Considering the extensive Austronesian migration eastwards into Oceania, it would have been most surprising if the Austronesians had not also ventured west, and indeed we find them as far west as Madagascar. The main reason why we do not also find them along the northern perimeter of the Indian Ocean between the Mergui Islands and Madagascar today is that, in my opinion, these regions were not as uninhabited as the islands of Oceania, so that Austronesian coastal settlers must, sooner or later, have been assimilated into the local mainland population.

In 'The Dispersal of Austronesian Boat Forms in the Indian Ocean' (Mahdi, Volume III, Ch. 5), I conclude that Austronesians with an already advanced tradition of long-distance navigation must have begun to settle in the South-Asian subcontinent between 1000 and 600 BC, having perhaps been preceded in this by Austronesian groups with less developed seafaring skills (also Mahdi 1994b: 470 n. 120). The dates are based on:

- the first settlement of Micronesia (the main locus of distribution of shunting single-outrigger boats also occurring in Sri Lanka) at around 1000 BC (Bonhomme and Craib 1987);
- the beginnings of the boat burial custom around 600 BC (Higham 1989: 195) marking the end of the period in which the double canoe and single-outrigger boat (the boat forms transmitted to India) were prevalent in Southeast Asia.

The period between 1000 and 600 BC was probably also the time of the introduction of sorghum (*Sorghum bicolor* [Linn.] Moench) from India to West Malayo–Indonesia (Mahdi 1994b: 433–434), indicating the existence of communications at that time. The dating required that sorghum must have been introduced from the west at about the same time that foxtail millet (*Setaria italica* Beauv.) was brought in from the north. Foxtail millet, already attested in Taiwan at 2800 BC, was apparently uncovered at a site in Timor in a layer just above 1000 BC (Bellwood 1985: 214, 227).

A much more recent period in history was involved in a controversial hypothesis suggesting a Malayan origin or affiliation of a ruling house of medieval Sri Lanka (Paranavitana 1960, 1966). This hypothesis met with vigorous rejection (e.g. Sastri 1962; Indrapala 1967; Sirisena 1971). The *Culavaṁśa* does record two Malay invasions against Sri Lanka in the thirteenth century (Kern 1896), and some indications seem to exist for obscure relations between Malays and South India. However, the invasions not only took place later than the founding of the allegedly Malayan dynasty, but also remained unsuccessful, the Malays gaining only a temporary foothold in Jaffna in the north of the island (de Silva 1981: 67).

LINGUISTIC EVIDENCE

Linguistic evidence for Southeast Asian elements in Dravidian and Indo-Aryan languages is in the main of either Austro-Asiatic or Austronesian origin. With regard to Austro-Asiatic contributions, the pioneering work of Przyluski (1921), Lévi (1923) and Gonda (1932) was followed up by the monograph of Kuiper (1948). The earliest indication of Austronesian borrowing into Tamil comes from Kern (1894), who showed that most of the numerals used in Tamil traders' slang reported by Sastri (1894: 49) originated from a language of West Indonesia. Kern (1897) was also the first to assume an Austronesian origin of Hindi forms for lime/lemon.

Another early assignment of a Malayo-Indonesian origin to a Sanskrit word is that of van der Tuuk (1901: 720) who, citing Old Javanese *wuŋa-lawaŋ* 'clove' (with Old Javanese *wuŋa* 'flower') and Toba Batak *labaŋ* 'nail', suggested that Sanskrit *lavaṅga* 'clove' could be a loan from a language of Indonesia. The same opinion was expressed by Gonda (1932: 326–329). I have elsewhere demonstrated that the word must indeed have been borrowed from Malay, in approximately the second century BC, probably through the intermediation of a Dravidian language (Mahdi 1994a: 188 and 215 n. 92; see below). Schoff (1922: 362–363) assumed Malay or Sumatran origin for Pali and Prakrit *kappūra* 'camphor' and the corresponding Sanskrit *karpūra* 'camphor' (probably a back formation from the former). Beside the word for 'clove' mentioned above, Gonda (1932) also suggested that Sanskrit *làśuna* 'leek, garlic' and *marīca* 'pepper' were of either Austronesian or Austro-Asiatic origin. However, this has meanwhile been shown not to be the case for *marīca* (Zide 1976: 420–421).

During the period of Dutch rule (1656–1795), Sri Lanka, like South Africa, served as a place of exile for war and political prisoners from Indonesia. During the same period, slaves from Indonesia were brought here, and Indonesian soldiers served in Dutch garrisons. Like the Cape Malays in South Africa who emerged under similar conditions (but speak a form of Afrikaans rather than Malay), an Indonesian community was formed in Sri Lanka speaking its own Malay vernacular (Adelaar 1991). Therefore, linguistic evidence of a Malay presence in Sri Lanka cannot be easily evaluated for preceding periods, given the greater possibility that a loanword originates from the recent Sri Lankan Malay vernacular. Also difficult to date are isolated borrowings in the language of the Maldives, reported by Gray (1878: 190–195), e.g. *timara* 'tin' (Malay *timah*, Minangkabau *timarah*), *kreis* 'dagger' (Malay *kəris*). These too are probably relatively recent, and would then fall outside the scope of this chapter. Additional linguistic evidence for Austronesian contacts and their dating will be proposed in this chapter.

THE SEVENTH CENTURY AS THE LIMIT OF THE PERIOD UNDER INVESTIGATION

The seventh century AD may perhaps serve as a convenient limit on the period discussed here. Although the oldest Sanskrit inscriptions in Indonesia, those of King Mūlavarman (Kern 1882; Chhabra 1945), have been dated at around 400 AD (Vogel 1918), the texts provide only rather marginal information about the king, and even less about his country and its inhabitants. In the seventh century we have, on the one hand, the emergence of the Buddhistic Malay thalassocracy *Srī Vijaya* (Cœdès 1918), about which we are much better informed (e.g. Wolters 1967; Hall 1985: 78–102; Manguin 1987, 1993). On the other hand, for the seventh century there are also reports from Persian historians about skilled seafarers, the *Sayābiğa*, who formed settlements in the Persian Gulf (de Goeje 1894, 1903: 18, 20, 86–91; Ferrand 1934). The name is the plural form of *Sābağ* which, like Arabic *Zābağ*, evidently reflects Pali *Jāvaka*, originally referring to Malays or countries they inhabited or ruled (Mahdi 1995 and Figure 8.1). Thus, by the seventh century AD, westward trade and colonizing activities of Malay-speaking seafarers had apparently already advanced beyond the South Asian subcontinent.

In the subsequent period, Malays are reported doing trade with the *Zanğ* and *Sofāla* in Africa (see Figure 8.1) by Arabian geographers, particularly al-Idrīsī (Ferrand 1910: 301, 1919: 63; Trimingham 1973: 125–126). The dating is further corroborated by Chinese reports of the Tang period (618–906 AD) mentioning *Sēng qí-nú* or *Sēngzhī-nú*, which denoted African slaves presented to the Chinese emperor by Malayo-Indonesian envoys (Chinese *nú* 'slave, bondsman'). The variants *Sēngqí* ~ *Sēngzhī* reflect Malay *jəŋgi* or *jaŋgi*, which in turn derive from Arabic *zanğī* 'pertaining to the *Zanğ*' (see Pelliot 1904: 231 fn. 4; Ferrand 1919: 211 fn. 3). The *Kitab ajā'ib al-Hind* ('Book of the Marvels of India'), credited to the tenth-century sea captain Buzurg ibn Shahriyār, actually records an invasion of the East African coast by a fleet of 1,000 ships from the Far Eastern *Wāqwāq* in the year 334 Hegira (= 945–946 AD) (Ferrand 1910: 324; Trimingham 1973: 133; Tibbetts 1979: 163). *Wāqwāq* was a geographical term covering a part of Malayo-Indonesia (usually Sumatra, sometimes perhaps also Kalimantan) in the east and Madagascar in the west, which Arab geographers believed, following Ptolemy, to be connected by a landmass bounding the Indian Ocean in the south (see Ferrand 1908: 480–506).

CLUES FROM LEGENDS AND LITERARY TRADITION

The Rāmāyaṇa: coconut-eating vassals of King Sugriva

If Austronesians had indeed been present in South Asia since 1000 BC or earlier, they could hardly have remained unmentioned in the rich literary tradition of the region. Indeed, the name of a place in Indonesia, *Yavadvīpa*, is already mentioned in the *Rāmāyaṇa* of Vālmīki (Kern 1869: 640; Lévi

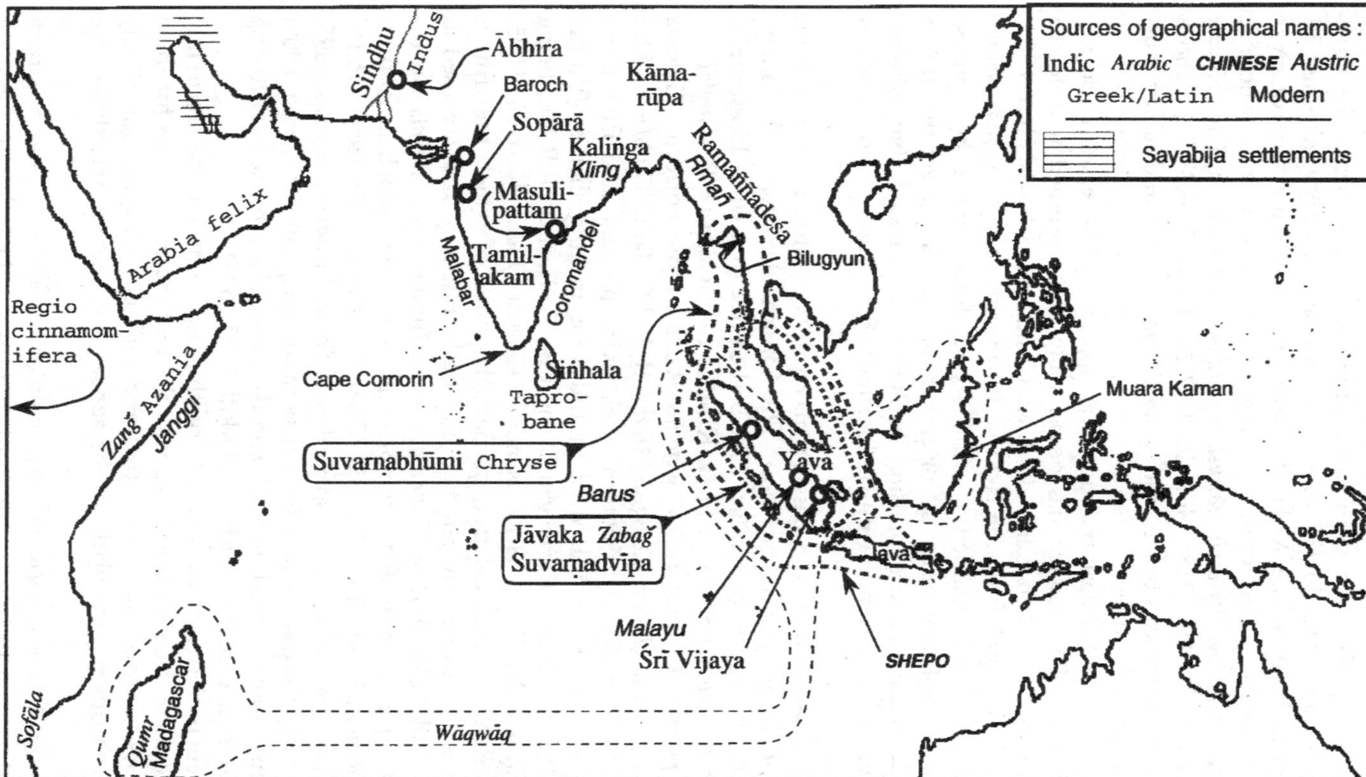

Figure 8.1 Some historical and traditional geographical names in and around the Indian Ocean from Indic (Sanskrit or Tamil), Arabic (and Persian), Chinese, Austric (Austro-Asiatic or Austronesian) and Graeco-Latin sources. Encirclements indicate a region within which countries referred to by a given name could be located.

Source: Mahdi

1918: 20; see Mahdi 1994a: 215 n. 93 and 1994b: 469–470 n. 111), referred to on Figure 8.1 as *Yava*. An even more important passage from the same epic work is one in canto 36 (37 of the Goresio and Bombay editions, 30 of the Lahore) of book 4, the *Kiṣkindhākāṇḍa*, in which King Sugriva gives the order to summon all his vassals, among which are also those who . . . *kṣīrodavelānilayāstamālavanavāsinaḥ nārikelāśanaścauva teṣaṁ saṁkhyā na vidyate* (4.36.25 in Mankad 1965: 218; Goresio 4.37.22, Bombay 4.37.21, Lahore 4.30.21.) 'on the milky sea's beach, and in *tamāla* woods live, and of coconuts eat, their number is countless'.

The 'Milky Sea' of Sanskrit tradition was apparently the eastern sea, i.e. the Bay of Bengal. The passage has occasionally been assumed to refer to the inhabitants of the Nicobar Islands (S.K. Gupta, cited by Mathur 1968: 230–232). However, apart from the fact that inhabitants of the Nicobar Islands would hardly have been regarded as a military reinforcement to be counted upon in India of the first century BC or earlier, one would expect the insular rather than the littoral nature of their habitat to be stressed. No allusion is made, however, to a remote insular location of the summoned vassals in the text.

The plants mentioned in the passage may provide some information on the identity of the peoples in question. The Sanskrit dictionary of Monier-Williams (1899: 438) gives for *tamāla*: 'dark-barked tree (but white blossomed), *Xanthochymus pictorius* L.; a sort of black Khadira tree; *Crataeva Roxburghii* L.'. *Xantochymus pictorius* L. must be *X. pictorius* Roxb., more correctly *Garcinia xanthochymus* Hook.f. (*Index Kewensis* 1895, II-1232), fam. *Guttiferae*. This tree has a dark bark that dyes cotton black, with pleasant acid fruit eaten fresh in the Malayan Peninsula, used in India for making sherbet (Burkill 1935: 1056–1057). *Crataeva Roxburghii* L. must be *C. Roxburghii* R.Br., being either *C. religiosa* Ainslie = *Aegle marmelos* Correa (*Index Kewensis* 1895, I-637), fam. *Rutaceae*, the bael tree or Bengal quince of which the fruit is either used fresh or made into sherbet, and both fruit and bark are used as medicine (Burkill 1935: 57), or *Crataeva religiosa* Blume = *C. nurvala* Buch.-Ham., fam. *Capparidaceae*, which resembles the former very much in its fruit, leaves and medicinal use (Burkill 1935: 676–677). *Crataeva Roxburghii* R.Br. could, however, also refer to *Crataeva religiosa* Forst.f. (*Index Kewensis* 1895, I-637), fam. *Capparidaceae*, which has on occasions (e.g. in de Clercq 1909: #893) been mistakenly united with *C. nurvala* Buch.-Ham. because of having many characters in common with it (Burkill 1935: 677).

Where even professional taxonomists of the last century were apparently confused, traditional Sanskrit botany apparently used the term *tamāla* for three or four species of trees, belonging to three different families, but closely resembling each other, particularly in the appearance, use and refreshing taste of their fruit.

From the lost *Nánzhōu yìwù zhì* ('Account of Curiosities of the Southern Islands'), written by Wan Zhen in the third century AD, the following fragment has been preserved in the *Tàipíng yùlǎn*:

> There is a tree styled *móchú* which grows in the country of *Sīdiào*. Its juice is fat and moist. It is glossy like grease. As to its odour, it is very fragrant and beautiful. It can be used for boiling and frying foods which thus assume a fine smell, in the same manner as oil is employed in China.
>
> (Ferrand 1916: 531)

The name *Sīdiào* apparently resulted through a scribal error from *Yèdiào* (Middle Chinese *i̯äp-dʿieu*, Karlgren 1940: #633d and #1083k) which was a Chinese rendering of Sanskrit *Yavadvīpa* (see Pelliot 1904: 266–268; Laufer 1915: 351; Ferrand 1916; Mahdi 1994a: 204–205 n. 25, and 1995; *Yava* in Figure 8.1). The contemporaneous Middle Chinese pronunciation of the Chinese rendering of the fruit was *muâ-d̂ʿi̯u* (Karlgren 1940: #17e and #127m). Ferrand (1916: 523, 531–532) has identified it as Javanese *mɔjɔ* (Old Javanese *maja*),[8] which refers to the bael fruit/tree or Bengal quince (*Aegle marmelos* Correa; de Clercq 1990: #58) already mentioned above, and to the elephant apple or wood apple (*Limonia acidissima* L. = *Feronia elephantum* Correa), also known as the 'false bael' (de Clercq 1909: #1428).

Evidently, the *tamāla* of the *Rāmāyaṇa* and the *móchú* of the *Nánzhōu yìwù zhì* referred either to the same tree, *Aegle marmelos* Correa, or to trees so similar that the respective local traditional botanical terminologies grouped them under the same name. That the tree proved noteworthy enough to be mentioned in an early Chinese reference to West Malayo-Indonesia suggests that it was a conspicuous feature of Austronesian villages of the region. The *Rāmāyaṇa* also notes either the same or a very similar tree as a characteristic feature of villages of the vassal peoples, thereby creating the impression that here too is a reference to Austronesians, an impression already suggested by the littoral habitat of these peoples.

The impression is further strengthened by the indication that coconuts were important in the diet of these peoples. The peoples whose use of coconut as foodstuff was still uncommon enough to serve as a distinctive characteristic in the *Rāmāyaṇa* must have been the very coastal settlers who had introduced the coconut to this region. As already noted by Hornell (1920: 222, 235), these were probably Austronesians. The distribution of *niəuR 'coconut' as one of the most widely represented protoforms in Austronesian languages throughout its distribution area (except Taiwan) shows that the plant already occupied an important place in Austronesian culture from a very early stage (Kern 1889).

Nāgas in the Mahābhārata

The *Rāmāyaṇa* does not give the name of the coconut-eating coast-dwellers, but simply includes them among the *Vānara*, the 'apes' making up the whole of King Sugriva's army. In both Sanskrit and Tamil literary tradition, relatively frequent mention is made of peoples referred to as *Nāga* (literally 'serpent, snake'). Various attempts have been made to identify them with historical peoples, but together with local traditions of serpent worship, this is a vast complex with elements of disparate origin (see Vogel 1926: 2–6). The modern *Nāgas* of north India are Tibeto-Burmans, but there have probably been shifts in the meaning of the ethnic term as a result of the gradual absorption of the original *Nāgas* into Dravidian, Indo-Aryan, Munda and Tibeto-Burmic language communities. Tibeto-Burmans were apparently referred to initially as *Yakṣas*. It has been occasionally assumed that the *Nāgas* were Mundas, or simply the autochthones. That the original *Nāgas* included Munda-speaking peoples is likely, but they were probably not exclusively Mundas. The *Nāgas* were often associated with the sea, piracy, a river, even fish odour, and Austronesians probably played a substantial part in the formation of the *Nāga* complex in Indian tradition.

In a passage in the first book of the *Mahābhārata*, the ocean is said to be the *Nāgānām ālayaṁ*, i.e. place/abode/country of the *Nāgas* (Sörensen 1904: 492 under *Nāga*). The excerpt from the first book of the epic, the *Adiparva* ('The Beginning'), in the translation of van Buitenen who renders *Nāga* as 'Snake', goes as follows: '21. . . . take me to lonely and lovely Ramaṇīyaka on the ocean bay that is the country of the Snakes' (van Buitenen 1973: 79). To the question of the possible significance of the toponym *Ramaṇīyaka* I shall return below. With regard to the association of the *Nāgas* with an aquatic or insular habitat, compare further:

> 22. . . . And as Indra rained, the Snakes were transported with joy, and Earth was everywhere filled with water. . . . they soon came to the island which was encompassed by the
> 23. waters of the ocean . . . . . . strew showers of blossoms on the Snakes that dwelt there.
>
> (van Buitenen 1973: 80).

All this suggests that the *Nāgas* were maritime peoples. The meaning of the word *nāga*, 'serpent, snake', is already a strong hint that it referred to Austronesians. The cult of the divine or sacred serpent or snake is widely distributed among Austronesian peoples. This may have led to the use of the Hindu term *Nāga* for the latter, and possibly influenced the Hindu mythological figure originally denoted by that name (a five-headed serpent, cf. Greek *hydra*). Kern (1916) indicated that, originally, water as an element of the mythological *nāga* was not the liquid element of rivers, lakes and seas, but atmospheric water, waters of the air (a frequently recurrent concept in Vedic literature). The author suggested that the *nāga* developed from a personification of rainclouds, the five heads representing forks of lightning. In later tradition, however, the *nāga* was often a single-headed serpent, and its element hydrological and maritime rather than atmospheric. This shift is not quite natural, because the lower world (i.e. not atmospheric) snake in various mythologies tends to dwell in the underbrush and holes in the earth, rather than water. Austronesian contacts may have contributed to this shift.

The sacred serpent cult in Indonesia

In West Austronesia, and particularly in West Malayo-Indonesia, where there has been considerable Indian and also some Chinese influence, the original serpent cult is permeated with conceptions of the Hindu-Buddhist *nāga* (usually symbolized as a sacred cobra) from India, and more recently also of the *lóng* ('dragon') of Chinese mythology. This has unfortunately led to a tradition of referring indiscriminately to all manifestations of the serpent in local mythology as *nāgas*. In Malayo-Indonesia, just as in Indochina, one can quite clearly distinguish an indigenous and an intrusive or foreign element in the divine serpent/*nāga/lóng* complex. The interactions of indigenous and Hindu-Buddhist serpent-and-water cult and symbolism in mainland and insular Southeast Asia is the subject of an interesting Buckminster Fuller-inspired study by Jumsai (1988), focusing particularly on its manifestations in architecture. A similar relationship between traditional and Indic Serpent cult in Burma was noted by Sörgel (1970).

Chinese influence in Malayo-Indonesia is particularly evident in Kalimantan (Borneo), where the import of large glazed earthware jars from China had the effect of firmly installing the Chinese *lóng* in local indigenous ornamental style (Kühr 1896: 234). Nevertheless, a distinction is maintained between the indigenous sacred serpent, a female deity associated with earth, water, storms, providing protection in daily life and ensuring safe passage of the dead to the next world on one side, and the Chinese *lóng* dragon on ceramic jars, which is male, but which did not occupy the position of male upper-world counterpart to the female lower-world serpent. That function in the mythology of the peoples of Kalimantan still remains a prerogative of the hornbill (Sellato 1989: 44–45).

In Sumatra and parts of Kalimantan, the indigenous sacred serpent manifests itself in the notion of a divine snake referred to in Malay as *ular saŋiaŋ* from *SulaR 'snake' (cf. Kongadavanu Rukai *sořaʔa* [Taiwan], Tombatu *ulah* [North Sulawesi], Lampung *ulay* [Sumatra] 'snake'.) and *sa-ŋ 'honorific personal article' + *qiaŋ 'ancestor, spirit, god'. In Malay, the qualifying word follows the head-noun, rather than preceding it as in English. The expression thus literally translates as 'snake of, or which is a venerable god', which is distinct from what is referred to as *ular naga*, the Indian *nāga*, any mythological magical serpent, (Chinese, European) dragon. The former (indigenous Malay) expression has also been loaned in other languages of the region. In some places, *ular saŋiaŋ* refers to the rainbow. The interpretation of the rainbow as a snake is apparently very widespread (Robert Blust, p.c.). This indigenous term in any case exhibits an older connotation than *naga*, the corresponding Sanskrit loan having cognates in most major languages of West and Central Malayo-Indonesia.

Further east, loaned cognates of *naga* are more sparsely represented, and probably recent. Manifestations of the snake cult are more markedly indigenous. Vatter (1932: 234–235) notes that the cult of the sacred serpent is the principal element in the indigenous religion of Alor (in the east of the Lesser

Sunda Islands), where it is called *haʔard*, which is its proper name, and is not derived from a word meaning 'snake'. Only in the east of Central Alor is the expression *ul-naŋ* known, a loaned cognate of Malay *ular naga* for any mythological magical serpent, dragon.

A Letinese legend holds that when *Tivurleti-Paisleti*, the progenitor who lived on the mountain *Mesmori* (< *ma-isa 'be one, alone' + *ma-quDip 'be alive'), scooped the sea dry and laid bare the island of Leti (in South Maluku), he came upon a rock named *Invatmuamra* under which lived a snake. The latter, regarded as the master or ruler of the original Letinese, is called *Ralieti*, literally 'king of Leti'. At the place of the rock now rises the mountain *Vuarlavna* on top of which the sacred snake *Ralieta* is believed to be enthroned. A similar legend exists on the nearby island of Kisar, where the sacred serpent is called *Nilavna*, literally 'big snake', as is the rock under which it originally dwelt. The population in Leti is divided into *Ornusa* 'of the land', who represent the original population and alone uphold the serpent cult, considering themselves subjects of *Ralieti*, on one side, and *Orleta* 'of a domain', descendants of peoples who had gold and came to Leti by boat, who do not worship the sacred serpent, but have their own totem-poles, on the other (Aone van Engelenhoven, p.c.). The snake cult thus not only dominates the origin myth and religion of the apparently autochthonous population, but seems not to be shared by a late-coming and technologically more advanced population, thus making its introduction from India or elsewhere seem very unlikely.

Andaya (1993: 29–30) notes that Antonio Galvão, in his sixteenth-century treatise on the Moluccas, retells a Moluccan tradition in which the hero, one of the principal men of the island of Bachan named *Biquociguara* (read *Bikusegara*), discovers four serpent eggs in an enchanted clump of rattan that gushed blood when cut. *Bikusegara* took the eggs home. From them hatched three boys and a girl who grew up to became kings of three countries and the wife of the king of a fourth in North Maluku (Jacobs 1971: 80–81). The tradition reveals several elements that have Hindu-Buddhist analogues, and the name of the hero appears to be of Malayo-Sanskrit origin (Malay *səgara* 'the sea' < Sanskrit *sāgara* 'id.'). However, Kim (1982: 182–187) has found that the oviparous myth in East and Southeast Asia and in the northeast of India generally coincides with megalithism. Its manifestation in the Moluccas could either result from Indian influence, or reflect an original feature of Austronesian megalithism. As the oviparous myth is also attested in Oceania (e.g. Fiji, see below), the latter seems more likely.

The blood-gushing rattan may be associated with the sacred red plant cult reported for Oceania by Riesenfeld (1950: 657). The cult of red-leaved plants appears to be connected with magical qualities ascribed to the colour red in general, the colour of blood symbolizing strength, vitality, fertility, life.

In the tradition of the Numfors of Cendrawasih Bay, Irian Jaya, the progenitor of the Burwos clan named *Mamori* (*-mori* presumably reflects *ma-quDip 'be alive') and his wife *Insrendi* had a child, *Abrakui*, who is not a human being but a snake. At the time of its initiation by the spirits of the dead, the latter bestowed rich gifts upon *Insrendi*. This made the other women so jealous that they tormented her till she ran away and disappeared with her snake-child. Since then, a snake or serpent dance is performed at every initiation ritual, commemorating the first novice, *Abrakui* (Held 1940: 140–143). The tradition does not appear to have come under Hindu–Buddhist influence.

Serpents in Oceanic folklore and mythology

In Oceania, Indian and Chinese components in manifestations of the cult of the sacred serpent are negligible. Heine-Geldern (1952) called attention to shared features in art style between China and Oceania, ascribing these to a movement from China over insular Southeast Asia to Oceania. These views have meanwhile been further substantiated (see Solheim 1980). However, I shall not regard such common features as intrusive Chinese elements in Oceania, because they were evidently introduced by the very Austronesian peoples who exhibit them. In other words, they are only 'intrusive' insofar as the corresponding ethnic groups themselves are intrusive. For example, the lizard and crocodile cult attested in the archaeology of Indonesia and Oceania, and reflected in ornamental style and mythology of Oceania (e.g. the Polynesian legendary *Moko*, usually depicted as a monster lizard), may ultimately originate from a lizard cult in the Proto-Austronesian homeland in southeast China, and may thus have a common source with the Chinese *lóng* which, unlike the Hindu-Buddhist *nāga*, has clawed feet, the most conspicuous feature distinguishing a lizard from a snake.

In Oceania, the cult of the sacred serpent is not as ubiquitous as in Malayo-Indonesia. Occupying a central position in the indigenous religion in some places, it plays a subordinated or peripheral role in others, being absent altogether in others still. Fiji tradition provides an example of the former. Here the great serpent *Ndeŋei* is involved in the emergence of the first man and woman, and provides these latter with the basic crafts and knowledge. Another tradition holds that *Ndeŋei* had created Viti Levu, Vanua Levu, and other Fijian islands, that he lived in a cave guarded at its mouth by a curtain of snakes hanging head down, and that he had three sons. In another tradition, the serpent god is called *Ratu Mai-mbulu*. The source collection of myths and legends includes not only the numerous communities of the various (relatively strongly divergent) Fiji dialects, but also those of neighbouring Rotuma (Reed and Hames 1967). *Ndeŋei* is the god of growth and fertility, and is responsible for the changing of seasons. There are also various legends involving a 'Snake Chief', and a speaking, children-devouring giant snake (Reed and Hames 1967: 13–15, 17, 69–79, 155–160).

In the mythology of the Papitalai (northeast coast of Manus, main island of the Eastern Admiralties, north Papua New Guinea), snakes occupy an important, though not central, position. In one legend, a snake called *Malai* caused the reefs to rise out of the sea and become the land, created two children who grew up to become the foreparents of all the people, and provided them with foodstuffs. In another legend, a snake called *Moat* approached a maiden who bore him two children, a son and a daughter. *Moat* fed them, and when they had grown up, it told the son to creep into its belly. From there he brought out fire, the coconut, taro, sugarcane and the banana. Keeping the fire and planting the foodcrops, the two made all of it available for all the land (Meier 1907: 650, 655–656, see also Nevermann 1934: 366, 369). A variant of this legend will be discussed in the section on sacred trees.

The mythology of the Tolai living around Rabaul in north New Britain is an example of eroded remnants of the sacred serpent cult. Two mythological brothers are the chief heroes of Tolai tradition, about whose origin divergent versions exist, including that they had as parents *To Lagulagu*, a male spirit residing in a volcano, and *Ia Kupia*, a half-snake half-woman. It is told that one of the brothers once met his mother, and failed to recognize her, because she had just changed her skin and looked like a young girl. To return her to the form he was familiar with, he forced her back into her old skin. Consequently, all humans became mortal, whereas snakes, which continue to cast their skin, never die (Janssen *et al.* 1973: xiv-xv, 90–91).

In Polynesian mythology, there is hardly any mention of prominent snakes; and indeed snakes are absent in the region, except in Futuna and, perhaps, Samoa (Matthew Spriggs p.c.). I am aware of only one example from Samoa, included in the collection of Hambruch (1979: 160–164): a woman gave birth to a male snake. When it grew up, a maiden came wanting to become its wife. Its wish to turn into a human in order to return the maiden's love was granted, and it became the handsomest youth in the land. The rest of the tale deals with the further fate of the couple. There are more stories involving a great sea-eel *Tuna* (< *tu[ñ]a 'eel') which, for example, bites *Tiki* the first man, e.g. in a tale from Tuamotu, or which turns into a handsome man and makes love with a fairy maid, e.g. in a legend from Mangaia (Alpers 1970: 71–75). These are perhaps remote echoes of earlier legends of serpents and the sea, as a comparison of the Mangaia and Samoa stories suggests. I shall come back to the fundamental role of the divine serpent in Austronesian cosmogony in the treatment of the sacred tree cult below.

Waterside people and urbanized Nāgas in Sanskrit and Tamil tradition

We return now to the *Nāgas* of Indian literary tradition. According to early Tamil tradition, Tamilakam is considered to have been initially inhabited by two primitive tribes, the *Villavar* (literally 'bowmen') who dwelt in the hills and jungles, and the *Minavar* (literally 'fishermen') who lived in the valleys and plains or on the sea coast. These two tribes were then conquered by the *Nāgas*, who were a very numerous and civilized race (Kanakasabhai 1904: 39). Thus we find a distinction between littoral fishermen and *Nāgas*. The latter brought the entire region, later to be inhabited by Tamils, under their rule, and even founded the later capital of the Cholas. It is from the *Nāgas*, depicted as ferocious warriors, that the Tamils reportedly wrested their presently inhabited country as a result of prolonged military strife (Kanakasabhai 1904: 40–43, 48).

The coastal fishing folk are obviously comparable to the coconut-eaters of the *Rāmāyaṇa*, presumably Austronesians. Their mention, side by side with the *Nāgas*, might preclude the latter term also referring to Austronesians. However, the westward movement of Austronesians along the northern perimeter of the Indian Ocean apparently proceeded in successive waves. It is possible that earlier contingents of Austronesian migrants were not yet referred to as *Nāgas*. It seems doubtful that Austronesians alone could have accounted for all the activity and influence ascribed to *Nāga* peoples by Sanskrit and Tamil tradition and these *Nāgas* probably encompassed more peoples than just Austronesians.

Tamil evidence for buffalo sacrifice, head hunting and megalithism among Nāgas

Early Tamil literature also provides some insight into certain features of the early religion of the *Nāgas*. The *Eyinar* are described in the Tamil *Chilappathikaram* as the most lawless of *Nāga* tribes. Their chief occupations are said to be cattle rustling, pillage and murder. They reportedly worshipped the dread goddess Kāli, and slaughtered buffaloes at her shrine (Kanakasabhai 1904: 43, 227–228). The buffalo sacrifice is one of the most widespread and persistent rituals in mainland and insular Southeast Asia, and has survived religious conversion to Hinduism, Buddhism and Islam. In places where it has retained much of its original pagan character, e.g. among the Ifugao of Luzon or the Toraja of Sulawesi (Celebes), it is a relatively gory performance. Although the name of the goddess Kāli is Sankrit, her cult apparently developed in post-Vedic times only as a result of male–female dichotomization of the cult of the god Rudra-Śiva. Noteworthy is the fact that Kāli is a female deity, a circumstance that may reflect matrilineal traditions widespread among Austronesians. In some areas of south India, stone monuments depict the serpent deity as half-snake, half-woman (Vogel 1926: 272). The female gender of the divine serpent, atypical in Vedic and Buddhist tradition, is a fundamental feature of early Austronesian cosmology.

Austronesian *Nāgas* perhaps contributed substantially to the bloodthirsty character of the cult of Kāli. It is tempting to connect the ritual murders associated with the cult – apparently originating, judging from the description of the murderous inclinations of the *Eyinar* in the Tamil source, in early *Nāga* custom – with well-known headhunting traditions of early Austronesians, expressed in the protoform *kaiau 'headhunting' (Dempwolff 1938: 72 under *kajav). Human sacrifice by decapitation may have been part of mainly non-Vedic Aryan worship of a goddess, but it was apparently part of a ritual in which horses were sacrificed, likewise by decapitation, and more often than humans (see Parpola 1988: 250, but see also ibid. pp. 258–259). The *Eyinar Nāgas* apparently sacrificed buffalo, and it is not said that the 'murdered' humans were offered in sacrifice as well.

The distribution area of the *Nāgas*, as far as one may judge from early Sanskrit and Tamil literature, tends to coincide with megalithism. The connection between *Nāgas* and megaliths is common in Tamil literature. Based on information from the Tamil *Chilapp-athikaram*, Kanakasabhai (1904: 228–229) suggested that *Nāgas* and the 'lower classes' (in the then Tamil-ruled country, perhaps descendants of autochthonous peoples with which the *Nāgas* had been in contact before the Tamil conquest) worshipped stones and springs that were believed to possess supernatural powers. One such object of worship was reportedly a large stone in the middle of an open square in a city named *Kavirip-paddinam*.

The peoples presently known as *Nāgas* in the northeast of India, though speaking Tibeto-Burmic languages, are the only ones presently still continuing a megalithic tradition on the subcontinent. In short, with reference to India of the present, 'Nagas' and peoples practising megalithism are in effect synonymous. The coincidence of buffalo sacrifice and megalith worship in the *Nāgas* corresponds to an observation by Münsterberger (1940: 737) that the buffalo sacrifice ritual is a feature of megalithism in Southeast Asia.

Nāga status transition from alien to associate ethnicity

The *Rāmāyaṇa* also mentions *Nāgas*. These are depicted as already urbanized peoples, which is in full agreement with the treatment in Tamil tradition. In the *Mahābhārata* we also find reference to *Nāga* kingdoms, with which the Vedic Aryans had to wage war to secure the land they wished to settle.

The *Nāgas* in the *Mahābhārata* probably also encompass a much larger group of ethnicities than only Austronesians. Kosambi (1964: 31, see also Vogel 1926: 66–71) called attention to the fact that the epic in its present state actually takes the form of a frame-story of a great *yajña* fire-sacrifice to encompass the destruction of the *Nāgas*. In this, the epic was anomalous in that the purpose of the sacrifice is not fulfilled. The *yajña* is invoked by Janamejaya to exterminate all *Nāgas* in vengeance for the death of his father Parikṣit at the hands of the *Nāga* Takṣaka. But the narrative lets the latter (and thus the main 'culprit') escape through being saved by the priest Āstika, himself son of the Brahman ascetic Jaratkāru and his *Nāgī* wife. This is perhaps an allusion to Hinduization of the *Nāgas*, which brought them into the category of non-barbarians (non-*mleccha*), not to be exterminated with impunity. Meanwhile, Jaratkāru's marriage with the *Nāgī* was matrilocal, and their son Āstika was brought up by his (*Nāga*) maternal uncle Vāsuki. This is another instance associating *Nāgas* with matrilineal customs. Among the Minangkabau of west Sumatra, for example, such a marriage and upbringing of the son would have been quite normal.

Marriage to a *Nāgī* is no rarity in the *Mahābhārata*. Arjuna, one of the Pāṇḍava brothers, is married twice to *Nāgī* princesses (Vogel 1926: 74–77). A similar liaison lies at the base of some dynastic lines, and, in a copper-plate charter, the Pallava king Skandhaśiṣya claims descent of the founder of his line from *Aśvatthāman* and a 'snake woman' (Kosambi 1964: 32). There are other examples of this kind in Tamil tradition, although here too, as in Hindu tradition in general, such a *mésalliance* would have originally been looked down upon. Thus, the Tamil *Mani-mekalai* describes the marriage of

Killi-Valavan, the first Chola king, with a beautiful *Nāgī* princess (Kanakasabhai 1904: 42). Such marriages are frequent in often legendary or mythological beginnings of dynastic pedigrees of royal houses in various places in South and Southeast Asia.

There is also a variant, in which the wife is not explicitly declared a *Nāgī*, but is associated with an aquatic habitat and means of transportation. In the first book of the *Mahābhārata*, the *Ādiparva*, the king of the Paravas, who lived on the banks of the Jumna, found an infant girl in the belly of a fish and adopted her as his own daughter. When she grew up, she ferried passengers over the river, as was the custom for Parava women. She meets the sage Parāsara, of whom she bears a son named Vyāsa, writer of the *Purāṇas*. She then shares the royal bed of King Santanu, bearing him a son, Vicitravīrya, the grandsire of the *Pāṇḍavas* and *Kauravas* (Thurston and Rangachari 1909: 142). As in the case of the distinction drawn between *Nāgas* and littoral coconut-eaters in the *Rāmāyaṇa*, or between the former and *Minavar* fishermen in Tamil tradition, this is another instance of a distinction between *Nāgas* and more ancient peoples associated with boats, fish, or the coast, the common denominator being contact with water.

Hornell (1920: 232, 235–237) made measurements of the cranial index of Paravas in the Tinnevely region, and found them to be distinctly brachycephalic, in contrast with the surrounding Tamil-speaking rural population. The author therefore grouped them with Indian peoples to which he ascribed 'Polynesian' racial affinities. Hornell believed that the Paravas represented a part of the *Nāgas*. Cranial measurements are treated with considerable scepticism today, but Hornell's data do indicate some underlying contrast in physical features, perhaps through Austronesian admixture.

Nāgas in Sri Lanka

The oldest Sri Lankan chronicles appear to be rather vague on the population of Sri Lanka before the first Indo-Aryans: the island is considered to have been inhabited by *Yakṣas* and *Nāgas* (de Silva 1981: 6). It is obviously this tradition that is retold in the travelogue of the Chinese pilgrim Faxian who visited Sri Lanka in 413 AD (see also Legge 1886: 101; and Giles 1923: 66):

> *Qí guó běn wú rénmín, zhèng yǒu guǐshén jí lóng jū zhī, zhū guó shāngrén gòng shìyì*
>
> (Legge 1886: p. 37 of Chin. suppl., cols. 2–3 right-to-left)
>
> 'This country originally was without people, there just were spirits and dragons living here, merchants of various countries carried on trade [with them]'

'People' (*rénmín*) here of course stands for Indo-Aryans (Sinhalese), whilst *lóng* 'dragon' is the word by which Sanskrit *nāga* is translated into Chinese throughout Faxian's report. The *Yakṣas*, at least insofar as reference is made to peoples in north India, were apparently Tibeto-Burmans, but probably not the ethnic group inhabiting legendary pre-Aryan Sri Lanka. As the Veddas of the island arrived here before the Indo-Aryans, they obviously were one of the pre-Aryan groups. As they could not have been the *Nāgas*, it follows that they must have been the 'spirits' (*Yakṣas*, rendered *guǐshén* in Faxian's Chinese text), provided of course that the tradition has indeed retained a relation to proto-historical reality. It is noteworthy that the tradition that Faxian retells in the quoted passage shows pre-Aryan Sri Lanka to have already been involved in maritime trade on the Indian Ocean, a circumstance that tends to lend it some credibility because merchants and trade would not normally be expected as a typical feature of fantastic tales about spirits and dragons. In another rendering of a Buddhist tradition, a *Nāgī* princess mediates in the recovery of a treasure from the *Nāga* king who lived at the bottom of the sea. The tale itself may be a distorted recount of a first acquaintance with Austronesian pearl-divers.

Fergusson (1873: 58–59) has called attention to the important role credited to local *Nāga* kingdoms in the conversion of Sri Lanka to Buddhism in the third century BC. The island itself, or part of it, is even called *Nāgadīpa* ('*Nāga* island') in corresponding passages of the *Mahāvaṁśa* (Vogel 1926: 118–120). It is interesting that one conflict, to which the introduction of Buddhism is said to have brought a peaceful solution, was between a *Nāga* king named Mahodara and his sister's son Chūlodara ruling over another *Nāga* kingdom. The maternal uncle of the former, King Maṇiakkhika, was also involved. It is perhaps significant that all three kings are related to each other by matrilineal line of descent.

Nāga pirates, Nāga cannibals and the Nāga homeland

The Buddhist treatise *Bodhisattvāvadāna Kalpalatā* of Kṣemendra, written in the tenth century AD, has perhaps preserved reports from the time of King Aśoka of serious depredations against maritime trade in the Gulf of Bengal by *Nāga* sea pirates. Seeing his income from trade revenues endangered, Aśoka issued an edict that was ignored by the *Nāga* pirates. Only when Aśoka became a devout Buddhist did the *Nāgas* respect his edict and turn in the captured booty (Mookerji 1912: 113–115). Freebooting Sea People (*Orang-lauts*) in the seas in and around West Malayo-Indonesia are very ancient, and are already mentioned in early Chinese reports on the region. It is through the command over their allegiance, based in part on the alleged magical power of the Malay ruler to bring great misfortune upon them in case of recalcitrance, that early Malay polities could apparently guarantee secure passage to merchant ships when these stopped to pay port dues, or threaten sure destruction to those ships that failed to pay. What Kṣemendra described appears to be a similar submisson to Aśoka, the latter's conversion to Buddhism supposedly bestowing additional magic upon him, and suggests that the *Nāga* pirates were such Sea People. Despite the transparent motive behind the tale, to demonstrate the beneficence of true religious devotion to the reader, there is no reason to disregard its relevance. King Aśoka did indeed convert to Buddhism, and such a political act would surely not have been contemplated if it had not promised to lead to a consolidation of power. It is in any case difficult to imagine any other freebooters in these waters, and at that time, than Austronesians or Austronesianized Negritos.

The *Mani-mekalai* mentions some islands lying to the east of Sri Lanka that were inhabited by a *Nāga* group called *Nakkasāranar* ('naked nomads') who were cannibals, perhaps referring to the Nicobars (Kanakasabhai 1904: 11). If the location is correct, this would serve as a further argument against placing the coconut-eaters of the *Rāmāyaṇa* in the Nicobars. After all, why would the epic highlight the coconut as an exotic item of their diet, but remain silent about so scandalous a one as human flesh? Cannibals in this region are also reported in the *Geōgraphikḗ Hyphḗgēsis* of Ptolemy, on ten islands named Maniola (see Figure 8.2). As the geographer of antiquity placed these immediately northwest of his Barusa Islands (the emporium of Barus on Sumatra's west coast, see Figure 8.1), the Maniola Islands should be the Nicobars. To the north of these are placed the islands *Chalínē* and *Bazakáta*, of which the inhabitants always go naked, perhaps the Andaman Islands. The *Nakkasāranar* of the *Maṇi-mekalai* could correspond to the inhabitants of either the ten islands of cannibals or the twin islands of the naked, or both at the same time. This would not be the only instance in which a geographical feature was reflected in Ptolemy's Geography, being compiled from a number of sources in which it might have figured under different names respectively, as two or more distinct objects (see below).

Ptolemy also located Bēsyngite cannibals (*bēsyngeití anthrōpophági*) on the Martaban coast between the mouths of the Irrawaddy (*Témala*) and Salween (*Bēsýnga*) (Figure 8.2). Luce (1965: 145–146, see also Forbes 1878: 234) has called attention to an ancient Mon tradition, according to which the Martaban coast was under constant threat of raids from the sea by what the author termed 'Malayan Vikings', referred to by the Mons as *Rakṣasas*, depicted as 'cannibal demons'. Luce regarded them as the ancestors of the Moken (so-called 'Sea gipsies') of the Mergui Islands. In Myanmar (Burmese) tradition, the *Rakṣasas* of the Mons were called *Bilù*, glossed in the dictionary of Judson (Stevenson and Eveleth 1921: 727) as: 'a kind of monster which eats human flesh and possesses super-human eyes'.

There still is an island in the Gulf of Martaban immediately before Moulmein in the mouth of the Salween, bearing the name *Bilù-gyùn* (*gyùn* 'island') (Figure 8.1). Grierson (1906: 14) proposed that the *Rakṣasas* or *Bilù* of the Martaban coast may have been Negritos. The Sea People, many of whom speak dialects of Malay (see Kähler 1960), and the Moken are indeed in the main either Negrito or exhibit a substantial Negrito admixture. Apparently, to Kṣemendra writing in Kashmir, as also to the writer of the *Mani-mekalai* in the south, *Nāga* was a cover term also applicable to *Oranglaut* freebooters and cannibalistic sea-nomads, both operating on the distant and close maritime approaches to the Strait of Malacca as well as within it. The *Mani-mekalai* even specifies quite explicitly that beyond the island with the cannibal *Nāgas* lies *Cāvakam* (Kanakasabhai 1904: 11). The latter term is

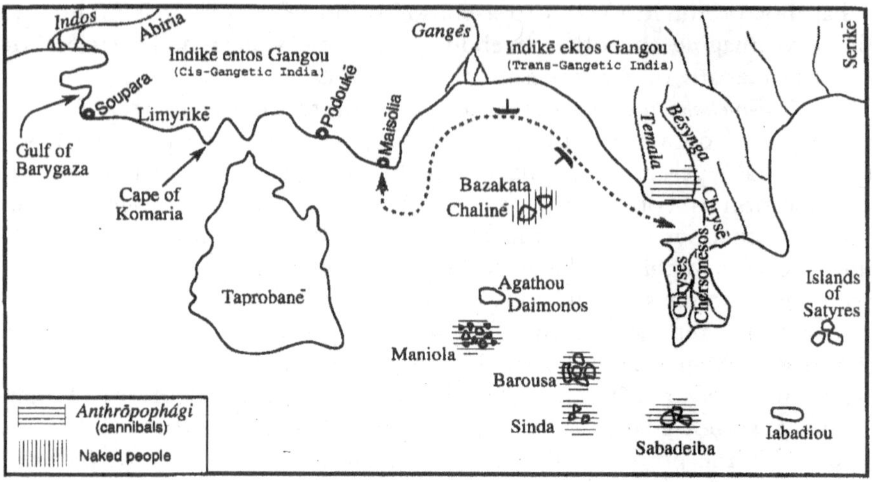

Figure 8.2 South and Southeast Asia as described by Ptolemy

Source: Mahdi

Note: Sea communication routes and – in Figure 8.3 – the name of the ships involved are given as described in the source. Only geographical items relevant to this discussion or for general orientation have been included.

commonly regarded as referring to Sumatra, Java, or even West Malayo-Indonesia in general (cf. Pillai *et al.* 1925–36: 1392). But, like Arabic *Zābağ*, it reflects Pali *Jāvaka*, which must have referred to Malays or their country (Mahdi 1995). The approaches to the Strait of Malacca would, therefore, be a likely place to locate the islands of the cannibal *Nāgas*.

Ptolemy's Barusa, Sinda and Sabadība – the latter probably an *alter ego* of Iabadiu, split as a result of failure to recognize that independent reports with divergent renderings of the original name referred to the same country, Sanskrit Yavadvīpa, that is *Yava* Island (*Yava* in Figure 8.1, see Mahdi 1994a: 215 n. 93, 1994b: 469–470 n. 111, 1995 and n.d.) – are apparently parts of the island of Sumatra, the interior of which was inhabitated by cannibals until around two centuries ago. It is less likely that these could have been the ones meant by naked cannibal *Nāgas* in the *Mani-mekalai*.

Ramaṇīyaka, appearing in the *Mahābhārata* excerpt cited above as the name of the ocean bay homeland of the *Nāgas*, can perhaps be equated with *Ramañña-deśa*, Rmañ-land, Land of the Mons (Figure 8.1). This would practically mean that the homeland (more correctly perhaps 'home-sea') of the *Nāgas* was located by the epic before the Martaban coast or, if one takes the expression ocean bay literally, in the approaches to the Strait of Malacca, i.e. at the same place as the *Nāga* naked cannibals and pirates of Tamil and Buddhist tradition, and the *anthrōpophági* and naked islanders of Ptolemy.

Foreign people or countries in ancient geographies are often associated with that part of their territory that lies closest to the observer. Thus Chins and

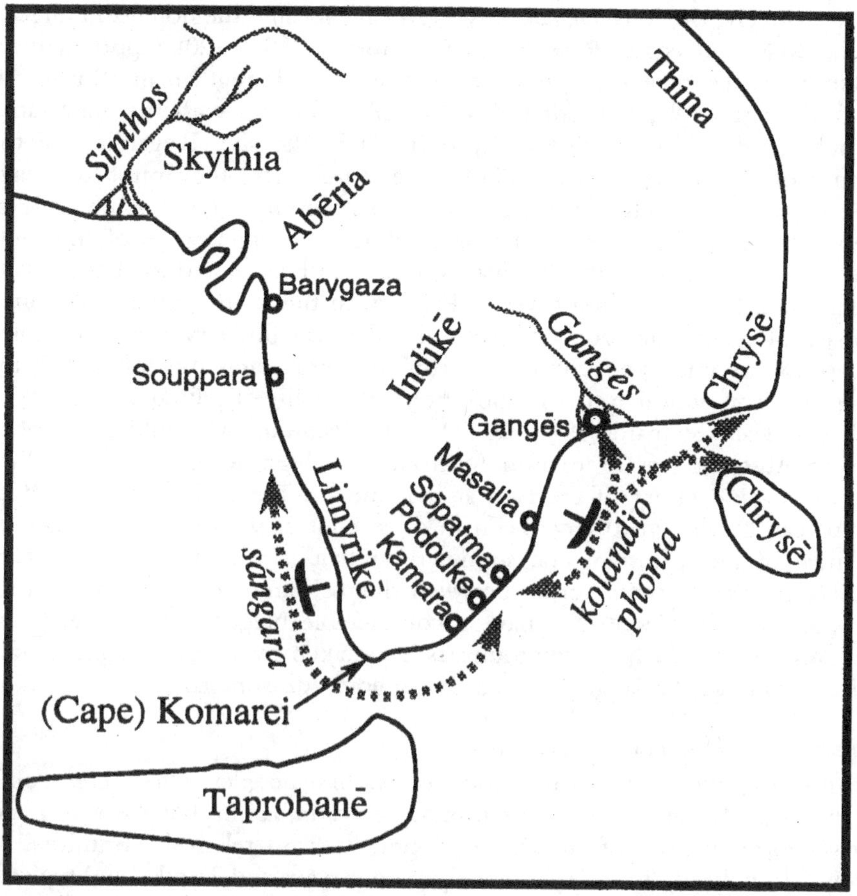

Figure 8.3 South and Southeast Asia as described by Periplus
Source: Mahdi
Note: See Figure 8.2

Kachins perhaps provided the name by which *Zhōngguó* (the 'Middle Kingdom') came to be known in India and the West, whereas *Bhārata* is known in the West by a name derived from that of the Indus river. Greece and the Graeco-Roman world were known to the orient by a name originating from the Old Ionian-Greek word for Ionian, whereas Asia has become known to the occident by the name of a Near-Eastern province of the Roman Empire, deriving from a Greek designation of the region encompassing Anatolia and the Levant already used by Herodotus. The *Nāga* infested waters, bound on three sides by Burma, the Malayan Peninsula and Sumatra, would in this context serve as the natural place for the writer of the *Adiparva* to locate the homeland of the Austronesians.

Alternatively, the *Mahābhārata* may even have located the homeland directly on the Martaban coast (*Ramañña-deśa*). Stargardt (1979: 20) reports a thermoluminescence date placing the earliest iron at Taungthaman, Burma, in the fifth century BC, indicating also that this coincides with the appearance of onyx beads (which originate from India) at the site. Regardless of the identity of the traders who brought these wares here, they must also have intermediated between this region and India, serving at the same time as transmitters of information about the region. The appearance of the same type of beads together with the first iron has also been noted for Ban Chiang in Thailand, and the Tabon caves of Palawan in the south of the Philippines. Remarkably, not only does the basically identical funerary rite at all three sites show an unbroken continuity from the preceding Neolithic, but the beads and iron appear only as innovations in the list of gifts buried together with the dead (Stargardt 1979: 20–22). The Palawan site could be ascribed only to Austronesians or to non-Austronesian Negritos, and it is not difficult to imagine either of the two at that time at the two other sites. Both groups can be shown to have been in close contact with each other around the turn of the Christian era, so much so that it would probably not be possible to draw a strict border between them. Many Austronesian communities are racially Negrito, or have a considerable Negrito admixture. Many non-Austronesian Negrito communities have taken over various culture traits and elements of the language from mongoloid Austronesians.

Some notes on identifications and chronology

In summary, the *Nāgas* were ferocious, warlike peoples who were mobile on the sea, were given to pillaging, piracy or cannibalism, but were also very non-uniform in level of culture development, from relatively 'primitive' to relatively 'civilized' (in the eyes of early Indo-Aryans and Tamils), and typically inhabited islands, the sea coast or the banks of rivers. Some of them worshipped megaliths and practised buffalo sacrifice and headhunting ('murder').

Of all the known peoples in the wider region, only Austronesians fit this description adequately. At the same time, *Nāga* apparently denoted a wider circle of ethnicities than only Austronesians. The possible identity of the other ethnicities will be explored in the section on megalithism below. On the other hand, Sanskrit and Tamil traditions seem to agree with each other in that they mention some presumably Austronesian peoples without explicitly classifying them as *Nāgas*. They are usually described as being culturally less advanced than those explicitly called *Nāgas*. Passages on the earlier littoral peoples possibly reflect traditions dating back to times when the term *Nāga* was not yet used for Austronesians.

The *Nāgas* are described as being already urbanized. One can only speculate upon what this meant in reality, and, considering that apparently not only Austronesians were included among the *Nāgas*, it might have been the non-Austronesian *Nāgas* who were urbanized. However, culture affinity to megalithist Micronesians and Polynesians lets the existence of ditched stone fortifications, mighty war fleets and a relatively sophisticated social stratification and administrative hierarchy appear plausible. On the other hand, borrowed Indic words originally meaning 'city', 'fortress' and 'citadel' are used in some languages of Sumatra and Kalimantan for 'village community' or 'village surrounded by wooden palisade'. For languages of West Malayo-Indonesia, one can reconstruct the protoform *bi(n)tiŋ 'fort' (Dempwolff 1938: 31), but in some places in Kalimantan the corresponding reflex denotes a stockaded settlement of about village size. Thus Schwaner (1854: engraving across p. 22) shows a stockaded hamlet in the interior of Kalimantan, providing it with the caption: *A Dayak village (Benting)*. Two further engravings (Schwaner 1854: across pp. 18 and 31) depict similar settlements, named *Kotta Baroe* (read: Kota Baru) and *Kotta Karingan* respectively. The first component in the names reflects Malay kota 'city' (Classical Malay 'fortification, citadel'), from Dravidian *koṭ(ṭ)a(y) 'fort, wall', perhaps via Sanskrit *koṭ(ṭ)a*.

A chronological evaluation of the data on Austronesians provided by early Sanskrit and Tamil literature and tradition is problematic for two reasons. Firstly there is the question of dating the sources. As the Tamil material is from a later date than the Sanskrit epics, I shall concentrate on the latter. Both the *Mahābhārata* and *Rāmāyaṇa* contain substantial later additions. The main parts, however, to which the citations made above belong, were probably completed in the first century BC. The earliest components of the epics may, however, have been inspired by happenings as early as the fifth or fourth century BC. Insofar as references to *Nāgas* reflect happenings contemporaneous to the writing down of the two epics, they must therefore date from the last half of the first millennium BC. The second problem is ascertaining the contemporaneity of the happenings and their notation in the sources, allowing for the possibility that parts of the narration reflect traditions already of undefined antiquity at time of notation. It seems safer tentatively to assume contemporaneity or at best relatively recent age of reports on peoples explicitly denoted as *Nāgas*. Older traditions, in which the term *Nāga* was not yet used, that were co-opted into the epics could have origins several centuries earlier. Speculations on the original homeland of the *Nāgas* were possibly inspired by the traditions of the Austronesians themselves, being again of an antiquity that is difficult to estimate. The earliest Tamil and Sinhalese sources are apparently from the first half of the first millennium AD, but legends about original populations perhaps also reflect traditions that were already of unknown antiquity at the time of their notation in the chronicles.

MEGALITHISM

Some common features in the megalithism of India and Indonesia

The nature of the affinities between megalithism in India and in Austronesia has been the subject of study and speculation since the first serious attempts at systematization of the knowledge about these regions. In India, one can discern three geographically separate areas of megaliths: the northwestern exhibiting westward affiliations, the southern, and the northeastern, the two latter having numerous elements in common with megalithic horizons in Malayo-Indonesia and Oceania (Sarkar 1982).

In an early work by Logan (1848), culture parallels were already drawn between the megalithist *Nāgas* of Assam and some peoples of Kalimantan. Important contributions in pre-war research on the subject were made by Heine-Geldern (1928, 1934). With regard to the megaliths of insular Southeast Asia, the author distinguished between an 'early' and a late 'megalithic' horizon. The former was characterized by menhirs, dolmens, stone seats, stone compositions (stone rows, circles, megalithical meeting places, etc.), platforms and terraces, pyramids, stone fortifications, stone-lined or paved wells, baths, stone stairways, etc.; these were assigned to what the author termed 'Proto-Austronesians', referring to neolithic peoples moving into the Southern Islands between 2000 and 1500 BC and not in general to speakers of Proto-Austronesian in the modern linguistic sense. The late megalithic horizon, characterized by stone cisterns or sarcophagi and presence of metal, is assigned by the author to a bronze age movement of the last two or three centuries BC. This corresponds reasonably well with modern estimates of the beginning of the metal age in the Philippines at around 500 BC (Bellwood 1980: 152). In conformance with terminology used in this chapter, I shall speak of neolithic and metal age 'megalithism' in place of the early and late megalithic periods respectively.

The relationship of northeast Indian megalithism to Malayo-Indonesia

An interesting common feature connecting megalithism in northeast India with that in Malayo-Indonesia needs to be touched upon here, because one relevant site has been misinterpreted. The stone monoliths at Nartiang in northeast India, shown on plate I*c* of Rao (1977: across p.194), are probably related to the so-called *pənji rəti* (literally 'flag of tomb') of east Sumba (see e.g. Hoskins 1988: 121 fig. 127, 124 fig. 130, 284 ill., 285 pl. 46). They are upright-standing, tall, flat stone slabs with rather thin, angular-edged, rectangular cross-section, and asymmetric contour of the widest surface profile, which are each placed close to a stone mound to which they are oriented like the rudder of an overturned boat towards the hull.

Similar flat and upright-standing monoliths with angular-edged asymmetrical contour and relief carvings, but not immediately adjacent to a mound or tomb, are known from the Sungai Udang site at Pangkalan Kempas in Negeri Sembilan, West Malaysia. Miller (1969: 264–266) regarded these as remains of visiting Phoenicians, but failed to provide convincing evidence for similar objects in the Phoenician homeland. The Phoenician stone anchor, proposed by Miller as a precursor, hardly resembles them, other than being carefully hewn and made of stone. It has a symmetrical (rectangular or trapezoidal) rather than an asymmetric main profile, which is generously rounded rather than angular-edged, has a round hole in it, and shows a considerably higher thickness-to-breadth ratio. Remains of stone anchors that at least resemble

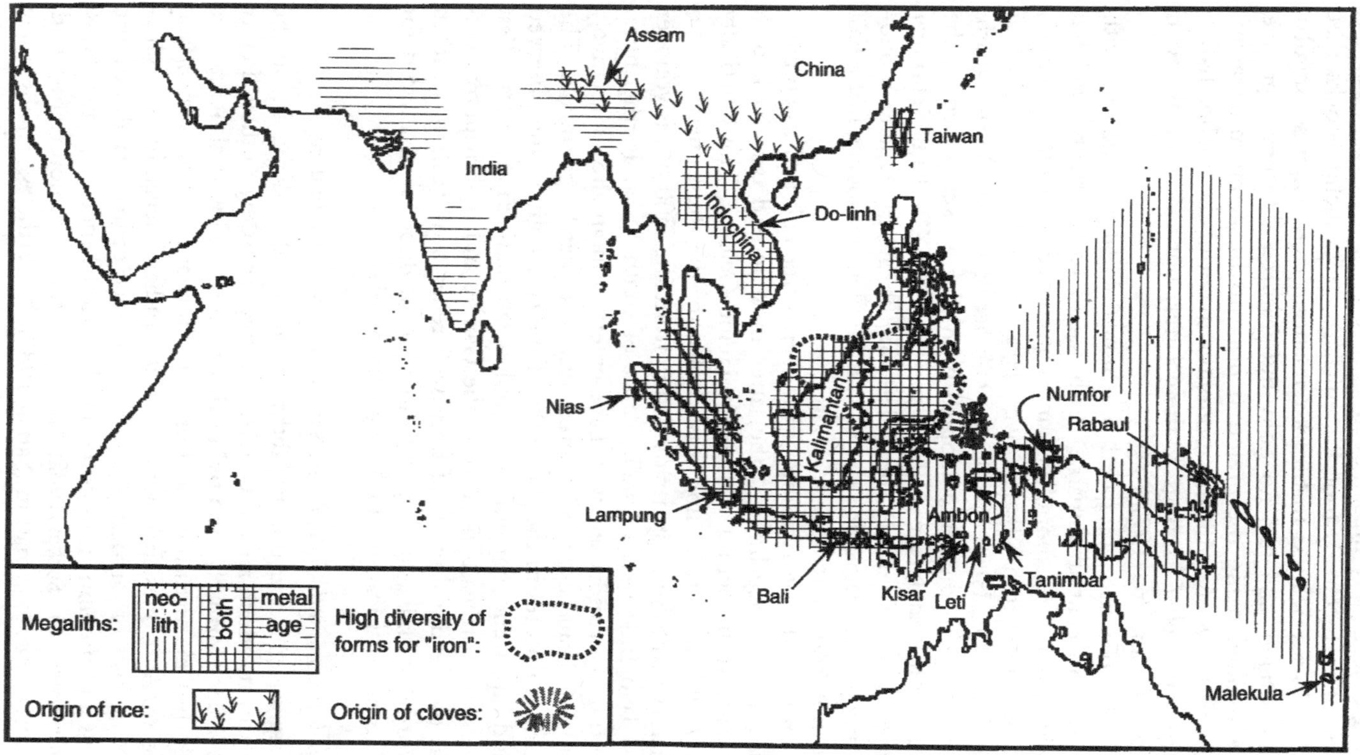

Figure 8.4 Neolithic and metal age megalithism in South and Southeast Asia

Source: (for high diversity of forms of 'iron') Loofs (1969: x); Sarkar (1982: 134); Kim (1982: 170, 186); Riesenfeld (1950: across pp. 118, 236, 538); Morgan (1988); (for the origin of rice) Nakao (1958: 399 fig.1); Chang (1976: 144 fig. 2); Watson (1983: 16 map 2); (for the origin of cloves) Burkill (1935: 961)

the Sungai Udang objects have indeed been uncovered in the Mediterranean. These are halves of early Greek (not Phoenician) stone anchor stocks that had a comparable low thickness-to-breadth relation, were likewise angular rather than round-edged, and had no hole, but featured a fastening groove or notch in the middle, being the preferred place to break up into halves. Though resembling the Sungai Udang monoliths in form, the anchor halves were not incised with relief carvings as these latter, but, when at all, with Greek inscriptions (Gianfrotta 1977).

Schnitger (1939: 167) had compared the Sungai Udang monoliths with those at Puar Datar in central Sumatra. With regard to the figure depicted on one of the Sungai Udang monoliths, and considered to depict a Semite, compare the stone statue from Sukaraja, west Java, shown by van Kinsbergen (reproduced in plate 13 of Heine-Geldern 1934: across p. 24). A particularly 'Semitic' looking figure in the sense adopted by Miller is exemplified by the right-most of the Nias sculptures shown in plate XXXIV-1 in Schnitger (1939). In artistic style, of course, the reliefs on the Sungai Udang monoliths resemble much more the *pənji rəti* of east Sumba. Analogues from East Austronesia are not known to me.

In conjunction with the distribution of the sacred tree cult and its connections with Austronesian megalithism, the latter must indeed date back to a time at least as early as the period assumed by Heine-Geldern. This practically excludes an introduction from northeast India to the Austronesian area. Northeast Indian megalithism, apparently dating from after 500 BC, is younger than that of Malayo-Indonesia as well as that of south India. Furthermore, megalithism in northeast India seems to have emerged in an already relatively developed state, without signs of a preceding pre-metal age phase. The numerous common features in the megaliths of the two regions, the observed distribution and apparent chronology, implies an extraction from West Malayo-Indonesia to northeast India along the Ganges and Brahmaputra. This is corroborated by the distribution of borrowed reflexes of *qaBaŋ 'boat' in Burma and Northeast India, and by the direction of dispersal of a word for 'ditch around stone fortification' in Malayo-Indonesia.

The relationship between south Indian and Southeast Asian megalithism

Of particular significance are the megaliths of south India, as this is, on the one hand, the area in which ancestors of the Tamils, according to their literary tradition, met *Nāga*s formerly occupying this country. This is also the area of early south Indian maritime tradition, the introduction of which was ascribed above to Austronesian colonists. South Indian megalithism is characterized by stone cists or sarcophagi and the presence of iron, as also by urn burials, dolmens and menhirs, thus corresponding to the metal age megalithist horizon of Southeast Asia (Sarkar 1982: 128). A further common feature with the latter is reburial after excarnation (Sarkar 1982). Nevertheless, an alignment with megalithism of Southeast Asia cannot be established as simply as in the case of northeast Indian megalithism.

In a 1973 publication I have not seen, Asko Parpola shows that megalithism in south India must have been introduced by Indo-Aryans (in *Arguments for an Aryan Origin of the South Indian Megaliths*, Madras. Lars Martin Fosse, p.c.). This proposition seems convincing, in view of the dispersal of Black and Red Ware pottery from north India around 2000 BC, proceeding down the Ganges valley, and then spreading throughout the south around 1000 BC (Parpola 1988: 248), where it is represented by the greater part of earthware vessels in megalithist sites (Wessels-Mevissen 1991: 15). The pottery was apparently in use both by autochthonous Dravidians as well as by non-Vedic Aryans, who were the first to move south through previously solely Dravidian-inhabited territory (Parpola 1988: 249). Another important indication is horse sacrifices, included among grave goods of the early phase of south Indian megalithism (see below), indicating that the initiators of south Indian megalithism were neither Dravidians nor Mundas, or Austronesians, but cattle-tending semi-nomadic horsemen like the Indo-Aryans. The unfavourable climate for this occupation in south India explains the gradual disappearance of the sacrificial custom (McIntosh 1985: 473). A particularity that likewise distinguishes south Indian megalithism from metal age megalithism of Southeast Asia are the portholes in the (often eastern) sidewall of many stone cists.

Against all this stands the evidence that links *Nāgas* with, on the one hand, worship of megaliths and, on the other, with the sea, cannibalism, piracy, buffalo sacrifice, matrilineal organization, which militate against an Aryan identity but fit well with an Austronesian one. However, the distribution area of the *Nāgas* considerably exceeds that which could comfortably be assigned to Austronesians. The term *Nāga* may have referred to non-Vedic Aryans practising megalithism as well as to Austronesians, since the former apparently descended the Ganges between 2000 and 1000 BC. Some time during the first half of the first millennium BC, Austronesians must have ascended the Salween, Irrawady, Brahmaputra and Ganges rivers, bringing the double canoe and a word for boat/coffin. By the time the earliest version of the *Mahābhārata* was being written in the fifth to fourth centuries BC, in which the conquest of *Nāga* territory along the Ganges by Vedic Aryans is extolled, Austronesian riverine fishermen and non-Vedic Aryan cattle-holding agriculturists must have formed a certain symbiosis, perhaps involving intermarriage and cultural exchange.[9]

It was probably this ethnically mixed population, which the Vedic Aryans encountered sometimes on horseback, at other times with a matrilocal family organization or having a homeland on a distant ocean bay, that formed what the *Mahābhārata* called *Nāgas*. This ethnic concept must then have been carried southwards by Vedic Aryans, to be adopted also by the Tamils in their epics.

Sarkar (1982: 153) dates the period of megalithism in south India with a relatively high level of confidence from around 1000 BC to 300 AD. McIntosh (1985: 469, 471, 473, 475–477, 482–489) suggested the four periods and

Table 8.1 Four periods and respective characterizations in the development of south India

No.	*Years* BC	*Features*
I:	1100–800	Earliest iron; simple oval pit grave with one or more capstones, sometimes surrounded by stone circle; vessels with perforated ringfoot.
II:	800–550	First truly megalithic graves, burials with grave goods, including horses, within a stone circle; also pits lined on some sides with slabs, and simple cists often with portholes.
III:	550–300	A. Continuation of the latter period, innovation in grave goods, fewer horses, abundant bronze and gold; first appearance of earthenware 'bathtub' sarcophagi in pits or simple cists (with portholes); first three-legged jars; end of early phase (lack of substantial stone architecture). B-C. Gradual disappearance in the north as a result of north Indian subjugation; massive settlement expansion assigned to introduction of irrigated rice agriculture; increased complexity of cist grave, subdivision of cists, often through increased use of containers (urns, sarcophagi); cist in cairn with leading slabs; rock-cut chamber tomb; urn burials, originally from the extreme south in IIIB, apparently introduced from Sri Lanka, became common all over the area in IIIC.
IV:	300–100	Fully fledged towns; urn burial prevails over megalithic forms.

respective characterizations in the development in South India, based on available radiocarbon dates, shown in Table 8.1. Limited radiocarbon data does not permit individual dating of subdivisions A, B, and C of period III.

Period I can still be aligned quite simply with iron age cultures featuring cairn circles and horse domestication in Baluchistan, and stone circles in the Ganges basin, whereas related sites in Kashmir have even produced good examples of menhirs (Sarkar 1982: 130). The subsequent development of a truly megalithist culture complex appears to be a local process in the peninsular part of the subcontinent during periods II and IIIA, to be subsequently pressed southwards and finally confined to the south through expansion of northern Vedic Indo-Aryan states into Maharashtra and Andhra Pradesh, reaching as far as Brahmagiri at its maximum extension in the third century BC under Aśoka (McIntosh 1985: 482).

With period IIIA, large three-legged jars appear that show remarkable common features with the same type of jars of the Shang period (between 2000 and 1000 BC) in China, where they are much more variegated in form (see e.g. Chang 1978: 231, 240; see Figure 8.5. The large jars in south India tend to belong to the red ware, rather than to the black and red (Wessels-Mevissen 1991: 15). In Southeast Asia, however, three-legged jars are relatively rare and are not in the same style. The jar from Leang Buidane on Talaud Island (Figure 8.5*C*) is a burial urn, that from West Malaysia (*D*) is analogous

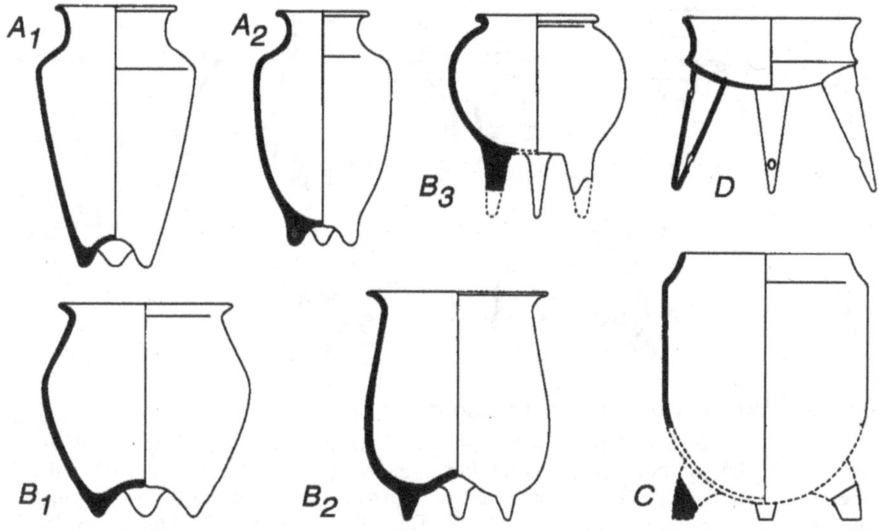

Figure 8.5 Earthware three-legged jars

Source: A1 Sanur based on McIntosh (1985: 476) and A2 Salem based on Wessels-Mevissen (1991: 56), both south India; B1 Western Zhou period I, B2 and B3 from Shangjie site in Zhengzhou, Henan, China, based on Chang (1978: 307, 231); C from Leang Buidane, Talaud, north of Sulawesi, and D Gua Berhala, Kedah, West Malaysia, based on Bellwood (1985: 386, 261)

Note: All drawings are out of scale

to one from Ban Kao, Thailand (see Bellwood 1985: 259). Perhaps the large three-legged jars of south India reflect earlier connections with Shang China by an overland route.

For vessels with perforated ringfoot, one can trace a continuous dispersal route from Lungshanoid sites on the Chinese mainland over Taiwan to metal age sites in the Philippines (see Figure 8.6, and Mahdi 1994a: 185). However, there is a gap between the Philippines and south India, because the ringfeet of analogous vessels from Indochina and Malayo-Indonesia are apparently never perforated (but the legs of the three-legged jars from Gua Berhala and Ban Kao are perforated).

Of crucial significance is the introduction of urn burial in period IIIB from the south, apparently from Sri Lanka, i.e. from the same direction as the coconut, and almost at the same time. Urn burials are well attested for West and Central Indonesia, the Philippines, and in Indochina along the Mekong Basin as far as the Plain of Jars in Laos. Cranial measurements at south Indian burials of different types reported by Sarkar (1982: 160) show that the peoples practising urn burial were quite different from the non-Vedic Aryans who buried their dead in stone cists (rather than burning them, as practised by Vedic Aryans). In period IIIC, the practice of urn burial seems to have spread rapidly, probably also to include Dravidians. By historic times, at least, urn

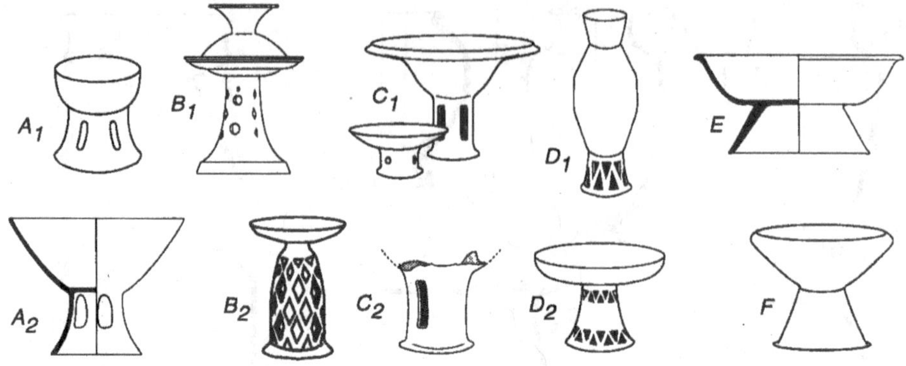

Figure 8.6 Vessels with perforated ringfoot

Source: A1 Vilpatti based on McIntosh (1985: 472) and A2 Perumal Malai, based on Wessels-Mevissen (1991: 53), both south India; B1 Huating, based on Chang (1978: 231) and B2 Ninyang, based on Foreign (1972: 28), mainland east/southeast China; C1 Fengbidou, C2 Dapenkeng, Taiwan, based on Chang (1969: 89); D1, D2 Novaliches, the Philippines, based on Sullivan (1956: 73); and with unperforated ringfeet from *E* Bukit Tengku, Lembu, West Malaysia, and *F* Ban Kao, Thailand, based on Bellwood (1985: 266, 259)

Note: All drawings are out of scale

burial seems to have been the common mode of burial among Tamils, as testified in early Tamil literature (Srinavasan 1946: 12–15)

The likeliest explanation of this constellation seems to be an introduction of the urn burial custom by Austronesians from West Malayo-Indonesia over Sri Lanka to south India, where it penetrated from the south around 450–400 BC, assuming period IIIB to be more or less in the middle of the entire period III. Occasional instances of urn burial have also been found in north India, but it is difficult to explain why the dispersal should take the roundabout route over Sri Lanka, particularly since the Aśokan expansion would have offered favourable conditions for a direct propagation by driving the putative original urn-buriers south. Apart from that, urn burial in the north is quite rare, whereas in Southeast Asia it was at one period a prevailing feature. Southeast Asian urn burial already begins towards the end of the Late Neolithic, and the Manunggul Cave site in Palawan produced a burial urn with a ship of the dead on its cover (Fox 1979: 233). Perhaps the instances in north India also resulted from Austronesian infiltrations.

Arriving on the scene from the third century BC onwards, the Tamils must have chanced upon an ethnic situation similar to that which had been encountered by Vedic Aryans in the Ganges basin: megalithist non-Vedic Aryans, but this time probably with Austronesian and other admixture partly already acquired in the Ganges basin, and urn-burying Austronesians. They apparently used the term *Nāga* in the same manner as the Vedic Aryans had in the Ganges basin, in its 'urban' aspect perhaps more in reference to non-Vedic Aryans, in its head-hunting and maritime aspect more to Austronesians. With regard to the dating of Austronesian activity in South Asia, the above implies Austronesian presence in Sri Lanka at least as early as 450–400 BC, and in the Ganges basin since sometime between the tenth and the fifth centuries BC. This agrees well with the interval of 1000–600 BC within which settlement of Austronesian long-distance navigators in India began.

Ditches around stone fortifications in Indonesia

Some information on the dispersal of megalithic traditions in the Philippine–Indonesian Archipelago can be extracted from the distribution of reflexes of the doublet protoforms *parij ~ *parigiʔ 'ditch around stone fortification'. The variety of meanings of the reflexes of the latter suggests a more advanced stage of development than those of the first. The distribution of reflexes of these protoforms and their respective meanings is explored in some detail in an earlier publication (Mahdi 1994b: 442–450). Reflexes of both forms are limited to the Philippines, Central and West Indonesia, and Madagascar.

The form *parij is mainly represented in Sumatra, where reflexes usually mean 'ditch', occasionally also 'fence' or 'earthen wall'. It is additionally represented in Madagascar with the meaning 'boundary', and in Cebuano (Central Philippines) as either 'stone wall enclosing or at ridge of area' or 'stone fish corral in tidal flats' (the same semantic shift is attested for a word for stockade in Fiji). Its association with stone as a material of the fortification wall is not very pronounced and could be a later development.

The form *parigiʔ is densely and evenly represented along a continuous strip leading from Sangir and the Sea of Sulawesi over Sulawesi and the whole of Java (with adjacent lesser islands), with occasional reflexes in Luzon and Sumatra. The meanings of the forms clearly indicate proximity to stone masonry, and the distribution generally coincides with remains of metal age megalithism in island Southeast Asia. Assuming that the latter doublet developed out of the former, its origin can be situated in the Sangir–North Sulawesi–Mindanao area, because only here do we find automatic post-fixation of a post-glottalized vowel to originally final-voiced consonants (Mahdi 1994b: 442–450).

The region (with the addition of the Sulu Islands and parts of East Malaysia) is also the place of convergence of the distribution areas of reflexes of various protoforms meaning 'iron' (*Bəsi ~ *[bB]asi, *bari[]) or of reflexes meaning 'iron' of protoforms otherwise referring to weapons, utensils or other metals (*Həuasəi axe, *maLat machete, *putəu iron or steel utensil, *bulau-an copper, *ntiti copper; see Mahdi 1994a: 173–184, and Figure 8.). The metal age megalithist horizon, as noted above and reflected in the name itself, is distinguished from the neolithic megalithist among others by the presence of metal, and particularly iron. It thus appears probable that it was a lexical item of the language(s) of peoples chiefly involved in the dissemination of metal age megalithism in the Archipelago. This implies a propagation from the border area between the south Philippines and north Sulawesi first southwards, then westwards into West Malayo-Indonesia.

If the metal age megalithist culture of south India and Southeast Asia are related, it must have originated in Southeast Asia, because otherwise one would have expected a propagation within Indonesia from west to east. However, the bronze–iron period in the Archipelago, and hence also the metal age megalithist period, must date from around 500 BC, which is too late to be the source of south Indian megalithism. The megaliths of period I and most of period II in south India correspond to those of neolithic megalithism in Austronesia. But in view of the megaliths of Baluchistan, an Austronesian influence is possible but not imperative, even on purely diffusionist assumptions. Some Austronesian contribution in the distinct (from Baluchistan) south Indian development from periods II and particularly IIIA onwards seems much likelier, and that apparently received substantial reinforcement from an Austronesian influx from Sri Lanka around 450–400 BC.

Activities ascribed in traditional sources to *Nāgas* suggested an area of operation exceeding that which one could confidently assign to Austronesian movements in India. When compared with the distribution of South Indian megaliths, however, covering a considerable portion of the peninsular part of the subcontinent (see Figure 8.4), the geographical agreement with *Nāga* movements improves. Whereas the term *Nāga* referred to Austronesians when connected with maritime activities, the coincidence of *Nāga* movement with mainland diffusion of megalithism should be assigned mainly to the activity of non-Vedic Aryans who experienced Austronesian admixture in the Ganges. In south Indian megalithism, it is not the stone cists and related objects but the introduction of the urn fields (not quite strictly included in megalithism) for which the Austronesians were responsible. If the *Minavar* fishermen of Tamil tradition were (non-urn-burying) Austronesians, and the conquest of the *Villavar* and *Minavar* by *Nāgas* referred to the southward migration of non-Vedic Aryans from the Ganges basin, Austronesians could already have been in littoral south India from a rather early date.

THE SACRED TREE CULT

Kalpavṛkṣa, the wishing tree, in Southeast Asia

The problem of megaliths in the cultural relations between Southeast Asia and India is also connected with that of the sacred tree cult in both regions. The problem arose for the first time in conjunction with the Muara Kaman *yūpa* inscriptions of King Mūlavarman of around 400 AD. Inscription C includes among the donations of the king one that was called *kalpavṛkṣa*, a ritual tree. Kern (1882: 187), who was the first to publish the Sanskrit text, believed it to be a mythical wonder tree that fulfils all wishes, possibly a tree laden with precious gifts to be grasped by the beneficiaries, like a kind of Christmas tree. Vogel (1918: 215 fn. 2), like Kern, was unaware of a comparable donation in India, noting that it was customary there for a royal benefactor to liken himself to a wishing tree that fulfils all desires, but not to donate a real tree laden with gifts.

Ironically, first analogous cases became known not from India, but from other countries of Southeast Asia. Blagden (1918: 615) called attention to a Mon inscription of around 1100 AD, containing a list of gifts of a Choli prince to the Mon king, including '*chu kalpabrikṣa ma tmūy na sattaratna*' – 'a *kalpavṛkṣa* adorned with seven [kinds of] jewels'. The same author pointed to a report in the Pegu chronicles *Slapat Rājāwaṇ*, that King Dhammacetī, who reigned in Pegu in the fifteenth century, ordered twenty-five trees, referred to as *kawslabamruik*, a corruption of *kalpavṛkṣa*, to be presented every year to the Shwedagon temple. Furthermore, the Mon version of the Buddhist *Abhidhānappadīpika*, published at Pak Lat, contains the less corrupted version *kalpabruik*. Finally, Blagden (1918: 616) quoted a report by Adhémard Leclère on the festivities at the consecration of the great Buddhist pagoda of Pnom Penh in February 1903, indicating that the gift-bearing ritual tree was still a familiar thing for Khmers: 'The three *kalbo-priks* or *kalpa* trees, in the tops of which sat three men who threw silver coins and ingots into the crowd'.

Vogel (1920: 432–434) published part of a personal letter by George Cœdès, addressed to him in 1919, informing that a Khmer inscription from Sukhothai (in Thailand) reported that the king '*pre laṁtap slā lāja dyaṅ dhūpa puspa kalpabṛkṣa*' – 'let be prepared areca, popped rice, candles, incense, flowers [laden, bedecked?] *kalpavṛkṣas*'. Cœdès noted further that the *kalpavṛkṣa* continued to be known to the Thais as *kala-* or *kara-phrŭk* with the meaning: 'a tree on which are attached lemon-like bundles containing money, which are thrown to the people on feast days'.

The first mentions of the *kalpavṛkṣa* in India itself did not seem to be adequate as precursors of the Southeast Asian variant. In an editor's note to Blagden (1918), Vogel indicated that a *kalpavṛkṣa* is mentioned in the Sanskrit *Mānasāra* among the royal insignia to be used at coronations, not as a gift, but as an emblem. Finot (1919) discovered a passage in the *Matsya-pūraṇa* enumerating sixteen great gifts, including a *kalpapādapa* which literally means the same thing (*vṛkṣa* and *pādapa* 'tree' are synonyms). In Hemādri's voluminous treatise on the subject, the *Caturvargacintāmaṇi* written around 1300 AD, it is described as a tree of gold and precious stones, having figurines of people and animals hanging like fruit from its branches.

By contrast, the Southeast Asian *kalpavṛkṣa* is a real tree. It was not itself the gift, but a ritual medium in which the gifts were placed, or from which these rained onto the beneficiaries, whereas examples of the Indian counterpart in Southeast Asia are obviously results of Indian influence. Could this be an instance of the phenomenon decribed by de Casparis (1986: 248) as the tendency in early civilizations of Southeast Asia to apply Sanskrit terminology to indigenous concepts and institutions that show only superficial likeness to those known from Indian texts? Before directly answering this question, we need to inspect the roots of the sacred tree cult in Southeast Asia.

The sacred benjamin tree and the bird–tree–serpent cosmological complex

The concept of the sacred tree, originally the benjamin tree, *Ficus benjamina* L. (often confused with the banyan tree *F. benghalensis* L. = *F. indica* L. which also has aerial roots, and more rarely with the peepul, bo, or bodhi tree *F. religiosa* L., which does not have them), is an immediate feature of megalithism in Western Austronesia (Mahdi 1994b: 478–479 n. 164). In the cosmological mythology of early Austronesians, the tree apparently had the function of connecting the upper-world with the lower-world. The paradoxical aerial roots of the benjamin tree, reaching down from the branches instead of being in the ground, apparently marked it as a 'heavenly' tree. These aerial roots reached all the way down, finally to penetrate into the earth, appearing to enact the unification of the upper- and lower-worlds. Marriage between heaven and earth was the subject of a dissertation by H.Th. Fischer (Rijksuniversiteit te Utrecht, 1929) cited in Münsterberger (1940). Furthermore, aerial roots that had reached the ground developed into secondary trunks, so that an older tree presented itself as a gigantic maze of entangled trunks, in which an eerie dark and moist atmosphere provided an ideal biotope for many insects and reptiles, supplying ample nourishment to the imagination.

The sacred benjamin tree cult not only coincides in its distribution area with megalithism at least in Western Austronesia, but actually forms an integral part of some megalithic monuments or ritual meeting places. Colani (1937) noted that the stone terraced construction at Do-linh in central Vietnam had a tree of the genus *Ficus* at its top. A benjamin tree is reported by Röder (1939) at the top of the megalithist hill-sanctuary and meeting place at Soya near Ambon, central Maluku. A sacred tree growing out of the top of a terraced pyramid is a well-known motif on the *kayɔn* or *gunuŋan* of the Javanese *wayaŋ* shadow-theatre (Stutterheim 1926a: 135 fig. 225; Bosch 1960: plates 66–67).

In variants of the sacred tree cult among peoples of Kalimantan, the unification of male upper-world (typically personified by the hornbill) and female lower-world (the divine serpent) through mediation of the benjamin tree results in the lower-world becoming pregnant with the further procreation. Unlike the Kayan, the peoples of the Barito reportedly believe that the fertilization act between upper-world and lower-world leads to the destruction of the tree. In a plea for prosperity, they stage ritual fights or contests, pitting upper-world and lower-world affiliated teams against each other in re-enaction of the primordial fertilizing destruction of the *bataŋ gariŋ* tree (Sellato 1989: 46).

In the myth of origin of the Tombra Lumuleli in Leti, these regard themselves as the offspring of Upu Nusa ('Mistress Island', the lower-world goddess). A yearly feast is held under a benjamin tree (*nunu*) standing in the middle of the village, with the aim of pleading to Upu Lero ('Master Sun', the upper-world god) for rain, abundance of food, cattle, children and wealth. It is believed that Upu Lero descends through the *nunu* tree to fertilize Upu Nusa. In earlier times, free sexual intercourse was allegedly permitted during these festivities (Riedel 1886: 372–377). From the nearby island of Kisar, van Hoëvell (1890: 204–205) has also reported that the sun god (here Upu Lera) descends through the sacred benjamin tree, at the foot of which there are often megalithist objects, to bring good fortune to the people. The significance of this belief is greatly enhanced by an analogous example in Taiwan. According to the Tsou, the god Hamo, a remote sky deity whose body exudes light, descends through the benjamin tree called *eono* (*rono* in the phonologically more conservative Duhtu dialect) in the same fecundating function (Ferrell 1969: 36).

The sacred tree cult thus presents us with what one may evidently see as several variations and developments of the same basic idea: the primordial connection of the upper-world and the lower-world through the benjamin tree, resulting in fertilization and procreation. In this, the lower-world of land and water is originally represented by the sacred or divine snake, and the upper-world by either a sacred bird or the sun. The latter may perhaps have been conceptualized as a fiery bird (because it also moved along across the sky?).

The best-known manifestation or remains of the cult is the depiction of what is often referred to as 'tree of life' (e.g. Steinmann 1939; Gittinger 1972; de Jonge and van Dijk 1978) on the *tampans* and *paləpays* (ritual ship cloths) of Lampung at the southeastern end of Sumatra. The tree, called *kayu hara*, is often depicted on both sides of the ship, or on the ship itself in the place of its mast. In spite of the complexity of mythological notions associated with the *kayu hara*, it is not difficult to discover its connection with the Southeast Asian *kalpavṛkṣa* on the one hand, and pre-Hindu Austronesian mythology on the other. At a wedding ceremony, sticks with branches are carried to the village of the bridegroom and set up as *kayu hara*. Gifts from the maternal grandparents are hung on the branches, and a mythological bird (referred to as *Garuda*, a word of Sanskrit origin, here perhaps a reinterpretation of the hornbill?) wrapped in a *tampan* is fastened at the top. At the end of the ceremony, boys of the bridegroom's clan climb the *kayu hara* and take out the gifts (Gittinger 1972: 157; cf. also Gittinger 1976: 215). Lampung *kayu hara* is also the word for *Ficus* spp. (de Clercq 1909: #1432) to which belongs the benjamin tree.

A similar reflection of the original cult with a light Hindu veneer can be seen in the Balinese correspondence to the *gunuŋan* or *kayɔn* of Javanese *wayaŋ* shadow-theatre. One such *gunuŋan*, reproduced in Kats (1923: across p. 24), shows a tree with a mythological bird perched on its crown, and an entangled pair of serpents depicted as *nāgas* under the tree. Considering that the Indian *kalpavṛkṣa* is essentially a feature of Buddhism, the circumstance that the Indic element in both the Lampung and the Balinese manifestations are Hindu confirms the secondary and superficial nature of that element, and the indigenous origin of the basic tree cult itself.

Depictions of the sacred tree with a bird, the hornbill, perched at its top, and either a single serpent or an entangled pair (but with traditional Dayak rather than Hindu stylization), can be seen on the carved ornament of a quiver of the Ot Danum of southeast Kalimantan described by Loebèr (1911: 49–52, fig. 5 between pp. 48 and 49). In the example with single serpent, the tree appears to be growing right out of its back. Stutterheim (1940) called attention to the ornamentation of the Ot Danum quiver in connection with an unusual Javanese *tahas* or *talam* – a circular, flat, stone, ornamental element decorated with carvings placed in concentric zones – found in East Java. One of the carved motifs was that of a serpent carrying a stylized tree on its back, which the author correlated with the sacred tree on a snake's back motif of the Ot Danum quiver. Significantly, the serpent on the *tahas* did not offer any likeness to characteristic Hindu–Javanese depictions of the *nāga*, but showed the typical indigenous ornamental style of depictions of the sacred serpent as is to be found all over Kalimantan (see e.g. Sellato 1989), and specifically on the Ot Danum quiver.

Another interesting relict of the primordial fertilization myth around the original sacred tree cult is reported from the Shan States of Burma. According to the origin myth of the Palaungs, they proceed from the liaison between the sun prince from the sky and a *Nāgī* princess from deep in the earth (see Milne 1924: 379–383). A noteworthy detail is that the bird, equated with a crow, is identified as the messenger of the sun prince and as originally having been gold-feathered and a dweller in the heavens, becoming black and banished to the world of people as punishment. It will be remembered that the hornbill too is black-feathered. The tree in the Palaung tale is where the crow alights with a gift of a precious gem from the sun prince for the *Nāgī* princess. It is in this tree that the gem is cunningly replaced by a worthless pebble by a fisherman (who thereby fetches his fortune), causing punishment of the crow. It is tempting to speculate that 'fisherman' here might refer to Austronesians who once ascended the Salween.

Preliminary linguistic aspects of the sacred tree cult

In one expansion or reinterpretation of the cosmic fertilization act, indicated above for the Letinese and Kisarese of south Maluku, and the Tsou of Taiwan, the upper-world deity descends through the sacred tree to bring the people, the children of the primordial creation, fulfilment of their plea for sufficiency and prosperity. This extension evidently underlies the Southeast Asian *kalpavṛkṣa* and, for example, the Lampung *kayu hara* cult. This extended feature still survives even in modern Indonesian symbolism, in which the benjamin tree represents a spiritual category to which expectations of fulfilment of the communities' collective aspirations are attached.

The same idea is apparently reflected in the etymology of the Malay word for the tree itself, (*pɔhɔn*) *bəriŋin* 'benjamin (tree)', from *bər-* a verbal prefix, and *iŋin* 'desire, want'. The literal meaning was thus 'wishing (tree)' (Aichele 1928: 28 fn. 4). This must have been a relatively early Malay innovation, because borrowings into other languages of West Malayo-Indonesia exhibit subsequent sound shifts, some of which took place at a relatively early time:

Balinese *baiŋin* *r* > *h*, ø (see Teeuw 1965)
Sundanese *cariŋin* *b* > *c* (see Nothofer 1975: 301–307)
Javanese *wariŋin* *b* > *w*

The shift in Balinese is relatively recent. In Sundanese it is probably somewhat older, but I have no data that would permit using it for dating. The shift in Old Javanese predates the earliest written records in the language, and thus points to a time prior to the seventh century AD. In loans from Malay, a *b* of the donor is usually retained in Old Javanese. Borrowings in which it has been shifted to *w* are very rare, e.g.: Old Javanese *wrāt* ~ *wərat* < Malay *bərat* < *BəRqa[tC] 'heavy', beside the regular: Old Javanese *uwat* < *BəRqa[tC] 'heavy'.

Malay loans having *w* in Old Javanese for the *b* of the original are restricted to the very earliest Malay stratum in Javanese, evidently dating from the beginnings of Malay–Javanese contacts. As I shall show below in the section on the spice trade, the earliest dispersal of Malay lexical items in the Archipelago may probably be dated to the second century BC. It is possible that the loan into Javanese took place at this early date or not very much later.

In the Old Malay inscriptions of Śrī Vijaya (late seventh century AD), the prefix *mar-* persists in place of the *bər-* of Late Medieval (e.g. *bər-bajik-i* in line A-4 of the Trengganu inscription, fourteenth century AD, Paterson 1924) until modern Malay. This would contradict the etymology of ***bəriŋin*** adopted above, as one would expect ***mariŋin*. However, *Śrī Vijaya* was Buddhist, and may have represented a different dialect than the Malay of Hindu *Malayu* which, as I indicate elsewhere (Mahdi 1995, and n.d.), must have been the continuation of legendary *Yavadvīpa*. In Hinduized Old Malay inscriptions of Java, dated only some 150 years later, one finds *var-* (read *bar-* or *bər-*) rather than *mar-* (cf. e.g. de Casparis 1956: 3 fn. 16).

The oldest Austronesian form for benjamin tree has been reconstructed by Blust (1983–84b: #242) as *nunuk,[10] for which the different cognates are listed in Table 8.2 (see also de Clercq 1909: #1442). I have not included in the table Kayan *lunuk*, which is probably cognate, but has irregular *l*, and Duhtu Tsou *rono*, which is probably not cognate.

With the possible exception of the Aklanon and dialectal Sundanese forms, all reflexes apparently refer to *Ficus* species, mostly with aerial roots and thus, in the Philipine–Indonesian Archipelago, typically the benjamin, although in many of the sources the respective form is glossed as banyan (*Ficus benghalensis* L.). Typically, Dutch sources use the term *waringin* or, less frequently, *beringin*, respectively the Javanese and Malay word for benjamin tree, whilst English sources employ the word banyan. In a Philippine or Malayo-Indonesian context, the term refers to the benjamin tree, in an Indian one exclusively to the actual banyan (*Ficus benghalensis* L.), and in Oceania to any large *Ficus* with aerial roots that develop to secondary trunks. In French sources, the benjamin is usually referred to as 'species of fig'. The meaning given thus more often depends upon the language of the respective author than the

Table 8.2 Cognates of *nunuk

	**nunuk*	*Meaning*
Philippines–north Sulawesi:		
Aklanon	*nunuk*	tree believed to be bewitched or enchanted
Tirurai	*nunuk*	strangler fig, *F. benjamina* L.
Mongondou	*nunuk*	*F. benjamina* L.
Gorontalo	*lulu?o*	*F. benjamina* L.
Maranao	*nonok*	*Ficus* sp.
Sangir	*nunu?*	*F. benjamina* L.
Uma	*nunu?*	*F. benjamina* L.
Marianas:		
Chamorro	*nunu*	*Ficus* sp.
West Indonesia–Madagascar:		
Sundanese	*nunuk*	*Ficus* sp.
Sundanese dialect	*nunuk*	ghost in form of hunchbacked beldam
Mbalo	*nunuk*	*Ficus* sp.
Malagasy	*nónoka*	*Ficus* sp.
East Indonesia:		
Manggarai	*nunu*	*F. benjamina* L.
Letinese	*nunu*	*F. benjamina* L.
Gorom	*nunu*	*F. benjamina* L.
Asilulu	*nunu*	*F. benjamina* L.
Haruku	*nunu*	*F. benjamina* L.
Roti	*nunu(k)*	*F. benjamina* L.
Hatué	*lulu?*	*F. benjamina* L.
Kisar	*nunu*	*F. benjamina* L.
Wetar	*nunu*	*F. benjamina* L.
Babar	*nunu*	*F. benjamina* L.
Batumerah	*nunu?u*	*F. benjamina* L.
Oceania:		
Lenkau	*nun*	*Ficus* sp.
Raluana	*nunu*	*Ficus* sp.
Mbau Fiji	*nunu*	*Ficus* sp.

language itself. The banyan does not occur in Southeast Asia, except in isolation in West Malayo-Indonesia where it is known under other names (see de Clercq 1909: #1476). Parham lists the following *Ficus* species that may be referred as *nunu* in the Fijis:

> *F. fulvo-pilosa* Summerhayes,
> *F. greenwoodii* Summerhayes,
> *F. masonii* Horne, F. *pritchardii* Seem.,
> *F. prolixa* Forst.f.,
> *F. storckii* Seem.,
> *F. tinctoria* Forst.f.
>
> (Parham 1972: 137–139)

The species, principally referred to as *nunu* in Mbau Fiji, appears to be *F. vitiensis* Seem. (Capell 1941: 184). The *nunu* species do not, however, include the one formerly regarded as sacred, *F. obliqua* Forst. f., and called *mbaka* (Parham 1972: 138). It is remarkable that, beside the latter and *F. tinctoria* Forst f., the only other species referred to in Fiji as *mbaka* is apparently *F. benjamina* L., a post-contact introduction commonly grown as an ornamental or shade tree (Parham 1972: 136).

The meanings of the Aklanon form and of that tentatively assigned above to a Sundanese dialect reflect a tendency in some regions of the Philippines and Malayo-Indonesia to regard benjamin trees of which the aerial roots had developed into a cluster of secondary trunks as being enchanted or haunted. The dictionary of Coolsma (1930: 405) gave for Sundanese *nunuk*: '1. name of a fruit tree, 2. ghost in the form of a hunchback beldam, P[leyte]'. C.M. Pleyte, cited as the source for the second gloss, is named in the introduction as one who had made an outstanding contribution to the knowledge of Sundanese in general, and Bantenese in particular (Coolsma 1930: vi). It was a widespread error at that time to regard Bantenese as a dialect of Sundanese rather than, correctly, of Javanese. There are, however, Sundanese dialects spoken in Banten, particularly by Baduis, a non-Islamic minority in the area, so the gloss ascribed to Pleyte may be from Badui Sundanese. In any case, the standard dictionary of Sundanese (LBSS 1976: 33) and the comprehensive dictionary of Eringa (1989: 519) do not mention the second meaning. The tree name is furthermore not listed in de Clercq (1909), so it remains unclear what species of *Ficus* is implied.

The association of the benjamin trees with the dark and sinister, expressed in the Aklanon and dialectal Sundanese reflexes, as we shall see below, has analogues in Oceania. On the other hand, it also serves as a reason for caution in treating the Oceanic forms included above as cognates. These could theoretically also reflect a Proto-Oceanic form for 'shadow, reflection' for which the semantic extension 'soul, dream' is attested. It was first reconstructed by Blust (1972: 14 #76), originally as *[nñ]u[nñ]u, later defined more precisely as *(qa-)nunu (Blust 1978a: 6, 44; Chowning 1991: 61). In view of the possibility of a final *k in the protoform (see Table 8.3), I tentatively write *(qa-)nunu[k].

As Chowning indicated, the Manam reflex suggests a final *-k in the protoform, which would have predictably dropped away in most Oceanic languages. Nevertheless, Kove, Molima and Bwaidoga also often retain a final consonant by post-fixation of a paragogic vowel (typically *-a* as in Manam), and as the cognates in these languages show no trace of the final consonant, the author concluded that the original form was probably without a final *-k. Oceanic languages that retain final consonants do not do so very consistently, so that it is also possible that the three cited languages lost the final consonant in the reflex of this protoform.

Regardless of whether the original form had this final consonant or not, Lenkau *nun*, Raluana and Fiji *nunu* 'a species of *Ficus*' could obviously also reflect the protoform for shadow with mystical semantic extension. The evidence for the inclusion of Eastern Austronesia in the distribution area of

Table 8.3 Cognates of *(qa-)nunu[k]

Language	*Protoform*	*Meaning*
Papua-New Guinea:	*(qa-)nunu[k]	
Kove	*anunu*	shadow
Lakalai	*ha lulu*	shadow. *anunu-Gu la* I dream (my *anunu* goes)
Manam	*anunuka*	shadow
Molima	*ʔanunu*	shadow
Wogeo	*v-anunu*	shadow, reflection, soul
Arosi	*nunu-na*	image, reflection, soul
Bwaidoga	*anunu*	shadow
Solomons: Nggela	*nunu*	shadow
Sa'a	*nunu*	shadow
Micronesia: Kiribati	*nu*	shadow

reflexes of *nunuk '*Ficus* sp. with aerial roots' is therefore not fully watertight. On the other hand, *nunuk '*Ficus* sp.' and *(qa-)nunu[k] 'shadow' perhaps ultimately derive from a common source, an older hypothetical *nunuk 'shadow', and the tree was already referred to as shadow tree before the east/west split. Further investigation will perhaps at least bring clarity in the question of a final *-k in the eastern protoform for 'shadow'.

Some languages of West and Central Malayo-Indonesia have reflexes of the protoform reconstructed by Verheijen (1984: #6.59) as *[q]aRa, usually referring to various species of, or functioning as generic term for, *Ficus*, including the benjamin (see also de Clercq 1909: #1432), the uncertainty of the initial *q being caused by the missing reflection as *h* in the Malay reflex. But as initial *h-* is often lost in Malay, Blust (1986: 15 #16) reconstructed *qaRaʔ, accounting at the same time for the final glottal in Iban (see Table 8.4).

Lampung *kayu hara*, already mentioned above (*kayu* tree, wood), and the Sundanese cognate *ki-ara* (*ki-* < *kai* wood, tree) must be borrowings (probably from Malay), because of the irregular *r* for expected *y*. Old Javanese also has *ara* and *hara* which are obvious loans from Malay. In view of the two protoforms *nunuk and *qaRaʔ, Malay *pɔhɔn bə riŋin* benjamin tree, literally wishing tree (*pɔhɔn* tree), must indeed have been an early Malay innovation. Its borrowing into many languages that experienced intensive influence of

Table 8.4 Protoforms referring to *Ficus* sp.

Language	**qaRaʔ*
Malay, Toba Batak, Sasak, Makassarese, Buginese, Manggarai	*ara*
Iban	*araʔ*
Old Javanese	*hā*
Balinese, Sangir, Ratahan	*aha*
Ngadha	*ara*
Sikka	*ʔara*

Malay since early times suggests that the wishing tree probably occupied an important position in pre-Hindu Malay culture. As we shall see below, this may have implications for dating the first Malay involvement in maritime communication with India.

The sacred tree cult in Oceania

Evidence of the same sacred tree cult in Oceania increases the reliability of an early date for its beginnings. I do not have a particularly detailed overview of the religious concepts and mythology of Oceania, also not with special regard to the sacred *Ficus* tree cult. However, the fragmentary information from the literature, enhanced substantially by information kindly made accessible to me by some ethnologists and archaeologists, suggests that the cult was not restricted to Western Austronesians. Manifestations of the tree cult may be divided into two categories: first, the appearance of an actual tree with aerial roots of the genus *Ficus* (usually referred to as *banyan* in the literature) bearing the corresponding ritual function in the life of the community, particularly in association with megalithic objects, and second, reflection of the concept of such a tree mediating between personifiers of the upper- and lower-worlds in mythology, abstracted from an actual material tree as ritual medium.

I shall begin with examples of the former. The ritual dancing ground of a village of the Lambumbu on the island of Malekula, Vanuatu, is described by Deacon (1934: 28–30, fig. 2 on p. 29; see also Riesenfeld 1950: 48–49 fig. 4) as a large circle surrounded by upright monoliths and with a tier of four stone slabs at the centre. Two avenues lined by pairs of stones lead in opposite directions from the circle. At another point on the perimeter, approximately equidistant to the exits to the avenues, stands what is described as a gigantic banyan tree. The entire circular dancing ground lies within the shadow of the giant tree, which is apparently *Ficus prolixa* Forst. f., for which Guillaumin (1932: 101) reported sightings in Erromango, Tanna and Aneityum by Summerhayes. The author also gave its vernacular name as *nepang*, cf.: Erromango *paŋ*, Mataso *nabaŋ*, Bislama *nabaŋa* which refer to the benjamin-type *Ficus* species regarded as sacred in many areas of Vanuatu (Ralph Regenvanu, p.c.). For *Ficus obliqua* Forst. f., on the other hand, which is or was regarded as sacred elsewhere in Oceania instead of *F. prolixa*, the vernacular name given by Guillaumin is *nar-evirepp*.

Ralph Regenvanu of the Vanuatu Cultural Centre, Port Vila, has provided me with copious information on the banyan (i.e. benjamin-type tree) in Vanuatu. The location of a ritual ground in the shade of a banyan tree seems to be quite common in the northeastern part of Malekula, as also in Tanna. The shade of the tree apparently also serves as a meeting place, e.g. in Malekula, Pentecost, Tanna and Erromango. A banyan tree often stands before the men's house, e.g. in Malekula. In Tanna, the tree itself is the actual men's house, there not being a special house for this purpose. Instead, the aerial roots are fashioned into the form of a house. In various places in Vanuatu, the banyan features in origin myths as the tree of origin. In some areas of

Malekula, the tree is seen as a place of good as well as evil spirits, and people are afraid to approach the tree at night. One of the mythological dwellers of the tree is a female spirit with long hair which she hides in the aerial roots. She is known as *Leplepsepsep* in Bislama (the principal communication language of Vanuatu), and in the northeast of Malekula, where she is called *Lesserkamp*, she is said to devour children (Ralph Regenvanu, p.c.).

Cristophe Sand (*Service des Musées et du Patrimoine de Nouvelle-Caledonie, Noumea*) informs me that in New Caledonia, the 'banyan' is the principal sacred plant, and is usually found at burial and other hallowed or sacred sites. There appear to be two species commonly referred to as banyan in the country, the 'true' one with sacred quality and having leaves with a drop-conducting tip (like the benjamin), and a 'false' or 'wrong' one having leaves with a blunt tip. Judging from the data provided by Moore (1921: 411–413), these are apparently *Ficus prolixa* Forst. f. and *F. oreadum* S. Moore respectively. In one burial custom, the corpse was placed in the banyan. After excarnation, the skeleton was taken down, and the skull was placed on the ancestral altar. There are various oral traditions about people living in a banyan tree (Cristophe Sand, p.c.), but some caution may be required in comparing these traditions with possible analogues from insular Southeast Asia, because of possible recent contamination. Large contingents of Javanese labourers were brought to New Caledonia in the late nineteenth century, and their descendants still remain here till this day.

In the Marquesas islands in Polynesia, tribal or chiefly centres had as a rule a *tohua*, an elevated terrace on which was a large rectangular gathering field surrounded by smaller stone terraces with houses of the chiefs and priests, temples and other important houses or galleries. At one end usually stood the main temple with a sacred banyan tree before it, in which the bones of particularly revered dead ancestors were hung (Suggs 1982: 779). This Marquesan sacred banyan obviously paralleled the sacred benjamin atop stone terraces or other megalithist complexes in Southeast Asia noted above. The exposure of the revered bones, as also the New Caledonian burial custom, may perhaps be associated with the custom of reburial after excarnation, attested for the megalithism of Southeast Asia and south India. In the Marquesas, the banyan was also one of two tree species regarded as sacred in general, and which were almost always associated with sacred mortuary sites called *meʔae* (Handy 1923: 213).

The bird–tree–serpent complex in Oceanic mythology

The more purely mythological reflection of the cult has various manifestations, some more, others less obviously associated with the ideological conceptions of the sacred tree cult noted for Western Austronesia. As tree cults are widespread across the world, one must always keep the possibility of an independent origin in mind. It is instructive, in this respect, to compare the sacred linden tree cult among Germanic and Slavic peoples in Central Europe. It involved the large-leaved linden (*Tilia platyphyllos* Scop.), to a lesser degree also the small-leaved linden (*T. cordata* Miller), European linden (*T. europea* L.), and perhaps also the hybrid vulgar linden (*T. vulgaris* Heyne). It was customary in some regions of Central Europe that a hallowed linden stood in the village square or the yard of a castle (not unlike the banyan in Java or the Marquesas), or in a graveyard (cf. the banyan in Kisar or New Caledonia), or that the shade of an old linden tree served as a permanent place of session of a court of justice (like the meeting place in Tanna). Feasts and weddings were also often held under a linden tree (see e.g. Brockhaus 1990: 409 under *Linde, Tilia*).

Affinities to Southeast Asia of manifestations of the cult in Oceanic mythology are, however, suggested by numerous subtle reflections of various aspects of the cult from the former geographical region. The Fijians who, as indicated above, preserved the cult of the serpent as a central element in their original religion, also present us with one of the clearest reflections of the ideological concepts of the cult of the sacred *Ficus* tree with aerial roots.

In one Fijian legend about the great serpent god *Ndeŋei*, there stood a banyan tree before the cave in which he slept. In this tree sat a black dove which called out every morning to wake the serpent. One day, the people who had learned many things from the serpent, including how to build boats, decided to make themselves independent, and killed the dove so that the serpent would never awaken again. But somehow the serpent woke up, and in his wrath let huge waves raze the village of the miscreants. The tree was also swept away and, having become grounded at some place in the sea, turned into an island which became a refuge for people (Reed and Hames 1967: 26–28). We have the same constellation of a serpent, a bird (black-feathered) and a connecting *Ficus* tree, but modified so that the patrilineal male principle has been relegated to the lower-world serpent as well, leaving the bird, and thus also the tree, without the crucially important functions they had in the original myth. The act of fertilization, by which the personified upper-world activates the lower-world, is replaced here by an awakening call. That this latter was nevertheless felt to be sufficiently decisive is evident from the idea of deactivating the serpent by killing the bird. The close of the story perhaps reflects reminiscences of the providential protection emanating from the sacred banyan tree onto the community.

In another Fijian legend, an origin myth, the tree is no longer mentioned, but the tale presents us with what appears to be a reinterpretation of the act of fertilization of a female lower-world serpent by a male upper-world bird, adapted to the inversion of gender connotations of the respective poles: the serpent being male, the bird female. In the beginning of time, the great serpent *Ndeŋei*, the male ever-living god, was alone without friends other than the female hawk *Turukawa*, which could not speak but only flew around *Ndeŋei*'s home. One day the little hawk disappeared, and there was a downpour of rain. For many days, the serpent god waited in vain for his lost companion, then one day when the sun was shining, his solitude was pleasantly interrupted by the return of *Turukawa* who silently built a nest, laid two eggs in it, and disappeared again. *Ndeŋei* took the eggs into his house where he kept them warm. After a long time, two human babies hatched, the first man and woman (Reed and Hames 1967: 13–14).

Noteworthy here is the correlation of rain and sunshine with the absence and reappearance of *Turukawa*, apparently reflecting an earlier identification of the bird with the sun. Second, the replacement of the fertilization act by the gift of eggs to be hatched is significant because the oviparous myth of East and Southeast Asia and northeast India was associated by Kim (1982: 182–187) with the megalithic culture of the region. The inclusion of Fiji into its distribution area indicates a dating of its origin far before the period between 800 BC and the turn of the Christian era assumed by Kim, unless one assumes an intrusion into Oceania subsequent to the split of Proto-Oceanic. The inclusion of the Moluccas and Fiji shows it to be represented outside the distribution area of metal age megalithism, implying an origin within the Neolithic. Third, practically all deviations of the Fijian versions of the basic myth from the Indonesian ones can be explained by the adaptation of the Fijian legends (by a different way in each of them) conditioned by the transition from their original matrilineal to the present patrilineal principles.

In the Admiralty Islands and Bismarck Archipelago, corresponding legends retain far fewer features of the original myth. In the section on the serpent cult in Oceania, a Papitalai legend from Manus in the Admiralties was mentioned, in which the snake *Moat* begot a son and a daughter by a human maiden, and then let them extricate the first fire and food plants from its belly. In a variant of that legend, the sun was shining strongly so the heroes, two brothers, went to fetch water. The *Tjauka*-bird (*Philemon coquerelli*) cried out to warn them of a snake. One of the brothers climbed a tree, and looked around in vain. They went for the water, and again the *Tjauka* cried, and the brother climbed a tree. This time he saw the snake. The other brother grasped it – it was *Moat* – and took it into the house. There the same things as in the other variant were extricated from its belly, and additionally an earthen pot. One of the brothers also tore off the snake's tailbone. It was later planted and grew into a *mbuahat* tree (Meier 1907: 656–658). Unfortunately, no further identification of the tree is given, so that it is not clear whether it is perhaps a *Ficus* species with aerial roots.

The personification of the sun as a particular bird seems to be the only original feature of the cult still preserved in the mythology of the Tolai in the area around Rabaul on the island of New Britain, PNG, beside the snake-aspect of the earliest female ancestor already discussed above. In one Tolai legend, a man who saw a cockatoo seated in a *marita* tree decided to come back with a sling to catch it. He came back and sat in the tree with his sling, waiting for the cockatoo to come, but in vain, wondering also why there was no sun to warm him although it had already been light for a long time. After he climbed down to sit under the tree, the cockatoo returned and came down to him. He caught it with the sling, but he burnt himself, because the cockatoo was the sun. In a variation of the legend, the two heroes of Tolai mythology had become the sun and the moon respectively.

The one who was the sun once watched a green parrot sitting in a *gogo* tree. There came a man with a sling who snared the parrot and killed it. This provoked the wrath of the sun-hero, who thereupon burnt the culprit to death (Janssen *et al.* 1973: 20–21).

The variant legends from Tolai seem to have brought together the two variant personifications of the upper-world principle, the mythological bird and the sun god, and the former appears to remember that they descend to the people through the tree. Tolai *marita* is apparently a pandanus tree (Peekel 1984: 38, and Matthew Spriggs p.c.), and thus not *Ficus*. I have not as yet managed to identify Tolai *gogo*.

An even further erosion of the original almost beyond recognition can perhaps still be seen in the Tahitian tradition (see Davies 1851: 25 under *aoa*), that the banyan tree (Tahiti *aoa*) first grew in the moon, whence the seed was brought to the earth by a bird. For Samoa *āoa* 'banyan' tree (*Ficus aoa* Warb.), the dictionary of Milner (1993: 22) provides one sentence as example of its use: '*E aitua le āoa* "The [that] banyan'tree is haunted" (*aitu* ghost)'.

Although the significance of the Oceanic comparative data on the sacred tree cult and the associated bird–tree–serpent complex should be treated with some reserve, not only do they exhibit too many common features with corresponding traditions of Southeast Asia to assign to coincidence with confidence, but variant versions of the same basic myth tend to complement each other in this (e.g. the Fijian serpent-and-bird myths). The Oceanic and Southeast Asian variants presumably developed from a common source, thereby confirming the ancient date of the sacred tree cult in Western Austronesia.

Further linguistic data on the benjamin-type tree

The words for the benjamin- or banyan-like tree in languages of Oceania differ from region to region. Most cognate sets appear to have a relatively limited distribution. For one such set, Geraghty (1983: 127) reconstructed Proto-East-Oceanic (PEO) *qayaoa, from which are derived Proto-Central-Pacific (PCP) *ʔayawa and Proto-Polynesian (PPN) *ʔawawa k.o. tree, *Ficus* Geraghty (1986: 304). Most of the reflexes refer to a benjamin-type tree, but it is unclear whether the tree denoted in many of the instances is regarded as sacred (see Table 8.5).

Particularly interesting are, however, some forms for the sacred benjamin-type tree in languages of Vanuatu and in Fiji, indicated by Ralph Regenvanu (p.c.) and Matthew Spriggs (p.c.), reflecting a form reconstructed by Lynch (1983: #132) as Proto-Oceanic *mpaka. Cognates apparently also occur in the Bismarck Archipelago which, though referring to other species of *Ficus*, considerably enlarges the overall distribution area. Some forms for 'root' in languages of Kalimantan and some dialects of Malagasy are possibly cognate. The Fiji and Vanuatu forms reflect a protoform with pre-nasalized initial, whereas the Tolai (and assumably also the Pala) cognate does not.

The regular sound correspondence is: *p- > Tolai *p-*; *mp- > Tolai *b-* (Ross 1988: 266). The identification of the *Ficus* spp. in Table 8.6 is based

Table 8.5 Cognates of *qayaoa

Language	**qayaoa*	*Meaning*	*Source*
PCP	**ʔayawa*		
Rotuma	*aeva*	Ficus sp.†	(Churchward 1940)
Mbau Fiji	*γ-acau*	k.o. tree	(Capell 1941)
Tubaniwai Fiji	*γ-ayawa*	*Ficus* sp.†	(Geraghty 1983: 127)
Eastern Fiji	*γ-acawa*	*Ficus* sp.†	(Geraghty 1983: 127)
PPN	**ʔawawa*		
Samoa	*āoa*	*F. aoa* Warb.†‡	(Milner 1993; Ind. Kew. [suppl.I])
Marquesas	*aoa*	*Ficus* sp.†	(Dordillon 1931)
Tahiti	*aoa*	Ficus sp.†§	(Davies 1851)
Rarotonga	*aoa*	*F. prolixa* Forst.f.†‡§	(Wilder 1931: 41)
Futuna	*ʔaoa*	*Ficus* sp.†	(Grézel 1878)
Nukuoro	*aoa*	*F. prolixa* Forst.f.†	(Carrol and Soulik 1973)
Tonga	*ʔōvava*	*F. obliqua* Forst.f.†	(Yuncker 1959: 99–101)
Niue	*ovava*	*F. prolixa* Forst.f.†	(Yuncker 1943: 48)
	ovava niukini	*F. obliqua* Forst.f.†	(Yuncker 1943: 47)

Key: † benjamin-type tree, ‡ sacred/haunted, § bark used for cloth

Table 8.6 Cognates of *pakat

Region	*Language*	**pakat*	*Meaning*
Kalimantan:	Rejang Kayan	*pakat*	root
	Murik	*pakat*	root
	Baluy Kayan	*pakat*	root
Madagascar:			
	Mérina	*fáhany*†	root
		*paka	
Bismarck Archipelago:			
	Tolai	*paka*	*Ficus nodosa* Teijsm and Binn., *F. paka* Peekel
	Pala	*paka*	
		*mpaka	
Fiji:	Mbau	*mbaka*	*Ficus obliqua* Forst.f.°
Vanuatu:			
	Erakor	*mbak*	*Ficus prolixa* Forst.f.
	Mota	*paka*	*Ficus prolixa* Forst.f.
	Uripiv	*nu-mbək*	*Ficus prolixa* Forst.f.
	Kwamera	*nə-pək*	*Ficus prolixa* Forst.f.
	Mataso	*na-baŋ*‡	*Ficus prolixa* Forst.f.
	Aneityum	*paŋ*‡	*Ficus prolixa* Forst.f.

Key: † irreg. *-ny* for *-tra* ‡ nasalization of the *k, ° bark used for cloth

on Peekel (1984: 142, the author uses the term Kuanua for Tolai) for the Bismarck Archipelago, Parham (1972: 138) for Fiji, and Guillaumin (1932: 101) for Vanuatu. The *Ficus obliqua* in the Fijis was once regarded as sacred (Parham 1972: 138), and the *F. prolixa* in Vanuatu still is, at least in many places (Ralph Regenvanu, p.c.).

A final *-t is not lost in some languages of Vanuatu, e.g.:'*(Sa)ʔəpat 'four' > Mota *vat*, Erakor *pat*, Uripiv *vij*, Eromanga *de-vat*' (Ray 1926: 201, 433, 261, 172). Therefore, if the forms are cognate with the Kalimantan forms for 'root', the loss of the final *-t in the Vanuatu reflexes would mean that these were borrowings, perhaps from Fiji or another of the Central Pacific languages, which regularly drop final consonants.

The protoform is perhaps a doublet of another one, hitherto reconstructed on the basis of reflexes of languages not further east than the Moluccas: *uakat 'vine, aerial root' (Blust 1973: #300 *waka[Ct]), first reconstructed by Stresemann (1927: 66, 128) for his 'Proto-Ambon' (corresponding with modifications to modern Proto-Central-Maluku, see Collins 1983a: 12–22), with the meaning 'mangrove'.

The association of root or aerial root with a mangrove tree, an inhabitant of the tidal zone that is bare at ebb and immersed at flood, is suggested by the tree's standing on a cluster of stilt roots, exposed supporting roots lifting the bottom end of the trunk above ground and flood level. The species in question is identified by de Clercq (1909: #2968) as *Rhizophora conjugata* L. (see also Dunnebier 1951: under *uakat*). We thus have an interesting parallelism of doublet forms for 'root', each being associated with a respective tree species, distinguished from others in having aerial roots. Cognates of *uakat apparently also occur in languages of Melanesia and Micronesia (see Table 8.7). Occasional seemingly cognate forms in languages of the Southern Philippines reflect a doublet *ua(ŋ)kət, e.g.:

Maranao	*oaket*	vine
West Bukidnon Manobo	*waŋkət*	to entwine or entangle (of vine, fibres).

All the protoforms for 'root' reconstructed above can, however, be seen as doublets of *uakaR 'root' (Dempwolff 1938: 164 under *vaka[ɭ]), which has a much wider distribution area, though it does not include Taiwan.

The existence of two so similar protoform doublets of course makes it difficult to assign reflexes to the one or the other of them in languages that have dropped the final *-R as well as *-t, as e.g. Selayar *aka*, Maori *waka* 'root'. With regard to Kusaie *oak-oak*, the assignment to *uakat above takes account of the meaning, being associated with mangroves, and of a competing form, Kusaie *okah* root, reflecting *uakaR (cf. Marshallese *okar* 'root').

None of the protoforms considered above appears to represent an original Austronesian protoform for 'benjamin tree'. Whereas *qaRaʔ and *qayaoa have very limited distribution areas, *(m)paka apparently derived from a form for 'root', and *nunuk, having the greatest distribution area, perhaps derived

Table 8.7 Cognates of *uakat

Language	**uakat*	*Meaning*
Philippines:		
Aklanon	*wakat*	roots of mangrove tree –
Cebuano	*wákat*	intertwined
Bikol	*wákat*	roots of mangrove tree
Kankanai	*wákat*	vine
Hanunoo	*wákat*	vine
Sulawesi:		
Mongondou	*uakat*	root
	uakat-an	*Rhizophora conjugata* L.
Gorontalo	*uaato*	root
Buol	*wakato*	root
	wakat-ono	stilt roots
Limboto	*wu-waʔato*	root
Kaidipan	*wakato*	root
Atinggola	*wu-waʔato*	root
Kalimantan:		
Ma'anyan	*wakat*	root
Taboyan	*wakat*	root
Paku	*wakat*	root
Madagascar:		
Sakaláva	*váhatse*	root
Bára	*váhany*†	root
Tsimihéty	*váhatra*	root
Sumatra:		
Lampung	*wakat*	root
Central and South Maluku:		
Asilulu	*waʔat-e*	*Rhizophora conjugata* L.
Rumahkai	*waʔat*	root
Kayali	*wāt*	*Rhizophora conjugata* L.
Kamarian	*waat*	*Rhizophora conjugata* L.
Piru	*waat*	*Rhizophora conjugata* L.
Hatusua	*aat*	*Rhizophora conjugata* L.
Fordata	*waʔat*	mangrove tree
Irian Jaya:		
Biga	*kawato*‡	root
Amber	*kawak*‡	root
North Papua New Guinea:		
Onank	*ʌwʌts*‡	root
Musom	*kwats*‡	root
Dangal	*ka-kwats*‡	root
Micronesia:		
Kusaie	*oak-oak*	mangrove root

Key: † irreg. -ny for -tra, ‡ metathesis *u/*k

Table 8.8 Cognates of *uakaR

Language	**uakaR 'root'*
Ibanag	*uacag*†
Malay	*akar*
Penihing	*akah*
Mori	*haka*
Kai	*waʔar*
Yamdena	*wakar*
Haruku	*waʔar-elle*
Onin	*wakir*
Tolai	*okor*
Woleai	*wegar*
Banoni	*(b)aɣara*
Mota	*ɣar-iw*

Key: † vine

from a word for 'shadow'. Not all reflexes of these protoforms refer to a benjamin or some other sacred benjamin-type species of *Ficus*.

The mythological data also offer at first glance a rather contradictory picture of the antiquity of the sacred benjamin cult. On the one hand, the relatively pure manifestation of the cult, in which the tree mediates between upper and lower worlds, among the Tsou in Taiwan suggests a southeast China origin from the very homeland of the Austronesians. On the other hand, reflections of the cult, often in an advanced stage of erosion, are very unevenly distributed.

The data as a whole may be interpretated as indicating that the Proto-Austronesians were not yet megalithist. They may have already held the shade of the benjamin tree as a hallowed place, but had apparently not yet developed the ideological concept of the sacred tree as mediator between upper-world divine bird and lower-world divine serpent. The first wave of the Austronesian migration (supposing there were more than one) could not yet have been the propagator of these features from the moment of departure, but must either have developed them on the way, thereby also contributing them as substratum to subsequent movements of Austronesians, or acquired the features as adstratum on being caught up by a following wave.

It is unclear whether the original Lapita culture in the Bismarck Archipelago was associated with megalithism and knew the bird–tree–serpent cosmogonic complex. If not, this would suggest a subsequent introduction into Oceania. This could then perhaps have been the same movement that brought reflexes of *p[aə]DaHu 'ship for long-distance navigation' (Mahdi, Volume III, Ch. 5). This would explain the coincidence of megalithism (particularly in the northeast), the sacred banyan wishing tree cult and reflexes of *p[aə]DaHu in India. The beginnings of the sacred tree cult in Taiwan must in any case have been of such antiquity that influence from India must be excluded.

Original and intrusive sacred tree cults in India

The original Indian sacred tree cult also involved a large *Ficus* species, the peepul, or bo, or bodhi tree (*Ficus religiosa* L., Sanskrit *aśvattha*, Hindi *pippala*). It is extolled in the Rigveda as the seat of the gods in the third (that is the highest) heaven, and was already depicted as an object of worship on an early clay tablet from Mohenjo-daro (Bosch 1960: 65–69; Gupta 1971: 50–52; see also Parpola 1994: 229). However, the sacred peepul tree does not mediate between an upper- and a lower-world, but itself serves as a model of the social hierarchy rooting from Brahma; it thus does not channel the fertilization act between the personifiers of the upper and lower worlds. Unlike the banyan and the benjamin, the peepul tree has no aerial roots and therefore does not form a complex trunk cluster, nor does it develop an expansive shade-giving crown, and could not have acquired its sacred status in the same way as the benjamin tree apparently did in Austronesia. The wood of the peepul was apparently used in the kindling of fires, thus bringing it into connection with the fire god Agni and the hearth fire (see Bosch 1960: 68–74; Gupta 1971: 53–55). It is clear from the above that the original Indic sacred tree tradition could not have inspired the Southeast Asian version of the *kalpavṛkṣa*. This concept perhaps appeared at the threshold of state formation in regions in which the Austronesian tree cult was known, as a result of the ruler's taking in the role of the upper-world god by extending gifts to his profane subjects through the sacred tree.

The banyan tree (*Ficus benghalensis* L. = *F. indica* L., Sanskrit *nyagrodha*), resembling the benjamin tree of Southeast Asia in also having aerial roots, is likewise already depicted in Harappan tablets, not as an object of worship, but as a graph of the Harappan script. It was apparently used to write the phoneme sequence *vaṭa* (Proto-Dravidian *vaṭa(m) ~ *vaṭi 'rope, cord', *vaṭa-mara(m) 'banyan tree', literally rope tree in allusion to the rope-like aerial roots), particularly to write the morph *vaṭa* 'north' in *vaṭa-mīṇ* 'north star' (Parpola 1994: 241–242). The mythologically relevant denotate to be expressed in writing with the help of the banyan symbol was thus not the banyan tree, but the north star.

The banyan is not mentioned anywhere in original Vedic literature as a religiously exceptional tree. Records of ritual or religious traditions involving the banyan as a sacred tree apparently do not date from earlier than the third century BC. In reports and rumours about the gigantic banyan tree reaching Europe as a result of the conquests of Alexander the Great, and reflected in works of Theophrastus and Pliny the Elder (see Noehden 1827: 121–124), among others, in some of which the tree is described quite accurately, no mention seems as yet to be made of its being worshipped or having any ritual function.

The banyan seems to have started to acquire a sacred aura during the last half of the first millennium BC, a process that had apparently not yet been registered by the commentators in Alexander's entourage. In this newer cult, in which the tree performs the same fecundatory function as the benjamin tree in Southeast Asia, it is also for the first time referred to in India in Sanskrit as *kalpavṛkṣa*, literally 'wishing tree'. Its long hanging roots are depicted as dropping gold pieces in such abundance that collecting vessels placed underneath overflow. The failure of early indologists to discover the usage of Sanskrit *kalpavṛkṣa* is perhaps a further indication that it was initially an 'unofficial' feature of Indian religion. Perhaps the cult was originally limited mainly to non-Indo-Aryan ethnic groups, for which reason it found no reflection in Hindu literature and escaped the attention of the Greeks.

An immediate association between banyan tree and *Nāga*s is provided by the *Mahāvānija-jātaka*, in a fable about greedy merchants: resting under a banyan tree, they cut its eastern branches to quench their thirst. The tree produced clear and limpid water in surplus. They cut the southern branches, and the tree produced meat, rice and other dishes. They cut the western branches, and beautiful women appeared; they cut the northern branches, and the tree provided them with jewels, precious metals and cloths. Their greed unsatiated, they decided to get everything out of the tree by cutting it down. The *Nāga* king, incensed with wrath, ordered an army of *Nāga*s to exterminate them (Vogel 1926: 139–140).

A particular role in the consolidation in India of the sacred banyan cult is played by Buddhism, with which the *Nāga*s were closely associated (see Vogel 1926: 93–96). The attainment of *nirvāṇa* by the Buddha proceeded in three sessions of meditation: the first under a peepul or bodhi tree, the second under a banyan tree, and the third under what was called the tree of the *Nāga*, because in this last session Muchalinda the theriomorphic *Nāga* king (a giant cobra) protects the meditating Buddha with his/its hood from a storm (Vogel 1926: 56–58, Gupta 1971: 102). It is apparently not by coincidence that the most prominent stone representation of the *kalpavṛkṣa* in India is a Buddhist sculpture, that from Besnagar of the second century BC (see Gupta 1971: plate X).

The *Nāga*s are known to have played an active role in the propagation of Buddhism in India and Sri Lanka. The emergence of Buddhism (and Jainism) had in common with that of Taoism, Islam and Protestantism that it took place at a time of social turmoil due to what was felt by an industrious or enterprising part of the third estate (particularly its professional and mercantile élite) – or by a commercially committed or for other reasons discontented part of the nobility or gentry – as the oppressive or stifling rule of a decadent aristocracy and clergy (the second and first estates respectively). This may have been the reason why India-based Austronesians – either through their role in seafaring and trade, or the perhaps handicapped position of their ruling élite within the Hindu caste hierarchy – tended to be attracted to Buddhism. In Indonesia, a similar polarization took place in the Late Middle Ages, when coastal cities involved in maritime trade adopted Islam in their opposition to a decadent land-bound feudal monarchy.

Apparently, both regions had developed independent cults of a sacred *Ficus* tree, involving the benjamin in Southeast Asia, and the peepul in India. The

West Austronesian cult from Southeast Asia was apparently introduced to India during the last half millennium BC, the benjamin being replaced by the banyan which likewise has aerial roots. *Nāga* involvement in the propagation of Buddhism was possibly one of the factors leading to the incorporation of the sacred banyan into Indo-Aryan religion. The adoption of the term *kalpavṛkṣa* in Sanskrit, seemingly a calque from Malay *pɔhɔn bəriŋin* benjamin tree, 'wishing tree', suggests contacts of Malay speakers with India since around the second century BC, the date of the Besnagar sculpture. As Austronesians were apparently already active on the subcontinent since much earlier times, the cult itself may have been introduced some centuries earlier, e.g. by megalithist or urn-burial practising Austronesians in south India, without this having any immediate repercussions in Indo-Aryan culture. After it was introduced into Indo-Aryan religious life, the Sanskrit term for it was re-exported to Southeast Asia where, as noted above, it appears at around 400 AD on inscription *C* of King Mūlavarman.

We can now return to the observation of de Casparis on Sanskrit terms for things that essentially did not originate from India. Enough evidence exists in support of de Casparis' observation, so that its validity is not dependent on the situation in the present concrete instance. Although the *kalpavṛkṣa* evidently existed in India not later than the second century BC, I am not aware of instances of its use as a ritual medium through which the king exercises the dispension of divine providence to his subjects. If that were the case, the Southeast Asian use of the term and ritual merely copies an Indian precedent, even though the underlying sacred tree cult itself was of Southeast Asian origin. If no examples can be found for India, then the Southeast Asian ritual indeed qualifies as an additional instance of use of a Sanskrit term to refer to something only superficially resembling its designate in India.

The Southeast Asian sacred benjamin tree cult inspired ornamental designs, e.g. in architecture, carving, textiles, tatooing etc., and the same can be said about both original sacred peepul and introduced sacred banyan cults in India. Having become locally established, traditions of ornamental design reflecting the intrusive sacred banyan cult in India underwent further development, in many ways surpassing the original Southeast Asian correspondents in fine quality. Together with other influences from India to Southeast Asia, there was also the influence in ornament design, including those inspired by the original and by the borrowed Indian tree cult traditions (see e.g. Stutterheim 1926b; Maxwell 1991), adding further spirals to the process of mutual influencing. Analysis of the so-called 'tree of heaven' or 'tree of life' motif in Indonesian ornamental design therefore requires particular care in distinguishing between indigenous, imported and re-imported elements.

TRADE IN SPICES AND AROMATICS

The clove trade

Also unanswered is the dating of earliest Malay sailings to India. The first Austronesians could have sailed there as early as the second millennium BC, and regular two-way maritime communication between Austronesian colonies on the Indian coast and West Malayo-Indonesia probably began at some time between 1000 and 600 BC. With regard to activities of the Malays, we have an indirect clue suggesting first sailings in the second century BC based on the assumed calque of Sanskrit *kalpavṛkṣa* from Malay *pɔhɔn bəriŋin*.

The *Rāmāyaṇa* provides another important piece of evidence, in containing the earliest instance of the use of Sanskrit *lavaṅga* 'clove' (Monier-Williams 1899: 898; Gonda 1932: 326–329). The clove (*Eugenia aromatica* Kuntze = *E. caryophyllata* Thunb. = *Syzygium aromaticum* L.) originally grew exclusively on some north Maluku islands (Burkill 1935: 961, see Figure 8.4). Any mention of it outside that region implies transportation from north Maluku. Assuming that the passage in the *Rāmāyaṇa* had already been established by the first century BC, this points to the second century BC as a probable date for transportation of the spice to India, as it would have taken some time before an entirely new word would find its way into a work of *kāvya* literature. Furthermore, the word probably entered Sanskrit by way of a Dravidian language, thus introducing additional delay.

Van der Tuuk (1901: 720) noted that the Sanskrit word must have been cognate with Toba Batak *labaŋ* 'nail', and Gonda (1932: 328) has likewise assumed an Indonesian origin of the Sanskrit word. In a more detailed study, I found that borrowed reflexes of Malay *buŋa-lawaŋ* (< **BuŋaH* 'flower' + **laBaŋ* 'nail', i.e. the 'nail flower') occur in languages extending from Sumatra to the Maluku Islands as a word for 'clove' (Mahdi 1994a: 188–189).

Reflexes in languages of West and Central Malayo-Indonesia answer to an effective *buŋalawaŋ. Malay as the donor language can be identified quite confidently on the basis of known sound laws. Phonologically regular reflexes of the original *BuŋaH + *laBaŋ, which underlies the above effective protoform, are attested in only two languages:

Old Javanese	*wuŋa-lawaŋ*	clove
Malay	*buŋa-lawaŋ*	mace (of nutmeg)

Malay and Old Javanese both reflect medial *B between two *a-s as *w*, whereas other possible donors have *b*. On the other hand, only Javanese reflects initial *B as *w* too, whereas Malay reflects it as *b*. The present meaning of the Malay reflex is not the original one, which obviously must have been 'clove'. The word has been replaced in this meaning by *cəŋkeh*, a Chinese loan. It probably entered Malay via Chinese settlers in West Malayo-Indonesia, who began playing an important role in interinsular trade with Maluku during the Ming period (Ptak 1992).

Reflexes in languages of East Central Maluku appear to reflect a **pugalawan* as protoform (see Mahdi 1994a: 189). The reflection of initial *b as *p, and of final *ŋ as *n in Proto East Central Maluku (PECM) is regular, whereas the reflection of medial *ŋ as *g is not. As Collins (1983b: 360–361) noted, the same irregular reflection was reported by Stresemann (1918: 157) in a single case of obvious borrowing: 'Sanskrit *siṁha* > (Old) Malay *siŋa* "lion" > as-if PECM *siga "cat".' As these are the only reported instances of the sound shift, which is not attested in later borrowings into East Central Maluku languages, the corresponding sound law must have been operative only during a short period. The word for clove in East Central Maluku was presumably borrowed at roughly the same time as the word for cat from a Sanskrit-Malay word for lion; in addition, the borrowing probably took place in the earliest period of borrowing of Sanskrit words via Malay.

The employment of a borrowing from Malay as a word for clove over the entire length of the transportation route of the clove from East to West Malayo-Indonesia suggests that Malay must have played a key role as language in the transportation of cloves within Malayo-Indonesia. It seems reasonable to assume therefore, that Malay and not Javanese was the donor language that contributed the word to a language of India.

With regard to Sanskrit *lavaṅga* 'clove' it is doubtful that Malay final *-ŋ* would have been reflected as Sanskrit *-ṅga*, and not *-ṁ* in case of a direct borrowing. At least, in the Indic Pallava-based so-called Old Sumatran script used in Old Malay epigraphy, final Old Malay *-ŋ* was spelled with the symbol for *-ṁ* (the *anusvāra*). For Tamil, on the other hand, we have evidence for the first mentioned sound correspondence. In the Tanjore inscription dedicated to the successful military expedition of King Rajendrachola against Malay Šrī Vijaya, the name of one of the captured Malayan cities is given as *Yiruḍiṅga(m)*. Wheatley (1961: 71) pointed out that this must be the same town as that referred to in the *Zhūfānzhì* of Zhao Rugua as *Rìluótíng* (see Hirth and Rockhill 1911: 62). Colless (1989), meanwhile, has identified it as the town of Sating-pra, in which the first component evidently reflects the original Malay name which may have been something like *Cərətiŋ. It seems probable, therefore, that the Malay word for clove was first borrowed into Tamil or some other Dravidian language, and only from there into Sanskrit, cf. Tamil *(i) lavaṅkam*, Kannaḍa *lavaṁga*, Malayalam *lavaṁgam*.

In postulating that Malay-speaking seafarers were sailing to India in the first millennium BC, Solheim (1980: 334) took for granted that the seafarers spoke Malay. This would have been a reasonable assumption in the second half of the first millennium AD, but there is no evidence for the validity of the assumption a millennium earlier. After all, not only does the earliest written record of Malay and the first mention of the language itself (as the *Kūnlún* language in Chinese sources) date from the last third of the seventh century AD, but even the name *Malayu* is attested for the first time in Chinese transcription only in sources of the Tang period reporting an embassy from that country in 644 AD. In other words, we have evidence of neither the existence of the language nor even the name Malay before the seventh century AD. The linguistic evidence presented above confirms that the language spoken by at least a substantial part of the seafarers involved in the clove trade between the Maluku and India was Malay. Assuming that Sanskrit *lavaṅga* found its way into the *Rāmāyaṇa* in the first century BC, and that at least a century must have passed between the introduction of the word into the subcontinent and its becoming a sufficiently established element of literary Sanskrit to appear in a *kāvya* work, we arrive at a dating of the second century BC for Malay-speaking seafarers in India. This date compares well with that of the probable calque of Malay *pɔhɔn bəriŋin* as Sanskrit *kalpavṛkṣa*, and there are other indications for the correctness of the date.

Burkill (1935: 961) called attention to a passage in the records of the Earlier or Western Han Dynasty (206 BC–24 AD), according to which courtiers had to place cloves in their mouth when speaking to the emperor. It is not clear how soon this custom was introduced under the Western Han, but it was apparently relatively early. This implies that the Malayo-Indonesian trade and export of cloves had also been making the clove available in China since approximately the second century BC. Significant in this respect is also the indication in the 'History of the Earlier Han' (*Qiánhànshū*) that first Chinese contacts with countries of South and Southeast Asia were made during the reign of Emperor Wu (140–86 BC) (see Pelliot 1912: 458; Ferrand 1919: 451–455, 45–46; Wang 1958: 19–20), and that Chinese envoys did not use their own vessels, but: '*Mán Yí gǔ chuán chuǎn sòng zhì zhī*' (South and East-Barbarian merchant ships take turns to convey [them] on [each successive lap of their] way) (Pelliot 1912: 458). To this may be added the fact that the Chinese term used for large ships for high-sea navigation was *Kūnlún bó*, literally 'Malay ship'. According to Christie (1957), this is the etymon of the term *kolandiophōnta* used in the Periplus for ships sailing between the Indian east coast, the Ganges and *Chrysē̃*, 'Gold-land', i.e. the Malay world, and which were said to be the biggest of all (*tà mégista*).

Whereas from Austronesian and Indian linguistic evidence we can trace the route of Malay-speaking traders carrying cloves to India early enough to be mentioned in *kāvya* literature, the Han records provide evidence of comparable time-depth for the existence of the same clove trade, adding furthermore that adequate shipping facilities of the 'barbarians' existed at that time. The existence of these ships for long-distance navigation is, in itself, evidence of maritime trade, without which there would not have been any incentive for developing and building the ships. Mention of approximately contemporaneous sea transport is also to be found in Indian literature. Thus, the great compendium of Buddhist tales, the Pali *Jātakaṭṭhavaṇṇanā*, of which at least the core of the corpus already existed in the last centuries BC, and a good deal of the material of which is apparently of even earlier date, presents among others the picture of an established pattern of trade relations between certain ports in India and *Suvaṇṇabhūmi* (Wheatley 1983: 265). This is the Sanskrit *Suvarṇabhūmi*, the Gold-land, roughly coinciding with the land of the Malays (in Sumatra and the Malayan Peninsula), the *Chrysē̃s* of the Periplus.

Lime, camphor and an Egyptian mummy

In an interdisciplinary investigation of the Egyptian mummy PUM II, dated by the radiocarbon method at 170 ± 70 BC (Cockburn *et al.* 1980: 67), a mass-spectroscopic analysis was made of the polymerized resin filling (Coughlin, cited in Cockburn *et al.* 1980: 57, 62). One of the constituents was identified as camphor, the chief component of the oil of the camphor laurel (*Cinnamomum camphora* Nees and Eb.), in pre-industrial times a product mainly of China and Japan. The analysts overlooked the circumstance that the technology of extracting this camphor from chips and twigs of the camphor laurel by steam distillation was probably not yet known in the early second century BC. It was developed as a result of efforts to find a cheaper substitute for the extremely expensive camphor of Baros, obtained from the *Dryobalanops aromatica* Gaertn. which grows in Sumatra, the Malayan Penunsula and Kalimantan. Even centuries after the discovery of the camphor substitute, however, it had not gained much significance on the markets of the Near East as a substitute for the some forty times more expensive camphor of Baros (see Grasmann 1895: 310; Burkill 1935: 546). The persistent belief in the latter's medicinal powers prevented its replacement by the cheaper alternative until the last half millennium.

The chief component of camphor of Baros is the chemical compound borneol. It is not only very close in molecular structure to camphor, but can be transformed into the latter by a relatively simple chemical reaction, oxidation. The process of ageing, which the mummy experienced, is basically also oxidation, so that borneol, had it been originally present, would have been converted into camphor by the time the analysts inspected it. The analysts could not have detected anything else. Therefore the presence of camphor in the mummy, which could not yet have been used in the mummification in the second century BC, must be seen as an unambiguous indication of the use of camphor of Baros (borneol) in Egypt at that time (Mahdi 1994a: 190–191). In other words, camphor of Baros was already being transported to the Near East in the second century BC.

The polity of Barus in the north of the west coast of Sumatra, from which camphor of Baros got its name, was known since antiquity as the source of the product (see van Vuuren 1908; Drakard 1989), and appears in Ptolemy's *Geōgraphikḗ Hyphḗgēsis* (VII.2.28) as the *Baroûsa* Islands. Prakrit and Pali *kappūra* 'camphor (of Baros)', and Sanskrit *karpūra*, which is perhaps a backformation from the former, apparently derive from Malay *kapur* 'lime (chalk)', *kapur barus* 'camphor of Baros'. The ultimate etymon is Proto-Austronesian *qapuR 'lime' (Dempwolff: 1938: 16 under *'apuɣ) (see Table 8.9.) But the word for camphor reflects a secondary doublet with initial *k- (Dempwolff 1938: 75 under *kapuɣ), represented in the east not further than Maluku (e.g. Piru), and in the north only as far as Sarawak (e.g. Kelabit) (see Table 8.10).

Table 8.9 Cognates of *qapuR

Language	**qapuR 'lime'*
Amis	*qapul*
Ifugao	*ʔapul*
W. Bukidnon	*ʔapug*
Ratahan	*apu*
Simalur	*aul*
Lampung	*hapuy*
Old Javanese	*apū*
Balinese	*hapuh*
Bintulu	*apu*
Lawangan	*apuy*
Muna	*ɣefi*
Selayar	*aporo*
Sikka	*apur*
Solor	*apu*
Kai	*y-afur*
Fordata	*y-afur*
Kayali	*ahul*†
Onin	*y-afur*
Lenkau	*kop*
Ponam	*af*
Kiribati	*au-a*
Mekeo	*apu*
Sa'a	*s-ɛhu*
Tonga	*n-avu*‡

Key: † k- < *q-, ‡ to treat (the hair) with lime

Table 8.10 Cognates of *kapuR

Language	**kapuR*
Cham	*kapu*
Karo	*kapur*
Malay	*kapur*
Kelabit	*kapor*
Ma'anyan	*kapuy*
Sa'dan	*kapuʔ*
Sumba	*kāpu*
Piru	*kapul-e*

The reflection of final *-R as *-y* in Ma'anyan, and as *-ø* in Sumba, indicates that the forms are not borrowings from Malay. There also exist cognates with the meaning 'lime' in some East Austro-Asiatic languages of Indochina (see Table 8.11).

The restricted distribution areas of both the East Austro-Asiatic and the Western Austronesian cognate protoforms make borrowing in any of the two directions imaginable. However, considering that the Western Austronesian

Table 8.11 Cognates of *ka(m)pur

Language	**ka(m)pur 'lime'*	
Mon	*gǎpɣw*	
Kha	*pun*	
Khmer	*kǎmpor*	
My-son Muong	*pol*	
Vietnamese	*voγ*$_{A1}$	< pre-Vietnamese *Kpol†

Key: † *K is any unvoiced obstruent.

form is a doublet of an older Austronesian form represented in Taiwan (e.g. Amis) as well as in Oceania (e.g. Kiribati, Tonga, etc.), a borrowing from Austronesian into Austro-Asiatic appears likelier. Proof is provided by the Ma'anyan reflex, which has regular *-y* for final *-R. Even in earliest loans from Austro-Asiatic (through Malay), Ma'anyan already reflects the protophoneme as r, as for example in: Ma'anyan *karɛwaw* carabao < **kɛrɛwaw* < pre-Ma'anyan **kərəbaw* < < Malay *kər(ə)baw* < *kəRbau. The protoform, in which the *R (as opposed to *r) is certified by Lampung *kibaw* and Javanese *kəbo*, is a loan from East Austro-Asiatic (cf. Samre *krapao*, Kancho *krəbao*, see Mahdi 1994a: 200). The early date of the Ma'anyan borrowing is indicated by the reflection of Malay *ə* and *b* as Ma'anyan *ɛ* and *w* respectively (overshadowed by the rather recent shift of any vowel in the antepenultimate to *a*). In most loans from Malay, being of later date, the regular reflexes in Ma'anyan are *a* and *b* respectively, which would have led in this example to ***karabaw* instead of the observed *karɛ waw*, cf. Ma'anyan *karasik* 'sand' < Malay *kər(ə)sik* < *kəRsik. Thus, even in the very early instance of borrowing, Maanyan reflects the *R as *r* rather than as *y* (in final position, or ø elsewhere) as in original, not loaned forms.

The Austronesian etymology of Sanskrit *karpūra* 'camphor' was first proposed by Schoff (1922: 362–363), who suggested that the word was first borrowed into Prakrit or Pali as *kappūra*, from which the Sanskrit cognate is a back formation. On the other hand, Mayrhofer (1953–56: 175) considered the word to be probably of Austrosiatic origin, citing Khmer *kāpōr*, Mon *khapuiw*, and a form in Cham (which is an Austronesian language), as did Turner (1966: #2880). However, I have failed to find the cited forms with the meaning 'camphor' in Khmer and Mon dictionaries, but found them instead (in alternative transcription) with the meaning 'lime, chalk', as in Table 8.11 above. Cognates with either meaning are not listed in the Old Mon dictionary of Shorto (1971), whereas earliest (pre-Angkorian) Old Khmer has only *karpura*, which must be a Sanskrit borrowing (Jenner 1981: 17). The Old Khmer dictionary of Pou (1993: 85–86), which also includes later periods, has *kaṁpur* ~ *kaṁpor* 'chalk, lime' (but not camphor). We must therefore assume that cognate forms in Mon-Khmer are relatively late borrowings, and even then only as an expression for 'lime', not for 'camphor', so that the Indic cognates must derive from Malay.

Oriental tradition places the origin of camphor at the Sumatran port of Barus, and this is reflected in its name. It seems significant that the term for it in languages in the close vicinity of the port (see de Clercq 1909: #1174) contains either authentic, that is to say non-borrowed, reflexes of *kapuR, or very early loans that do not copy the Malay term. Cognates in languages spoken further away from Barus generally show sound correspondences that are regular for borrowings from Malay, and usually copy the Malay expression (see Table 8.12).

Sanskrit *karpūra* is first attested in the *Suśruta-saṁhita* and in the *Pañcatantra* (Monier-Williams 1899: 258); the former is also indicated as its earliest source in Turner (1966: #2880). The medical treatise of Suśruta probably dates from the last centuries BC, although the present version is apparently from the seventh century AD (Richardson 1982: 824). The lost original of Vidyāpati's collection of fables, the *Pañcatantra*, is considered to date from between 100 BC and 500 AD. Assuming that camphor of Baros was already mentioned in the earliest versions of the two sources, the dates would agree well with the transportation of the aromatic to Egypt at the beginning of the second century BC. It is perhaps noteworthy that another medical treatise, the *Caraka-saṁhita*, apparently a little older than the *Suśruta-saṁhita*, does not mention *karpūra* yet. Although the surviving version dates from the first century AD, there were older versions (Richardson 1982: 824). The 'land of camphor', *Karpūradvīpa*, first appears only in the compendium of tales *Kathāsaritsāgara* (Monier-Williams 1899: 258), compiled by Somadeva in the eleventh century AD.

For dating the first sailings to India of spice traders using Malay as principal language of communication in the second century BC, we thus have as the main evidence the etymology and distribution of the word for clove, supported

Table 8.12a Expressions for 'Camphor of Baros'

Near Barus:	*camphor of Baros*	*literal meaning*	*expected orig.*†	*for loan*‡
Nias	*fombõra gafu*	lime grain	***(h)afu*	***kafu*
Toba Batak	*hayu hapur*	lime wood	*hapur*	*hapur*
Malay	*kapur barus*	lime of Barus	kapur	

Table 8.12b Expressions for 'Camphor of Baros'

Far from Barus:	*camphor of Baros*	*expected orig.*†	*for loan*‡
Acheh	*kaphõ° barõh*	***kapõ*	***kapõ*
Old Javanese	*kapur (barus)*	***kapū*	*kapur*
Sundanese	*kapur barus*	***kapi?*	*kapur*
Madurese	*kapor bhʌrus*	*kapor*	*kapor*
Balinese	*kapur barus*	***kapuh*	*kapur*
Makassarese	*kapuru? barusu?*	*kapuru?*	*kapuru?*
Buginese	*kaporo baroso*	***kapo?*	*kaporo*

Key: † reflex of *kapuR, ‡ from Malay, ° influenced by Arabic *kāfūr*

by evidence from the camphor trade and the etymology of the word for camphor, 'camphor of Baros', and by the possible calque of the word for 'wishing tree' from Malay. This is flanked by information from Chinese sources that indicates access to the clove from around the second century BC, and availability of local Southeast Asian means of transportation at about the same time. It is remarkable, in this context, that the earliest iron as well as the first appearance of onyx beads in archaeological sites along the spice-trade route between Maluku and the Strait of Malacca has also been radiocarbon dated to the second century BC, e.g. in the Tabon Caves site in Palawan (Fox 1967: 13–14, see also Stargardt 1979: 20) and in north Maluku (Peter Bellwood, p.c.). The Malay-speaking seafarers who played a major role in the westward transportation of spice along this route apparently also occasioned the oppositely directed flow of goods from India, specifically onyx beads.

AUSTRONESIANS AND THE EASTWARD SPREAD OF HINDU-BUDDHIST CULTURE

Tamilization of Austronesian seafarers across the Bay of Bengal

The picture we arrived at above is one of major involvement and even principal responsibility of Austronesians in general since some time between 1000 and 600 BC, and of Malays in particular since the second century BC in maritime communication between India and Malayo-Indonesia. This now raises the question of the role of Austronesians, including Malays, in the transfer of Hindu-Buddhist culture to mainland and insular Southeast Asia. It seems likely that the role of the Austronesians, in which I shall here include linguistically Dravidianized and religiously Indianized peoples of Austronesian origin, was a very important one.

Austronesians in India, particularly on the Coromandel coast, had apparently been providing sea communication between the two regions for many centuries already, before elements of Hindu-Buddhist ideology began appearing in Southeast Asia. In a study on the introduction of cereal cultigens into Western Austronesia, I came to the conclusion that the transfer of sorghum from India into West Malayo-Indonesia must have taken place at about the same time as foxtail millet from China through the Philippines, fanning out in Central Indonesia to East and West Indonesia, i.e. at a time initially estimated at between 1500 and 700 BC (Mahdi 1994b: 431–434) but now corrected on the basis of more precise data to 1000 and 600 BC (see below). The reason was that the protoform for 'foxtail millet', *[bB]ətəŋ, and its secondary doublet, *bətəm, were distributed from Taiwan over the Philippines and Sulawesi all over East Indonesia as far as Cendrawasih Bay in Irian Jaya, but were not represented in West Malayo-Indonesia. On the other hand, there is a distribution of reflexes of the as-if protoform *zawaʔ, derived from Pali *java* (Sanskrit *yava*) 'barley'. As Yule and Burnell (1903: under *jowaur*) indicated, derived words of the same origin exist in languages of India, meaning 'sorghum'. The reflexes of *zawaʔ can be divided semantically into three groups: those meaning 'sorghum' are restricted to West Malayo-Indonesia, those meaning 'grain', 'ear of grain', occur in the Philippines, and those meaning 'foxtail millet' are spread over West and Central Indonesia, and the Philippines.

Apparently, when foxtail millet was introduced to West Indonesia from the Philippine-Sulawesi region, sorghum from India, called by the borrowed name of *zawaʔ, was already established. Instead of taking up the form *[bB]ətəŋ for 'foxtail millet', the meaning of already established *zawaʔ was generalized to mean 'grain'. Foxtail millet must have tended to replace sorghum as the main staple, so that the further spread of sorghum over the rest of the Archipelago was halted, whereas in West Indonesia itself *zawaʔ, in its generalized meaning of 'grain', came to refer to foxtail millet more frequently than to sorghum. Intensified Malay sailings through Central Indonesia and the Philippines to China in a roundabout route to avoid the Funan-controlled South China Sea from the third or fourth century AD onwards caused the form to be carried into the Philippines, but with the meaning of 'grain' or 'foxtail millet', and no longer 'sorghum'. The find of a grain of foxtail millet in Timor, dated at almost 1000 BC, and some other considerations lead to dating the propagation of foxtail millet through the Archipelago between 1000 and 600 BC (correcting a previous estimate of 1500–700 BC in Mahdi 1994b: 431–434). One implication of this dating is that occasional adoption of isolated elements of Indian culture in Malayo-Indonesia, even together with the Indian word for it, already took place many centuries before the beginning of massive transfer of Hindu-Buddhist ideology.

The beginning of this last-mentioned transfer, which was to play a decisive role in the formation of the later picture of Southeast Asia, is generally placed (quite correctly) in the first century AD. The question is, why not earlier, in view of the dates for regular Austronesian navigation in the Bay of Bengal elicited above? Two areas in India, it has been found, played important roles as places of immediate origin of the culture influence: Kalinga, located at present-day Orissa, and particularly the Coromandel coast which soon came under the rule of the Pallavas. The latter region, together with the hinterland behind it, is the core land of the Tamils. It is not a coincidence that the beginning of the Indian culture transfer to Southeast Asia, having the Tamil lands on the Coromandel coast as a principal point of departure, coincides with the defeat of the *Nāgas* by the Tamils as reflected in early Tamil literary tradition.

If the subjugation of the *Nāgas* by the Tamils resulted in the cultural and linguistic Tamilization of the Austronesians of the Coromandel, this would mean that the seafarers keeping up maritime communication across the Bay of Bengal, having originally been of a basically Austronesian culture, now converted to a Tamil one. Not a change in the frequency of sea communication between India and Southeast Asia, or the appearance of a new ethnic group on the scene, but a cultural change in the seafarers themselves, who had since centuries been traversing these seas, apparently started the cultural transfer.

In a contribution of 1932, further developed later in Sastri (1949), the author had called attention to an early Tamil inscription at Takuapa on the Malayan Peninsula. The inscription dates from the ninth century, however, which is on the late side for our purposes here. Other Tamil inscriptions have been found on the Peninsula and in Sumatra, but of comparable or even later dating (see Boeles 1966 and bibliography therein). The Tamil *Manimekelai* located *Cāvakam*, 'land of Malays', immediately to the east of the islands of cannibal *Nāga*s on the western approaches to the Strait of Malacca. The source continues, however, that the capital of Cāvakam was called *Nāgapuram*, literally 'Naga-burg, Naga City'. Most remarkable, however, is the information that the language spoken there was Tamil (Kanakasabhai 1904: 11). As I shall show elsewhere (Mahdi n.d.), the term *Yava/Jawa*, and consequently *Jāvaka/Cāvaka*, did not necessarily point to the Malay realm as a whole, but could also refer to any Malay or Malay-ruled country in West Malayo-Indonesia. The *Manimekelai* is probably referring here to a Tamilized Coromandel-Austronesian (*Nāga*) trading outpost, depot or emporium on the Malayan Peninsula.

That *Nāga*s apparently played a role in the introduction of state-legitimizing Hindu-Buddhist ideology appears to be borne out by widespread *Nāga* legends in Southeast Asia, particularly such involving marriage to a *Nāgī* princess at the beginning of royal pedigrees (see Cœdès 1911: 391–393; Przyluski 1925; Porée-Maspero 1950).[11]

In the *Pañcatantra*, there is a symmetrically complementary tale of a (human) maiden who wedded a snake (see Vogel 1926: 174–175), which may be associated with the Malay Peninsula: the wife of a Brahmin who had been promised the handsomest son in the land gave birth to a snake as an only child. When the snake attained maturity, the father embarked on a long journey in search of a bride for it, and found a beautiful maid in *Kukkuṭanagara*, which is said to be situated in a remote country (there was a wedding, the snake transformed into a handsome young man, and they lived happily ever after). Interestingly, Ptolemy in his Geography places in the Golden Chersonese, i.e. the Malayan Peninsula, a town he named *Kokkonágara* (7.2.25; see Cœdès 1910: 60) which is a corruption of a Sanskrit name. There is no such name known that would fit here more adequately than *Kukkuṭanagara*.

Kalinga (*Kaliṅga*) presents more of a problem than the Coromandel. Lévi (1923) says that the name is neither Indo-Aryan nor Dravidian, but of indigenous origin. It is doubtless related with Malay *Kəliŋ*, Javanese *Kliŋ*, which had come to be used until relatively recent times in the meaning of 'Indian' in general, and Tamil in particular (Dutch nineteenth and early twentieth century writers with Indonesian experience coined from this the term *Klingaleesch* 'Klingalese' with reference to the Indian immigrant population). In Malay of the New and Modern historical periods, a distinction was made between Hindu-Tamils, *Kəliŋ*s, and Muslim-Tamils, *Culia*s (see McPherson 1990: 44 n. 2).

The loss of final *-ga* would have been very unusual for Malay as well as for Javanese. On the other hand, the accretion of *-ga* in the Sanskrit form is easily accounted for by assuming the very likely borrowing of the word into Sanskrit via Tamil or some other Dravidian language. 'Klings', as the peoples of the country may be called, were apparently neither Indo-Aryans nor Dravidians.

There appear to be indications in *kāvya* literature associating Kalinga with *Nāgas*, but the exact connection remains unclear. Banerji (1930: 16) explains that the rulers of Vishṇupur belong to the *Nāgavaṁśa* line, which is apparently not of Rajput but of 'aboriginal' (*sic*) descent. The author expressed the opinion that a consideration of all available data tends to prove that most chiefs of Kalinga are of indigenous descent. Whether this could also include Austronesians I cannot judge.

One piece of evidence may possibly imply involvement of Austronesians in Kalinga. The Old Javanese Kaladi inscription dated 831 Šaka (909 AD) contains on faces 7a-b a list of foreign traders by ethnic group of origin. In this, the inscription, which in the surviving copper-plate copy unfortunately shows numerous spelling mistakes, makes the following distinctions:

> *lawan saŋ banigrāma ityaiwamadi tan tumana irikang śima muaŋ surā niŋ kilalān [wārgga kilalān] kli [kliŋ] arja [āryya] siŋhal [siŋhala] drawila [drawiḍa] banyaga 4 paṇḍikir [paṇḍikira] campa rammān [rmən] kismmira [kmir] . . .*
>
> (Barrett Jones 1984: 186)
>
> and the traders who are not allowed to enter the freehold, [subject to the consequences of being] recipients of provision: Klings, Aryans, Sinhalese, Dravidians, merchants of the four Paṇḍikira, Chams, Mons, Khmers . . .

Besides Indo-Aryans and Dravidians, and also Sinhalese who are on their own island, the text distinguishes traders from the Paṇḍikira territories (on the Malabar coast) and Klings. The Old Javanese inscription is possibly distinguishing merchants of the Malabar coast and Klings from Indo-Aryans and Dravidians. Considering that the Mundas are not known to have been active as seafaring traders at any time, it is unlikely that Munda traders were meant here. Perhaps the Paṇḍikirans and Klings were still felt to be Austronesian, or at least noticeably different from authentic Dravidians and Indo-Aryans.

Early dissemination of Sanskrit words by Malays

In spite of the important role that Dravidianized or Tamilized Coromandel Austronesians played in the transfer of Hindu-Buddhist ideology to Malayo-Indonesia, a key role in its dissemination within the Archipelago appears to have been played by Malays. There are numerous examples of Sanskrit borrowings, in which phonological or semantic deviations incurred upon assimilation into Malay are found again in cognate forms throughout the Archipelago (Mahdi 1994b: 483–484 n. 199; Adelaar 1994: 55, 63–64) (see Table 8.13).

Subsequently, some of the other language communities in the Archipelago developed their own direct relationship to Hindu-Buddhist culture and thus also to Sanskrit. This is particularly true for Old Javanese and Balinese, for which parallel borrowings of the same Sanskrit etyma are attested, which do

Table 8.13 Malay borrowings from Sanskrit and their cognate forms

Language	*Word*	*Meaning*	*Language*	*Word*	*Alteration*
Sanskrit	***upavāsa***	**'fast (not eat)'** >	Malay	*puasa*	(*u-* >ø; *-av-* > *-əw-* > *-u-*)
			Toba and Karo	*puasa*	
			Lampung	*puasa*	
			Sundanese	*puasaʔ*	
			Javanese	*p(u) ɔsɔ*	
			Madurese	*puwasa(h)*	
			Balinese	*puasʌ*	
			Makassarese	*puasaʔ*	
			Buginese	*puasaʔ*	
			Tausug	*puasa*	
			Cebuano	*puʔása*	'fast (not eat)'
Sanskrit	***cukra***	**'vinegar'** >	Malay	*cuka*	(*r* > ø
			Sundanese	*cukaʔ*	
			Javanese	*cokaʔ*	
			Madurese	*cokka(h)*	
			Balinese	*cukʌ*	
			Buginese	*cuka*	
			Maranao	*sokaʔ*	
			Tausug	*sukaʔ*	
			Cebuano	*sukaʔ*	
			Tagalog	*sukaʔ*	
Sanskrit	***jāgṛ***	**'wake', *jāgara* 'be awake, on guard'** >	Malay	*jaga*	(*r* > ø)
			Toba and Karo	*jaga*	
			Lampung	*jaga*	
			Sundanese	*jagaʔ*	
			Makassarese	*jaga*	
			Buginese	*jaga*	
			Tiruray	*diyaga*	
Sanskrit	***janma***	**'creature'** >	Malay	*jəlma* 'manifestation'	(*a* > ə; *n* > *l*)
			Toba	*jolma*	
			Lampung	*jəlma*	
			Sundanese	*jələmaʔ*	
			Madurese	*jhʌləma(h)*	
			Balinese	*jəl(ə)mʌ*	'person'
Sanskrit	***koṭi***	**'10,000,000'** >	Malay	*kəti* '100,000'	(*o* > ə)
			Sundanese	*kətiʔ*	
			Javanese	*kəṭi*	
			Balinese	*kəṭi*	
			Madurese	*kəṭṭɛ(h)*	
			Makassarese	*katti*	
			Buginese	*kətti*	
			Tiruray	*kati*	
			Maranao	*kati*	'100,000'

not exhibit the same phonological deviations. They thus demonstrate that the phonological conditions leading to the deviations in Malay loans were not also compelling for borrowing into other languages of the Archipelago. Compare the two languages in Table 8.14. The cognates that exhibit the same deviations as in Malay are thus loans from Malay.

The role of Malays in disseminating Hindu-Buddhist culture and ideology in the Archipelago may also help to clarify some unsolved problems of the Muara Kaman epigraphy of King Mūlawarman of *c.*400 AD, the oldest extant inscriptions in Malayo-Indonesia, all in Sanskrit. Inscriptions B and G indicate that the priests at Muara Kaman were from abroad:

> B. *yūpo-yam | kṛto viprair-ihāgataiḥ*
> 'this *yūpa* | was made by the priests who hither have come'
>
> G. *yūpo-yaṁ sth[āpito] viprair-nnānā . . . ih-ā[gataiḥ]*
> 'this *yūpa* was er[ected] by the priests (who from) different . . . hither [have come]'
>
> (Chhabra 1965: 86, 91)

[A vertical line indicates change of strophe, three dots represent an illegible part of the text, whereas uncertainly read ones are in square brackets.]

The most straightforward interpretation would have been that the priests had come from India. There were two circumstances, however, that require explanation. The first is numerous spelling and grammatical mistakes in all the texts. The most unpardonable mistake for a priest was the word *yaṣṭvā* in line 10 of inscription *A* – a non-existent form of the (irregular) verb *yaj* 'offer, sacrifice, donate'. The correct form fitting in the place would have been *iṣṭvā* 'having offered'. The second circumstance is that the priests very carefully but persistently avoided identifying themselves as *Brāhmaṇas*, representing the only *varṇa* whose members were allowed to practise the office of a priest. Most frequently, i.e. in inscriptions *B*, *C*, *E* and *G*, they refer to themselves by the word *vipra* 'the wise one, sage', and in two, *B* and *C*, as *dvijāti* 'the twice-born one' (with reference to rebirth after a former life), an expression normally referring to Aryans, but probably extended abroad to all bearers of the true (i.e. Hindu) faith and civilization. The expression *dvijendra*, used only once, in inscription *A*, comes the nearest to identifying the priests as *Brāhmaṇas*. The term, composed of *dvija* 'twice-born' and *indra*

Table 8.14 Parallel borrowings of Old Javanese and Balinese

Language	*Borrowing*	*Meaning*	*Language*	*Borrowing*	*Meaning*
Old Javanese	*upawāsa*		Balinese	*upawasʌ*	'fast (not eat)'
Old Javanese	*jagra*	'wake, guard'			
Old Javanese	*janma*	'reincarnation, person'	Balinese	*jadmʌ* ~ *janmʌ*	'person'
Old Javanese	*koṭi*	'100,000' (beside *kəṭi*)			

'the god Indra, supreme king, lord', thus seems to mean 'foremost among the twice-born', the foremost *varṇa* being that of the *Brāhmaṇas* (compare *dvijeśvara* 'a *Brāhmaṇa*', with *īśvara* 'lord, ruler' as second component). Nevertheless, it is still a circumscription that may technically also be understood literally as 'foremost among those who exercised the true faith', which is exactly what the priests at Muara Kaman claimed to be.

One explanation could be that the priests, though not *Brāhmaṇas*, were nevertheless Indian merchants. It is possible that they did not master the sacred language. By the beginning of the first millennium AD, living Indo-Aryan languages had already diverged considerably from Sanskrit. However, in the light of the linguistic evidence and also the coincidence of the date of the inscriptions (*c*.400 AD) with the beginning of intensive Malay sailings to China via the Strait of Makassar (on which Muara Kaman is situated) and the Philippines (third–fourth centuries AD, Mahdi 1994a: 187–188), the simplest explanation for the language mistakes and the restraint in the reference to themselves would be that the priests who had come hither from other parts were Malays.

The Old Malay word for 'shipmaster' was *puhawaŋ*, a word of Austronesian provenance (< *-pu 'elder, master' + *qaBaŋ 'boat'), showing regular Malay reflexion of *q as *h*, and of *B between two *a-s as *w*. The word for 'merchant', *baniaga*, on the other hand, was of Indic origin. This suggests a picture of Indic merchants travelling on Malay ships. The merchants could, of course, have been Tamilized India-based Austronesians, who would refer to their occupation by an Indic word. It is not unlikely, however, that, after the subjugation of Kalinga by King Aśoka, Indic merchants too were crossing the Bay of Bengal on Malay or other Austronesian ships. Nevertheless, Indian knowledge of Malayo-Indonesian geography, as e.g. reflected in the *Mahāniddesa* (a part of the Buddhist Pali Canon), the *Kathāsaritsāgara*, and other contemporary works, seems to be limited to the Malayan Peninsula (*Suvarṇakuḍya, Takkola, Tambraliṅga, Kadaraṁ*) and Sumatra (*Yavadvīpa, Yavakoṭi, Karpūradvīpa, Barusa*) or both as a whole (*Suvarṇabhūmi, Suvarṇadvīpa, Dvīpāntara*). Before the seventh century AD, merchants from India and countries to the west do not seem to have come further into insular Southeast Asia. I therefore consider it unlikely that Indian merchants could have been at Muara Kaman by around 400 AD.

Of course, replacing Indian colonists by Malays does not make the model of foreign newcomers 'planting the seed of civilization among the savages' any more plausible. It must have been the shift of the Malay trade route to China from the South China Sea to the Strait of Makassar (between Kalimantan and Sulawesi) as a result of Funan maritime control over the former, leading to involvement of peoples inhabiting the coasts facing the new seaways into the trade activities (the east coast of Kalimantan soon became well known in China as a source of expensive luxury articles, including camphor of Baros, eaglewood, etc.), that brought about profound changes in the economical, social, cultural and political life in the area.

CONCLUSIONS

With regard to Austronesian activity in India and Sri Lanka, there now appears to be more data available than has probably been realized. The main difficulty in identifying indications of Austronesian activity in South Asia is that its effect on the culture of the recipient region dates to an earlier period, and represents a less sophisticated level of cultural development than reciprocal culture influence on Southeast Asia. Consequently, features of an early culture horizon in Southeast Asia, having been exported to South Asia and then receiving a distinct Indic further development, tend to become superimposed over the original feature upon being reimported to Southeast Asia. This creates the impression that the feature is Indic in origin, and allows remnants of earliest manifestations to be reinterpreted as adaptations to local, less sophisticated levels of development. Examples of superimposition of the reimported version are the serpent cult and the tree cult, but these should probably be seen as the most visible tip of an iceberg.

From the material assembled above, it appears reasonable to assume four main periods in the activity of Austronesians in India and Sri Lanka up to the seventh century AD. The first, representing a gradual expansion along the coastline (and up major river arteries) of relatively small groups at still relatively early stages of culture–economic development, probably using relatively simple rafts and double canoes, cannot yet be dated more precisely than that it preceded the second period.

Beginning at some time between 1000 and 600 BC was the movement of more advanced groups with larger and seaworthier rafts, double canoes and single outrigger boats capable of long-distance navigation. The same craft apparently reached Micronesia around 1000 BC, and were probably finally displaced from Malayo-Indonesia at around 600 BC (when the eastward diffusion of boat-burial began, followed soon afterwards by the outriggerless, keeled, plank-hull boat). Some of the Austronesian groups involved probably practised megalithism, but manifestations in south India since around 1000 BC must be principally assigned to activity of non-Vedic Aryans, whereas the extent of Austronesian contribution, particularly in the early phase, is unclear. The period saw the beginning of regular maritime communication between India and Malayo-Indonesia, bringing sorghum to west Indonesia at about the same time as the dispersal of foxtail millet, attested in Timor shortly after 1000 BC. Sometime during this period, in any case before 450–400 BC, the custom of urn burial was apparently brought in from Southeast Asia to some places in South Asia, particularly Sri Lanka, from where it then spread into south India.

In the third period, beginning around 200 BC, sea communication between the Malay lands and Maluku made the clove available in West Malayo-Indonesia, from where it could be carried on to India by already established communication routes, as also to China. Malay speakers played an increasing role on the westward routes as well, evidenced by the terms for clove and camphor, and perhaps also wishing tree, in Sanskrit. This is probably also the period of increased exchange between Semitic and Austronesian seafarers.

The fourth period, beginning in the first century AD, is the period of the transmission of Hindu-Buddhist state-ideological concepts from India to mainland and insular Southeast Asia, in which the already established Austronesian sea routes play a decisive role. The process is perhaps due on the one hand to a rapid Tamilization of the Austronesians in India as a result of final political subjugation by the Tamils, and on the other hand to the advancement of politico-economic consolidation in some regions of Southeast Asia to a level that made the involved communities more receptive to sophisticated religious ideologies from India. Of particular importance was perhaps the establishment of principles of allegiance and loyalty of local rulers to the paramount king, who was likened to *Rāma* of the *Rāmāyaṇa* or *Yudhiṣṭira* of the *Mahābhārata*. If formerly only the superior prowess of the overlord kept his vassals in check, loyalty was now a virtue rather than a sign of inferiority, whereas challenging the overlord was immoral and disruptive to the cosmic order (except, of course, in case of success). The role of Indian epics here can be compared with the role of legends of King Arthur and the knights of the Round Table in the canonization of principles of knightly chivalry in West Europe. And just as all but one of the chivalrous knights are deprived of the reward of beholding the Holy Grail, in token of a realization that adversities of life and the all but too human qualities that enable the heroes to cope with them also distance these from ideals of the pure faith, so also are all but one of the *Pāṇḍavas* (the 'good guys' of the *Mahābhārata*) refused access to *Mahāmeru*. The oldest direct evidence of identification of a ruler in Indonesia with an epic overlord is inscription G of King Mūlavarman (around 400 AD), in which the latter is likened to *Yudhiṣṭira*.

The involvement of Indicized Austronesian colonies in the propagation of Indic spiritual culture to peoples of their former homeland was probably also an important factor behind the Hindu-Buddhistic peaceful cultural conquest of Southeast Asia, explaining also why it triumphed in the face of competing Chinese culture influence. Hindu-Buddhist culture practically entered through the backdoor along the trade-routes, carried by Austronesian seafarers into Austronesian emporiums. The propagation of Islam through the Archipelago about one millennium later took place not much differently, except that locally settled foreign traders (particularly Chinese) apparently played a greater role.

SOURCES

The chapter makes use of a large number of dictionaries of individual languages, only some of which have been explicitly referred to in this text. The others are referred to separately in Mahdi (1994a: 196–199 n. 1), to which add: Sundermann (1905) for Nias, and Meier (1906) for Papitalai.

Further, I used collections of wordlists, particularly Ferrell (1969) for Taiwan, Reid (1971) for the Philippines, Ray (1913) for Kalimantan with East Malaysia, Vérin *et al.* (1969) for Malagasy dialects, Lee (1966) for Austronesian Indochina, the Holle lists of Stokhoff (1980–87) for Indonesia as a whole, Smits and Voorhoeve (1992) for Irian Jaya, Dempwolff (1905) for the Admiralties and the Bismarck Archipelago, Hooley (1971) for the Morobe District and north Papua New Guinea, Capell (1943) for southeastern Papua New Guinea, Cochran (1983) for Milne Bay languages of southeast Papua New Guinea, Cashmore (1969) for Meso-Melanesian and Central–East Oceanic languages of Papua New Guinea, the Solomons, Vanuatu and Central Pacific, Ray (1926) for the Solomons, Vanuatu, Loyalties and New Caledonia, Lynch (1983) for Vanuatu, and Haudricourt (1971) for New Caledonia.

NOTES

1 Acknowledgements: I sincerely thank the World Archaeological Congress for a generous grant permitting me to attend its Third Congress in New Delhi, 4–11 December 1994, where I presented a preliminary version of this chapter, and to Roger Blench for his continued support. Particular thanks are due to Matthew Spriggs who opened my eyes to several important points of Oceanic archaeology, and helped me with some difficult botanical identifications. This chapter has also profited considerably from instructive and constructive comments from Gerd Mevissen, Lars Martin Fosse, David Zorc, Peter Bellwood, Lotika Varadarajan and S. Nagaraju. I am indebted to Aone van Engelenhoven, Ralph Regenvanu and Cristophe Sand for sacrificing their time to provide me with copious material on the folklore and ethnology of Leti, Vanuatu and New Caledonia respectively. I am grateful to the Department of Physical Chemistry of the Fritz-Haber-Institut, led by Gerhard Ertl, where I am employed, for the generous use of institute facilities in my independent linguistic studies. The presentation of this chapter nevertheless remains outside the responsibility of the Institute. Michael Wesemann proved indispensable in helping me with various software packages, particularly in preparing the figures and maps. I also owe gratitude to Marian Oort of the library of the Instituut Kern in Leiden, who unbureaucratically permitted me to use the library which was closed whilst moving to a new address, half of the books from the shelves being already packed in containers (I urgently needed the *Mahābhārata* and *Rāmāyaṇa* concordances and indexes). Thanks are also due to library staff members of the Institut für Indische Philologie und Kunstgeschichte of the Freie Universität Berlin for repeatedly letting me in outside opening hours (which coincided with my own work hours) and helping me locate publications.

2 With the inadequate tʃ and dʒ replaced by c and j respectively, and IPA j for palatal glide replaced by y. Furthermore, ʃ is replaced by š, and retroflex articulation is generally indicated by a dot under the main symbol.

3 The principal particularities are that i and y both represent /i/, o is read /u/, whereas ng, dr and tr are digraphs, respectively spelling /ŋ/, and a voiced and unvoiced retroflex affricate somewhat resembling /j/ and /c/.

4 This uses the digraphs dj and tj for the d and t with *cedille* (the tail under French ç) of some other sources.

5 For Vietnamese I use a phonemic transcription, with the toneme indicated as subscript in the notation of Fang-kuei Li, from which that of Chang Kun and Herbert Purnell differs only in that the B and C basic tones are interchanged.

6 Reviewed in Blust (1978b). Successive steps in the development of the transcription, departing from that of Dempwolff (1934) by a one-to-one replacement of symbols and a reallocation of reflexes defining *q and *h (the latter presently *S) in Dyen (1947a), are documented in Dyen (1947b, 1951, 1953, 1962 and 1965).

7 Whereas 'Neolithic', 'metal age', etc., serve to characterize a community in its material (economic, technological) culture development, megaliths and related artefacts are features of spiritual culture, testifying primarily, in spite of their impressive size, not to a community's technological capacities, but to its religious beliefs. To avoid the impression of its inclusion in one category with 'Paleolithic', 'Neolithic', 'metal age', and to accentuate its essentially more ideological than technological implications, I have added the suffix '-ism' to the term.

8 The Chinese transcription suggests that the immediate etymon may have been something like **majə*, indicating that knowledge of the fruit was transmitted by speakers of a language that shifted final *a* to ə, such as some early languages of southeast Kalimantan (e.g. Proto-Ngaju or Proto-East-Barito).

9 The actual ethnic situation would have been more complex. The non-Vedic Aryans, being the first Aryans on the scene, first encountered Dravidian peoples, from whose language they apparently made numerous borrowings, and perhaps also met Munda peoples. The Austronesians too must have encountered and partly intermingled with Mundas and Dravidians.

10 See also Stresemann (1927: 48), Mills (1981: #106) and Verheijen (1984: #6.57). Mills (1981) cites #2518 in David Zorc's 1971 unpublished Proto-Philippine Finder List.

11 Against the background of the sun/bird–tree–serpent cosmology, of course, marriage to a serpent princess identifies the founder of the dynasty with the sun prince.

REFERENCES

Adelaar, K.A. 1991. Some notes on the origin of Sri Lanka Malay. In *Papers in Austronesian Linguistics No. 1*, H. Steinhauer (ed.), 23–37 Pacific Linguistics A-81. Canberra: Australian National University.

Adelaar, K.A. 1994. Malay and Javanese loanwords in Malagasy, Tagalog and Siraya (Formosa), *Bijdragen tot de taal-, land- en volkenkunde* 150, 50–65.

Aichele, W. 1928. Oudjavaansche bijdragen tot de geschiedenis van den wenschboom, *Djåwå* 8, 28–40.

Alpers, A. 1970. *Legends of the South Sea: the world of the Polynesians seen through their myths and legends, poetry and art*. London: John Murray.

Andaya, L.Y. 1993. Cultural state formation in eastern Indonesia. In *Southeast Asia in the Early Modern Era: trade, power, and belief*, A. Reid (ed.), 23–41. Ithaca/London: Cornell University Press.

Barrett Jones A.M. 1984. *Early Tenth Century Java from the Inscriptions: a study of economic, social and administrative conditions in the first quarter of the century*. Verhandelingen van het Koninklijk Instituut voor Taal-, Land- en Volkenkunde 107. Dordrecht/Cinnaminson: Foris.

Bayard, D.T. 1972. Early Thai Bronze: analysis and new dates. *Science* 176, 1411–1412.

Bellwood, P. 1980. Indonesia, the Philippines and Oceanic prehistory. *Journal de la Société des Océanistes* 36, 148–155.

Bellwood, P. 1985. *Prehistory of the Indo-Malaysian Archipelago*. Orlando, Fla.: Academic Press.

Blagden, C.O. 1918. Kalpavṛkṣa, *Bijdragen to de taal-, land- en volkenkunde van Nederlandsch Indië* 74, 615–617.

Blust, R.A. 1972. Proto-Oceanic addenda with cognates in non-Oceanic Austronesian languages: a preliminary list, *University of Hawaii Working Papers in Linguistics* 4(1), 1–43.

Blust, R.A. 1973. Additions to 'Proto-Austronesian addenda' and 'Proto-Oceanic addenda with cognates in non-Oceanic Austronesian languages', *University of Hawaii Working Papers in Linguistics* 5(3), 33–61.

Blust, R.A. 1978a. *The Proto-Oceanic Palatals*, Memoir 43. Wellington: Polynesian Society.

Blust, R.A. 1978b. *Review of Current Trends in Linguistics 8: linguistics in Oceania* (ed. by Thomas A. Sebeok 1971. Mouton, The Hague/Paris), *Language* 54, 467–480.

Blust, R.A. 1983–84. Austronesian etymologies – II. *Oceanic Linguistics* 22–23, 29–149.

Blust, R.A. 1984–85. The Austronesian homeland: a linguistic perspective. *Asian Perspectives* 26, 45–67.

Blust, R.A. 1986. Austronesian etymologies – III. *Oceanic Linguistics* 25, 1–123.

Boeles, J.J. 1966. A note on Tamil relations with south Thailand and the identification of Ptolemy's Tacola, *Journal of the Siam Society* 54, 221–230.

Bonhomme, T. and J. J. Craib, 1987. Radiocarbon dates from Unai Bapot, Saipan – implications for the prehistory of the Mariana Islands. *Journal of the Polynesian Society* 96, 95–106.

Bosch, F.D.K. 1960. *The Golden Germ: an introduction to Indian symbolism*, Indo-Iranian Monographs 2. 's-Gravenhage: Mouton.

[Brockhaus] 1990. *Brockhaus Enzyklopädie*, 19. Aufl. Bd. 13. Mannheim: Brockhaus.

Buitenen, J.A.B. van (transl. and ed.) 1973. *The Mahābhārata, 1. The Book of the Beginning*. Chicago/London: University of Chicago Press.

Burkill, I.H. 1935. *A Dictionary of the Economic Products of the Malay Peninsula*, with contributions by W. Birtwistle, F.W. Foxworthy, J.B. Scrivenor and J.G. Watson. London: Crown Agents for the Colonies.

Capell, A. 1941. *A New Fijian Dictionary*. Sydney: Australasian Medical Press.

Capell, A. 1943. The linguistic position of South-Eastern Papua. Ph.D. dissertation, University of London.

Carrol, V. and T. Soulik, 1973. *Nukuoro Lexicon*. Honolulu: University of Hawaii Press.

Cashmore, C. 1969. Some Proto-Eastern Oceanic reconstructions with reflexes in Southeast Solomon Islands languages. *Oceanic Linguistics* 8, 1–25.

Casparis, J.G. de 1956. *Prasasti Indonesia, II. Selected inscriptions from the 7th to the 9th Century A.D.*. Bandung: Masa Baru.

Casparis, J.G. de 1986. Some notes on the oldest inscriptions of Indonesia. In *A Man of Indonesian Letters: essays in honour of Professor A. Teeuw*, C.M.S. Hellwig and S.O. Robson (eds) 242–256. Verhandelingen van het Koninklijk Instituut voor Taal-, Land- en Volkenkunde 121. Dordrecht/Cinnaminson: Foris.

Chang, K. 1969. *Fengpitou, Tapenkeng and the Prehistory of Taiwan*. Yale University Publications in Anthropology 73. Department of Anthropology, New Haven: Yale University Press.

Chang, K. 1970. The beginnings of agriculture in the Far East. *Antiquity* 44, 175–185.

Chang, K. 1978. *The Archaeology of Ancient China*. third ed. rev. and enlarged. New Haven/London: Yale University Press.

Chang, T. 1976. The rice cultures. In *The Early History of Agriculture: A discussion organized jointly for the Royal Society and the British Academy*, J. Hutchinson, G. Clark, E.M. Jope and R. Riley (organizers), 143–157. Philosophical Transactions of the Royal Society of London B 275. London: Royal Society.

Charoenwongsa, P. and D. Bayard, 1983. Non Chai: new dates on metalworking and trade from northeast Thailand, *Current Anthropology* 24, 521–523.

Chhabra, B.Ch. 1945. Three more Yūpa inscriptions of King Mūlavarman from Koetei (East Borneo). *Journal of the Greater India Society* 12, 14–17.

Chhabra, B.Ch. 1965. *Expansion of Indo-Aryan Culture During Pallava Rule (as evidenced by inscriptions)*. New Delhi: Munshi Ram Manohar Lal.

Chowning, A. 1991. Proto Oceanic Culture: the evidence from Melanesia. In *Currents in Pacific Linguistics: papers on Austronesian languages and ethnolinguistics in honour of George W. Grace* R. Blust (ed.), 43–75. Pacific Linguistics C-117. Canberra: Australian National University.

Christie, A. 1957. An obscure passage from the Periplus: Κολανδιοφωτα τα μὲγιστα. *Bulletin of the School of Oriental and African Studies* 19, 345–353.

Churchward, C.M. 1940, *Rotuman Grammar and Dictionary*. Sydney: Australasian Medical.

Clercq, F.S.A. de, 1909. *Nieuw plantkundig woordenboek voor Nederlandsch Indië*. Amsterdam: de Bussy.

Cochran, A. 1983. A comparative study of Milne Bay phonology. In *Second International Conference on Austronesian Linguistics: proceedings*, S.A. Wurm and L. Carrington (eds), pp. 851–866. Pacific Linguistics C-61. Canberra: Australian National University.

Cockburn, A., R.A. Barraco, W.H. Peck and T.A. Reyman 1980. A classic mummy: PUM II. In *Mummies, Disease, and Ancient Cultures*, A. Cockburn and E. Cockburn (eds), 52–70. Cambridge: Cambridge University Press.

Cœdès, G. 1910. *Textes d'auteurs grecs et latins relatifs à l'Extrême-Orient depui le IVe siècle av. J.-C. jusqu'au XIVe siècle*. Paris: Leroux (facsim. repr. 1977. Hildesheim/New York: Olms.

Cœdès, G. 1911. Études cambodgiennes. *Bulletin de l'École Française de l'Extrême-Orient* 11: 391–406.

Cœdès, G. 1918. Le royaume de Çrīvijaya. *Bulletin de l'École Française d'Extrême-Orient* 18(6), 1–36.

Colani, M. 1937. Anciennes irrigations et bassins dans le Do-linh (Quảng-trị). *Cahiers de l'École Française d'Extrême-Orient* 10, 18–19.

Colless, B. 1989. Satingpra in Sung Dynasty Records. *Archipel* 37, 31–42.

Collins, J.T. 1983a. *The Historical Relationships of the Languages of Central Maluku, Indonesia,* Pacific Linguistics D-47. Canberra: Australian National University.

Collins, J.T. 1983b. A note on cultural vocabulary in the Moluccan Islands. *Bijdragen tot de taal-, land- en volkenkunde* 139, 357–363.

Coolsma, S. 1930. *Soendasch-Hollandsch woordenboek*, 3e druck. Leiden: Sijthoff.

Dahl, O.C. 1977. *Proto-Austronesian*. 2nd rev. ed. Scandinavian Institute of Asian Studies Monograph Series 15. Lund: Curzon.

[Davies, John] 1851. *A Tahitian and English Dictionary with introductory remarks on the Polynesian language and a short grammar of the Tahitian dialect: with an appendix*. Tahiti: London Missionary Society (facsim. repr. 1991. Haere Po no Tahiti, Papeete).

Deacon, A.B. 1934. *Malekula: a vanishing people in the New Hebrides*. C.H. Wedgewood (ed.), with a preface by A.C. Haddon. London: Routledge.

Dempwolff, O. 1905. Beiträge zur Kenntnis der Sprachen von Deutsch-Neuguinea. *Mitteilungen des Seminars für Orientalische Sprachen an der Königlichen Friedrich-Wilhelm-Universität zu Berlin* 8, 182–254.

Dempwolff, O. 1934. *Vergleichende Lautlehre des austronesischen Wortschatzes, 1. Band, Induktiver Aufbau einer indonesischen Ursprache*. Beihefte zur Zeitschrift für Eingeborenen-Sprachen 15. Berlin: Reimer (Vohsen)/Hamburg: Friedrichsen & De Gruyter (facsim. repr. 1969. Nendeln: Kraus).

Dempwolff, O. 1938. *Vergleichende Lautlehre des austronesischen Wortschatzes, 3. Band. Austronesisches Wörterverzeichnis*, Beihefte zur Zeitschrift für Eingeborenen-Sprachen 19. Berlin: Reimer (Vohsen)/Hamburg: Friedrichsen & De Gruyter, (facsim. repr. 1969. Nendeln: Kraus).

Dordillon, R.I. 1931. *Grammaire et Dictionaire de la Langue des Iles Marquises: Marquisien-Français*. Travaux de l'Institut d'Ethnologie 18. Paris: Institut d'Ethnologie.

Drakard, J. 1989. An Indian Ocean Port: sources for the earlier history of Barus. *Archipel* 37, 53–82.

Dunnebier, W. 1951. *Bolaang Mongondousch-Nederlandsch woordenboek*. 's-Gravenhage: Nijhoff.

Dyen, I. 1947a. The Malayo-Polynesian word for 'two', *Language* 23, 50–55.
Dyen, I. 1947b. The Tagalog reflexes of Malayo-Polynesian *D. *Language* 23, 227–238.
Dyen, I. 1951. Proto-Malayo-Polynesian *Z. *Language* 27, 534–540.
Dyen, I. 1953. *The Proto-Malayo-Polynesian Laryngeals*. Baltimore: Linguistic Society of America.
Dyen, I. 1962. Some new Proto-Malayopolynesian initial phonemes. *Journal of the American Oriental Society* 82, 214–215.
Dyen, I. 1965. Formosan evidence for some new Proto-Austronesian phonemes, *Lingua* 14, 285–305.
Dyen, I. 1971. The Austronesian languages and Proto-Austronesian. In *Current Trends in Linguistics, vol. 8, Linguistics in Oceania,* T.A. Sebeok (ed.), 5–54. The Hague/Paris: Mouton.
Eringa, F.S. 1989. *Soendaas-Nederlands woordenboek*. Dordrecht/Cinnaminson: Foris.
Fergusson, J. 1873. *Tree and Serpent Worship: or illustrations of mythology and art in India in the first and fourth centuries after Christ. From the sculptures of the Buddhist topes at Sanchi and Amravati*, 2nd edn. London: India Museum/Allen.
Ferrand, G. 1908. Les îles Râmny, Lâmery, Wâķwâķ, Ķomor des géographes Arabes, et Madagascar. *Journal asiatique* 10e série 10, 433–566.
Ferrand, G. 1910. Les voyages des Javanais à Madagascar, *Journal asiatique* 10e série 15, 281–330.
Ferrand, G. 1916. Ye-tiao, Sseu-tiao et Java, *Journal asiatique* 11e série 8, 521–532.
Ferrand, G. 1919. Le K'ouen-Louen et les anciennes navigations interocéaniques dans les mers du Sud. *Journal asiatique* 11e série 13, 239–333, 431–492, 14, 5–68, 201–241.
Ferrand, G. 1934. Sayābidja. In *The Encyclopaedia of Islām*, M.Th. Houtsma *et al.* (eds) vol. 4. 200–202. Leyden: Brill/London: Luzac.
Ferrell, R. 1969. *Taiwan Aboriginal Groups: problems in cultural and linguistic classification.* Institute of Ethnology Academia Sinica Monograph 17. Taipei.
Ferrell, R. 1982. *Paiwan Dictionary*. Pacific Linguistics C-73. Canberra: Australian National University.
Finot, L. 1919. Review of Vogel (1918). *Journal asiatique* 11e série 13, 342–346.
Forbes, C.J.F.S. 1878. On the connexions of the Mons of Pegu with the Koles of central India. *Journal of the Royal Asiatic Society of Great Britain and Ireland* new series 10, 234–243.
[Foreign Languages Press] 1972. *Historical Relics Unearthed in New China/Archäologische Funde im neuen China.* Peking: Foreign Languages Press.
Fox, R.B. 1967. *Pre-History of the Philippines*. Manilla: The National Museum.
Fox, R.B. 1979. The Philippines during the first millenium B.C. In *Early South East Asia: Essays in archaeology, history and historical geography*. R.B. Smith and W. Watson (eds), 227–241. Oxford/London/New York: Oxford University Press.
Geraghty, P. 1983. *The History of the Fijian Languages*. Oceanic Linguistics Special Publication 19. Honolulu: University of Hawaii Press.
Geraghty, P. 1986. The sound system of Proto-Central-Pacific. In *FOCAL II: papers from the Fourth International Conference in Austronesian Linguistics*, P. Geraghty, L. Carrington, and S.A. Wurm (eds), 289–312, Pacific Linguistics C-94. Canberra: Australian National University.
Gianfrotta, P.A. 1977. First elements for the dating of anchor stocks. *International Journal of Nautical Archaeology and Underwater Exploration* 6, 285–292.
Giles, H.A. 1923. *The Travels of Fa-hsien (399–414 A.D.), or Record of the Buddhistic Kingdoms*. London: Routledge & Kegan Paul.
Gittinger, M.S. 1972. A study of the ship cloths of south Sumatra: their design and usage, unpublished Ph.D. dissertation, Columbia University, New York.
Gittinger, M.S. 1976. The ship textiles of south Sumatra: functions and design system. *Bijdragen tot de taal-, land- en volkenkunde* 132, 207–227.

Goeje, M.J. de 1894. De Sajâbidja. In *Feestbundel van taal-, letter-, geschied-, en aardrijkskundige bijdragen ter gelegenheid van zijn tachtigsten geboortedag aan Dr. P.J. Veth oud-hoogleraar door eeinige vrienden en oud-leerlingen aangeboden*, 10–12. Leiden: Brill.

Goeje, M.J. de 1903. *Mémoire sur les migrations des Tsiganes à traverse l'Asie*. Leiden: Brill.

Goloubew, V. 1929. L'âge du bronze au Tonkin et dans le Nord-Annam. *Bulletin de l'École Française d'Extrême-Orient* 29, 1–46.

Golson, J. 1977. No room at the top: agriculture intensification in the New Guinea Highlands. In *Sunda and Sahul: prehistoric studies in Southeast Asia, Melanesia and Australia*, J. Allen, J. Golson and R. Jones (eds), 601–638. London/New York/San Fransisco: Academic Press.

Golson, J. and P.J. Hughes 1980. The appearance of plant and animal domestication in New Guinea. *Journal de la Société des Océanistes* 36, 294–303.

Gonda, J. 1932. Etymologica. *Acta orientalia* 10, 326–335.

Gonda, J. 1952. *Sanskrit in Indonesia*, Sarasvati Vihara Series 28. Nagpur: International Academy of Indian Culture.

Gorman, C.F. 1969. Hoabinhian: a pebble-tool complex with early plant associations in Southeast Asia. *Science* 163, 671–673.

Grasmann, E. 1895. Der Kampferbaum. *Mittheilungen der Deutschen Gesellschaft für Natur- und Völkerkunde Ostasiens in Tokyo* 6, 277–315.

Gray, A. 1878. The Maldive Islands: with a vocabulary taken from François Pyrard de Laval, 1602–1607. *Journal of the Royal Asiatic Society of Great Britain and Ireland* new series 10, 173–209.

Grézel, 1878. *Dictionaire Futunien-Français avec notes grammaticales*. Paris: Maisonneuve.

Grierson, G.A. 1906. *Linguistic Survey of India, vol. 4, Muṇḍā and Dravidian Languages*. Calcutta: Office of the Superintendent of Gov.

Guillaumin, A. 1932. Contribution to the flora of the New Hebrides. *Journal of the Arnold Arboretum* 13, 81–126.

Gupta, S.M. 1971. *Plant Myths and Traditions in India*. Leiden: Brill.

Hall, K.R. 1985. *Maritime Trade and State Development in Early Southeast Asia*. Honolulu: University of Hawaii Press.

Hambruch, P. 1979. *Südsee-Märchen*. Düsseldorf/Köln: Diederichs.

Handy, E.S.C. 1923. *The Native Culture in the Marquesas*, Bernice P. Bishop Museum Bulletin 9/Bayard Dominick Expedition Publication 9. Honolulu: Bishop Museum (facsim. repr. 1971. New York: Kraus).

Haudricourt, A.G. 1971. New Caledonia and the Loyalty Islands. In *Current Trends in Linguistics, vol. 8, Linguistics in Oceania*, T.A. Sebeok (ed.), 357–396. The Hague/Paris: Mouton.

Heine-Geldern, R. 1928. Die Megalithen Südostasiens und ihre Bedeutung für die Klärung der Megalithenfrage in Europa und Polynesien. *Anthropos* 23, 276–315.

Heine-Geldern, R. 1932. Urheimat und früheste Wanderung der Austronesier. *Anthropos* 27, 543–619.

Heine-Geldern, R. 1934. Vorgeschichtliche Grundlagen der Kolonialindischen Kunst. *Wiener Beiträge zur Kunst- und Kulturgeschichte Asiens* 8, 5–40.

Heine-Geldern, R. 1952. Some problems of migration in the Pacific. *Wiener Beiträge zur Kulturgeschichte und Linguistik* 9, 313–362.

Held, G.J. 1940. Slangenfiguren in het initiatie-ritueel in de Geelvinkbaai. *Cultureel Indie* 2, 138–147.

Higham, C. 1989. *The Archaeology of Mainland Southeast Asia from 10,000 B.C. to the Fall of Angkor*. Cambridge: Cambridge University Press.

Hirth, F. and W.W. Rockhill, 1911. *Chau Ju-kua: his work on the Chinese and Arab trade in the twelfth and thirteenth centuries*. St. Petersburg: Imperial Academy of Sciences.

Hoëvell, G.W.W.C. Baron van 1890. Leti-Eilanden. *Tijdschrift voor Indische taal-, land- en volkenkunde* 33, 200–232.

Hooley, B.A. 1971. Austronesian languages of the Morobe District, Papua New Guinea. *Oceanic Linguistics* 10, 79–151.

Hornell, J. 1920. The origins and ethnological significance of Indian boat designs. *Memoirs of the Asiatic Society of Bengal* 7, 139–256.

Hoskins, J. 1988. Arts and cultures of Sumba. In *Islands and Ancestors: indigenous styles of Southeast Asia*. J.P. Barbier and D. Newton (eds), 120–137 and 284–295. München: Prestel.

Hutterer, K.L. 1983. The natural and cultural history of Southeast Asian agriculture: ecological and evolutionary considerations, *Anthropos* 78, 169–212.

Indrapala, K. 1967. Review of Paranavitana (1966). *Journal of the Ceylon Branch of the Royal Asiatic Society* new series 11, 101–106.

Jacobs, H.Th.Th.M. (ed. and transl.), 1971. *A Treatise on the Moluccas (c. 1544): Probably the preliminary version of Antonio Galvão's História das Molucas, edited, annotated translated from the Portuguese manuscript in the Archivo General de Indias, Seville.* Sources and Studies for the History of the Jesuits 3. Città di Castello (Perugia): Arti Grafiche.

Janssen, H., M. Mennis and B. Skinner (eds) 1973. *Tolai Myths of Origin*. Jacaranda Press, Milton (Queensland)/Ryde (N.S.W.)/Melbourne/Perth/West Richmond/Auckland/Port Moresby/London/Singapore/Rizal.

Jenner, P.N. 1981. *A Chrestomathy of pre-Ankorian Khmer, II. Lexicon of the Dated Inscriptions*, Southeast Asia Paper 20(2). Honolulu: University of Hawaii Center of Asian and Pacific Studies.

Jonge, N. de, and T. van Dijk 1978. Het Oud-Indonesische wereldbeeld (van scheepjesdoeken – de dingen die voorbijgaan), unpublished Ph.D. dissertation, Koninklijk Instituut voor de Tropen, Amsterdam.

Jumsai, S. 1988. *Naga: culture origins in Siam and the West Pacific*, with contributions by R. Buckminster Fuller. Singapore/Oxford/New York: Oxford University Press.

Kähler, H. 1960. *Ethnographische und linguistische Studien über die Orang darat, Orang akit, Orang laut und Orang utan im Riau-Archipel und auf den Inseln an der Ostküste von Sumatra*. Veröffentlichungen des Seminars für Indonesische und Südseesprachen der Universität Hamburg 2. Berlin: Reimer.

Kanakasabhai, V. 1904. *The Tamils Eighteen Hundred Years Ago*. Madras: South India Saiva Siddhanta Works (facsim. repr. 1989. New Delhi/Madras: Asian Educational Services; first appeared as articles in the *Madras Review*).

Karlgren, B. 1940. *Grammata serica: script and phonetics in Chinese and Sino-Japanese*. Bulletin of the Museum of Far Eastern Antiquity 12. Stockholm. [repr. 1940. Gothenburg: Elanders].

Kats, J. 1923. *Het Javaansche toneel, I.* Wajang Poerwa, [Balai Poestaka] serie 387. Weltevreden: Commissie voor de Volkslectuur.

Kern, H. 1869. Java en het goudeiland volgens de oudste berichten. *Bijdragen tot de taal-, land- en volkenkunde van Nederlandsch-Indië* 3e volgereeks 4, 638–648.

Kern, H. 1882. Over de opschriften uit Koetei in verband met de geschiedenis van het schrift in den Indischen Archipel. *Verslagen en Mededeelingen der Koninklijke Academie van Wetenschappen, Afdeeling Letterkunde*, 2e reeks 11, 182–203.

Kern, H. 1889. Taalkundige gegevens ter bepaling van het stamland der Maleisch-Polynesische volken. *Verslagen en Mededelingen der Koninklijk Akademie van Wetenschappen, Afdeeling Letterkunde* 3e reeks 6, 270–287.

Kern, H. 1894. Over de herkomst van eenige telwoorden in 't bargoensch der Zuidindische kooplieden. In *Feestbundel van taal-, letter-, geschied- en aardrijkskundige bijdragen ter gelegenheid van zijn tachtigsten geboortedag aan P.J. Veth, oud-hoogleraar, door eenige vrienden en oud-leerlingen aangeboden*, 199–200. Leiden: Brill.

Kern, H. 1896. Twee krijgstochten uit den Indischen Archipel. *Bijdragen tot de taal-, land- en volkenkunde van Nederlandsch-Indië* 46, 240–245.

Kern, H. 1897. Limoen. *Tijdschrift voor Nederlandsche taal- en Letterkunde* 16, 271–273.
Kern, H. 1916. Over den vermoedelijken oorsprong der nāgavereering. *Bijdragen tot de taal-, land- en volkenkunde van Nederlandsch-Indië* 71, 395–398.
Kim, B. 1982. A new interpretation of megalithic monuments in Korea. In *Megalithic Cultures in Asia*, Byung-mo Kim (ed.), 164–189. Seoul: Hanyang University Monograph 2. Pyung-hee Lee.
Kosambi, D.D. 1964. The autochthonous element in the Mahābhārata. *Journal of the American Oriental Society* 84, 31–44.
Kühr, E.L.M. 1896. Schetsen uit Borneo's Westerafdeeling. *Bijdragen tot de taal-, land- en volkenkunde van Nederlandsch-Indië* 46, 214–239.
Kuiper, F.J.B. 1948. *Proto-Munda Words in Sanskrit*. Verhandelingen der Koninklijke Nederlandsche Akademie van Wetenschappen, Afd. Letterkunde, nieuwe reeks 51(3). Amsterdam: Noordhollandsche Uitgevers.
Laufer, B. 1915. Asbestos and salamander. *T'oung pao* 16, 299–373.
LBSS 1976. *Kamus Umum Basa Sunda*, ed. by a dictionary commission of the Lembaga Basa & Sastra Sunda. Bandung: Tarate.
Lee, E.W. 1966. Proto-Chamic phonology and vocabulary, unpublished Ph.D. dissertation, University of Indiana. University Microfilms International #67–3690, Ann Arbor.
Legge, J. 1886. *A Record of Buddhist Kingdoms, being an Account by the Chinese Monk Fa-Hien of Travels in India and Ceylon (AD 399–414) in Search of the Buddhist Books of Discipline, translated and annotated with a Corean recension of the Chinese text*. Oxford: Clarendon (facsim. repr. 1991. New Delhi: Munshiram Manoharlal).
Lévi, S. 1918. Pour l'histoire de Rāmāyaṇa. *Journal asiatique* 11e série 11, 5–160.
Lévi, S. 1923. Pré-aryen et pré-dravidien dans l'Inde. *Journal asiatique* 203, 1–57.
Loebèr, J.A. Jnr 1911. Merkwaardige kokerversieringen uit de Zuider- en oosterafdeelingen van Borneo. *Bijdragen tot de taal-, land- en volkenkunde van Nederlandsch-Indië* 65, 40–52.
Logan, J.R. 1848. Customs common to the hill tribes bordering on Assam and those of the Indian Archipelago. *Journal of the Indian Archipelago and Eastern Asia* 2, 229–236.
Loofs, H.H.E. 1969. *Elements of the Megalithic Complex in Southeast Asia: an annotated bibliography*. Oriental Monographs Series 3. Canberra: Australian National University Press.
Luce, G.H. 1965. Rice and Religion, a study of Old Mon-Khmer evolution and culture. *Journal of the Siam Society* 53, 139–152.
Lynch, J. 1983. Proto-South Hebridean and Proto-Oceanic. In *Second International Conference on Austronesian Linguistics: Proceedings*, S.A. Wurm and L. Carrington (eds), 717–776. Pacific Linguistics C-61. Canberra: Australian National University.
McIntosh, J.R. 1985. Dating the South Indian megaliths. In *South Asian Archaeology 1983*, J. Schotsmans and M. Taddei (eds), 467–493. Papers from the 7th International Conference of the Association of South Asian Archaeologists in Western Europe, Istituto Universitario Orientale, Dipartimento di Studi Asiatici, Series Minor 23. Naples: Istituto Universitario Orientale.
McPherson, K. 1990. Chulias and Klings: indigenous trade diasporas and European penetration of the Indian Ocean littoral. In *Trade and Politics in the Indian Ocean: Historical and Contemporary Perspectives*, G. Borsa (ed.), 33–46. Proceedings of the 10th European Conference on Modern South Asian Studies, Venice 1988, vol. 10. New Delhi: Manohar.
Mahdi, W. 1988. *Morphophonologische Besonderheiten und historische Phonologie des Malagasy*. Veröffentlichungen des Seminars für Indonesische und Südseesprachen der Universität Hamburg 20. Berlin/Hamburg: Reimer.
Mahdi, W. 1994a. Some Austronesian maverick protoforms with culture-historical implications-I. *Oceanic Linguistics* 33, 167–229.

Mahdi, W. 1994b. Some Austronesian maverick protoforms with culture-historical implications-II. *Oceanic Linguistics* 33, 421–490.

Mahdi, W. 1995. Wie hießen die Malaien, bevor sie 'Malaien' hieß en? In *Südostasien und Wir: Grundsatzdiskussionen und Fachbeiträge. Tagung des Arbeitskreises Südostasien und Ozeanien Hamburg 1993*, A. Bormann, A. Graf and M. Voss (eds), 162–176. Austronesiana. Studien zum austronesischen Südostasien und Ozeanien 1. Hamburg: LIT-Verlag.

Mahdi, W. n.d. *The Hindu-Malay Yava Realm*, in preparation.

Manguin, P.-Y. 1987. Etudes sumatranaises, I. Palembang et Sriwijaya: Anciennes hypotheses, recherches nouvelles. *Bulletin de l'École Française de l'Extrême-Orient* 76, 337–401.

Manguin, P.-Y. 1993. Palembang and Sriwijaya: an early Malay harbour-city rediscovered. *Journal of the Malaysian Branch of the Royal Asiatic Society* 66(1), 23–46.

Mankad, D.R. 1965. The Kiṣkindhākāṇḍa: the fourth book of the Vālmīki Rāmāyaṇa, the national epic of India. Baroda: Maharaja Sayajirao University of Baroda Press.

Mathur, L.P. 1968. *History of the Andaman and Nicobar Islands (1756–1966)*. Delhi-6 – Jullundur City: Sterling.

Maxwell, R.J. 1991. The tree of life in Indonesian textiles: ancient iconography or imported Chinoiserie? In *Indonesian Textiles: Symposium 1985*, Gisela Völger and Karin v. Welck (eds), with a Supplement by Brigitte Khan Majlis, Ethnologica (neue Folge) 14. Cologne: Verein zur Förderung des Rautenstrauch-Joest-Museums der Stadt Köln.

Mayrhofer, M. 1953–56. *Kurzgefaßtes etymologisches Wörterbuch des Altindischen/A Concise Etymological Sanskrit Dictionary, vol. 1*. Heidelberg: Winter.

Meier, J. 1906. Berichtigungen zu Dr. Schnee's Mitteilungen über die Sprache der Moánus (Admiralitäts-Inseln). *Anthropos* 1, 210–228, 472–482.

Meier, J. 1907. Mythen und Sagen der Admiralitätsinsulaner. *Anthropos* 2, 646–667, 933–941 (continued 1908–09 in *Anthropos* 3, 193–206, 4, 354–374).

Mills, R.F. 1981. Additional addenda. In *Historical Linguistics in Indonesia Part I*. R.A. Blust (ed.), 59–82. NUSA 10. Jakarta: Badan Penyelenggara Seri NUSA.

Miller, J.I. 1969. *The Spice Trade of the Roman Empire 29 B.C. to A.D. 641*. Oxford: Clarendon.

Milne, L. 1924. *The Home of an Eastern Clan: a study of the Palaungs of the Shan State*. Oxford: Clarendon.

Milner, G.B. 1993. *Samoan Dictionary*. Aotearoa (N.Z.): Polynesian Press.

Monier-Williams, M. 1899. *A Sanskrit-English Dictionary, etymologically and philologically arranged, with special reference to cognate Indo-European languages*. New edition, greatly enlarged and improved, with collaboration of E. Leumann, C. Cappeller, and other scholars. Oxford: Clarendon (facsim. repr. 1988).

Mookerji, R. 1912. *Indian Shipping: a history of the sea-borne trade and maritime activity of the Indians from the earliest times*. Bombay/Calcutta/London/New York: Longmans.

Moore, S. le M. 1921. Gamopetalæ. In A.B. Rendle, E.G. Baker and Spencer Le M. Moore pp. 325–415, A Systematic Account of the Plants collected in New Caledonia and the Isle of Pines by Prof. R.H. Compton, M.A. in 1914. *Journal of the Linnean Society* 45, 245–417.

Morgan, W.N. 1988. *Prehistoric Architecture in Micronesia*. Photographs by Newton Morgan assisted by Dylan Morgan. Austin: University of Texas Press.

Münsterberger, W. 1940. Nias en Marquesas: Proeve van een ethnologisch onderzoek over het cultuurcontact (naar een voordracht, gehouden op den XXIII-sten Ethnologendag te Utrecht). *Tijdschrift van het Koninklijk Nederlandsch Aardrijkskundig Genootschap* 2e reeks 62, 728–739.

Nakao, S. 1958. Transmittance of cultivated plants through the Sino-Himalayan route. In *Peoples of Nepal Himalaya: scientific results of the Japanese expeditions to Nepal Himalaya 1952–1953*, H. Kihara (ed.), vol. 3, 397–420. Kyoto: Fauna and Flora Research Society, Kyoto University.

Nevermann, H. 1934. *Admiralitäts-Inseln.* Hamburg: Friedrichsen, de Gruyter.

Noehden, G.H. 1827. Account of the Banyan-tree, or *Ficus* Indica, as found in the ancient Greek and Roman authors. *Transactions of the Royal Asiatic Society of Great Britain and Ireland* 1, 119–132.

Nothofer, B. 1975. *The Reconstruction of Proto-Malayo-Javanic.* Verhandelingen van het Koninklijk Instituut voor Taal-, Land- en Volkenkunde 73. 's-Gravenhage: Nijhoff.

Paranavitana, S. 1960. Ceylon and Malaysia in mediaeval times. *Journal of the Ceylon Branch of the Royal Asiatic Society* new series 7, 1–42.

Paranavitana, S. 1966. *Ceylon and Malaysia.* Colombo: Lake House.

Parham, J.W. 1972. *Plants of the Fiji Islands*, rev. edn. Suva: Authority of the Government Printer.

Parpola, A. 1988. The coming of the Aryans to Iran and India and the cultural and ethnic identity of the Dāsas. *Studia Orientalia* (Helsinki) 64, 195–302.

Parpola, A. 1994. *Deciphering the Indus Script.* Cambridge/New York/Melbourne: Cambridge University Press.

Paterson, H.S. 1924. An early Malay inscription from Trengganu. *Journal of the Malayan Branch of the Royal Asiatic Society* 2, 252–258 (see also comment by C.O. Blagden on pp. 258–263).

Pawley, A. and R. Green 1973. Dating the dispersal of the Oceanic Languages. *Oceanic Linguistics* 12: 1–67.

Peekel, P.G. 1984. *Flora of the Bismarck Archipelago for Naturalists.* Transl. by E.E. Henty. Lae: Office of Forests Division of Botany (original: 1945, Illustrierte Flora des Bismarck-Archipels für Naturfreunde. n.n. n.pl.).

Pelliot, P. 1904. Deux itinéraires de Chine en Inde à la fin du VIIIe siècle. *Bulletin de l'École Française d'Extrême-Orient* 4: 131–413.

Pelliot, P. 1912. Review of Hirth and Rockhill (1911). *T'oung pao* 13, 446–481.

Pillai, V. *et al.* (eds) 1925–36. *Tamil Lexicon, University of Madras.* Mylapore: Madras Law Journal Press.

Porée-Maspero, É. 1950. Nouvelle étude sur la nāgī Somā. *Journal Asiatique* 238, 237–267.

Pou, S. 1993. *Dictionaire Vieux Khmer–Français–Anglais/An Old Khmer–French–English Dictionary.* Bibliotheque Khmère série B, Travaux et Recherches 3. Centre de Documentation et de Récherche sur la Civilisation Khmère, Paris.

Prentice, D.J. 1974. Yet another PAN Phoneme? *Oceanic Linguistics* 13, 33–75.

Przyluski, J. 1921. De quelques noms anaryens en Indo-Aryen. *Mémoires de la Société de Linguistique de Paris* 22, 205–210.

Przyluski, J. 1925. La princess à l'odeur de poisson et la nāgī dans les traditions de l'Asie Orientale. pp. 265–284. In *Études asiatiques, publiée à l'occasion du 25e anniversaire de l'École Française d'Extrême-Orient,* 2, Publication de l'École Française d'Extrême-Orient 20. Hanoi: EFEO.

Ptak, R. 1992. The northern trade route to the Spice Islands: South China Sea-Sulu zone-North Moluccas (14th to early 16th century). *Archipel* 43, 27–56.

Rao, S.N. 1977. Continuity and survival of Neolithic traditions in Northeast India. *Asian Perspectives* 20(2), 191–207.

Ray, S.H. 1913. The languages of Borneo. *Sarawak Museum Journal* 1(4). London: Kelly and Walsh.

Ray, S.H. 1926. *A Comparative Study of the Melanesian Island Languages.* Cambridge: Cambridge University Press.

Reed, A.W. and I. Hames 1967. *Myths and Legends of Fiji and Rotuma.* Wellington/Auckland/Sydney: Reed & Reed.

Reid, L.A. 1971. *Philippine Minor Languages: word lists and phonologies.* Oceanic Linguistics Special Publication 8. Honolulu: University of Hawaii Press.

Richardson, R.G. 1982. Medicine, History of. In 15th edn *Encyclopædia Britannica: Macropædia.* P.W. Goetz *et al.* (eds), vol. 11, 823–841. Chicago/Geneva/London/Manila/Paris/Rome/Seoul/Sydney/Tokyo/Toronto: Encyclopœdia Britannica.

Riedel, J.G.F. 1886. *De sluik- en kroeshaarige rassen tussen Selebes en Papua.* 's-Gravenhage: Nijhoff.

Riesenfeld, A. 1950. *The Megalithic Culture of Melanesia.* Leiden: Brill.

Röder, J. 1939. Levende oudheden op Ambon. *Cultureel Indië* 1, 97–105.

Ronkel, P.S. van 1902. Het Tamil-element in het Maleisch. *Tijdschrift voor Indische taal-, land- en volkenkunde* 45, 97–119.

Ronkel, P.S. van 1903a. De oorsprong van hat Maleische woord tjoema, Over den oorsprong van het Maleische woord binara, Over de herkomst van het Maleische woord djodo, De oorsprong van het Maleische woord bagai, Over den oorsprong van het Maleische woord pawai, De oorsprong van het Maleische woord Kamandikai. *Tijdschrift voor Indische taal-, land- en volkenkunde* 46, 71–72, 92–94, 128–131, 241–242, 469–471, 476–477.

Ronkel, P.S. van 1903b. Tamilwoorden in Maleisch gewaad. *Tijdschrift voor Indische taal-, land- en volkenkunde* 46, 532–557.

Ronkel, P.S. van 1918. Drawiḍische volksnamen op Sumatra. *Bijdragen tot de taal-, land- en volkenkunde van Nederlandsch-Indië* 74, 263–266.

Ross, M. 1988. *Proto-Oceanic and the Austronesian Languages of Western Melanesia.* Pacific Linguistics C-98. Canberra: Australian National University.

Sarkar, H. 1982. Megalithic culture in India. In *Megalithic Cultures in Asia*, Byung-mo Kim (ed.), 127–163. Hanyang University Monograph 2. Seoul: Pyung-hee Lee.

Sastri, K.A.N. 1949. Takuapa and its Tamil Inscription. *Journal of the Malayan Branch of the Royal Asiatic Society* 22(1), 25–30.

Sastri, K.A.N. 1962. Ceylon and Sri Vijaya. *Journal of the Ceylon Branch of the Royal Asiatic Society* new series 8, 125–140.

Sastri, S.M.N. 1894. Trader's slang in southern India. *Indian Antiquary* 23, 49–52.

Schnitger, F.M. 1939. *Forgotten Kingdoms in Sumatra.* Leiden: Brill.

Schoff, W.H. 1922. Camphor. *Journal of the American Oriental Society* 42, 355–370.

Schwaner, C.A.L.M. 1854. *Borneo: Beschrijving van het stromgebied van den Barito en langs eenige voorname rivieren van het zuid-oostelijk gedeelte van dat eiland,* 2e deel. Amsterdam: Van Kampen.

Sellato, B. 1989. *Hornbill and Dragon – Naga dan Burung Enggang: Kalimantan – Sarawak – Sabah – Brunei*, [bilingual: Indonesian] translated from English by Winarsih Arifin, photographs by G. Perret, W.P. Dicky, G. Houlbrèque and B. Sellato. Jakarta: Elf Aquitaine Indonésie/Kuala Lumpur: Elf Aquitaine Malaysia.

Shorto, H.L. 1971. *A Dictionary of the Mon Inscriptions from the Sixth to the Sixteenth Centuries.* London Oriental Series 24. London/New York/Toronto: Oxford University Press.

Silva, K.M. de 1981. *A History of Sri Lanka.* London: Hurst/Berkeley: University of California Press.

Sirisena, W.M. 1971. The Kalinga Dynasty of Ceylon and the theory of its South-east Asian origin. *Ceylon Journal of Historical and Social Studies* new series 1, 11–47.

Smits, L. and C.L. Voorhoeve, 1992. *The J.C. Anceaux Collection of Wordlists of Irian Jaya Languages, A. Austronesian Languages, Parts I-II, Irian Jaya.* Source Materials No. 4 Series B-No. 1, and No. 5 Series B-No. 2. Leiden: Department of Languages and Cultures of South East Asia/Jakarta: Irian Jaya Studies Project.

Solheim, W.G. II 1967. Southeast Asia and the West. *Science* 157, 896–902.

Solheim, W.G. II 1968, Early bronze in northeast Thailand. *Current Anthropology* 9, 59–62.

Solheim, W.G. II 1969. Reworking Southeast Asian prehistory. *Paideuma* 15, 125–139.

Solheim, W.G. II 1970. Northern Thailand, Southeast Asia, and world prehistory. *Asian Perspectives* 13, 145–162.

Solheim, W.G. II 1972. An earlier agricultural revolution, *Scientific American* 226(4), 34–41.

Solheim, W.G. II 1980. Neue Befunde zur späten Prähistorie Südostasiens und ihre Interpretation. *Saeculum* 31, 275–317, 319–344.

Sörensen, S. 1904. *An Index to the Names in the Mahabharata, with short explanations and a concordance to the Bombay and Calcutta editions and P.C. Roy's translation*. London: Williams & Norgate.

Sörgel, D. 1970. The influence of Buddhism on snake and Nāga cults in Burma. In *Buddhist Yearly 1970/Jahrbuch für Buddhistische Forschungen*, H. Mode, J. Mehlig, H.-J. Peuke and B. Thaler (eds), 13–23. Halle: Buddhist Centre.

Spriggs, M. 1989. The dating of the Island Southeast Asian Neolithic: an attempt at chronometric hygiene and linguistic correlation. *Antiquity* 63, 597–613.

Spriggs, M. 1991. Lapita origins, distributions, contemporaries and successors revisited. In *Indo-Pacific Prehistory 1990*, P. Bellwood (ed.), 306–312. Bulletin of the Indo-Pacific Prehistory Association 11. Canberra: Indo-Pacific Prehistory Association/ Jakarta: Asosiasi Prehistorisi Indonesia.

Srinavasan, K.R. 1946. The megalithic burials and urn-fields of south India in the light of Tamil literature and tradition. *Ancient India* 2, 9–16.

Stargardt, J. 1979. L'isthme de la Péninsule Malaise dans les navigations au long cours: nouvelle données archéologiques. *Archipel* 18, 15–39.

Steinmann, A. 1939. Enkele opmerkingen aangaande de z.g. scheepjesdoeken van Zuid-Sumatra. *Cultureel Indië* 1, 252–256.

Stevenson, R.C. and F.H. Eveleth (eds), 1921. *The Judson Burmese–English Dictionary*. Rangoon: American Baptist Mission Press.

Stokhoff, W.A.L. 1980–87. *Holle Lists: vocabularies in languages of Indonesia*, vol. 2, in cooperation with L. Saleh-Bronkhorst, Pacific Linguistics D-28; vols. 3(1–4), 4, 5(1), 6 and 7(1–2), in coop w. L. Saleh Bronkhorst and A.E. Almanar, Pacific Linguistics Series D-35, 44, 49, 50–52, 59, 61 and 60; vols 7(3–4), 8, 10(2–3) and 11, in cooperation with A.E. Almanar, Pacific Linguistics D-62, 66, 69, 75, 76, and 81. Canberra: Australian National University.

Stresemann, E. 1918. *Die Paulohi-Sprache*. 's-Gravenhage: Nijhof.

Stresemann, E. 1927. *Die Lauterscheinungen in den ambonischen Sprachen*. Beihefte zur Zeitschrift für Eingeborenen-Sprachen 10. Berlin: Reimer.

Stutterheim, W.F. 1926a. *Cultuurgeschiedenis van Java in beeld*. Weltevreden: Java-Instituut en Kolff.

Stutterheim, W.F 1926b. Oost-Java en de hemelberg. *Djåwå* 6, 333–349.

Stutterheim, W.F 1940. Een merkwaardige talam-versiering, *Cultureel Indië* 2, 200–204.

Suggs, R.C. 1982. Polynesian cultures. In *15th edn of Encyclopœdia Britannica: Macropœdia*. P.W. Goetz *et al.* (eds), vol. 14, 777–784. Chicago/Geneva/London/ Manila/Paris/Rome/Seoul/Sydney/Tokyo/Toronto: Encyclopœdia Britannica.

Sullivan, M. 1956. Archaeology in the Philippines. *Antiquity* 30, 68–79.

Sundermann, H. 1905. *Niassisch-Deutsches Wörterbuch*. Moers: Bataviaasch Genootschap van Kunsten en Wetenschappen.

Teeuw, A. 1965. Old Balinese and comparative Indonesian linguistics. *Lingua* 14, 271–284.

Thurston, E. and K. Rangachari. 1909. *Castes and tribes of southern India, vol. 6*. Madras: Government Press.

Tibbetts, G.R. 1979. *A Study of the Arabic Texts containing Material on South-East Asia*, Oriental Translation Fund n.s. 44. Leiden/London: Brill.

Trimingham, J.S. 1973. The Arab geographers and the East African coast. In *East Africa and the Orient: cultural synthesis in pre-colonial times*, by H. Nevill Chittick and R.I. Rotberg (eds), 115–146. New York/London: Africana (a division of Holmes & Meier).

Tsukada, M. 1966. Late pleistocene vegetation and climate in Taiwan (Formosa). *Proceedings of the National Academy of Sciences of the United States of America* 55, 543–548.

Turner, R.L. 1966. *A Comparative Dictionary of the Indo-Aryan Languages*. London: Oxford University Press.

Tuuk, H.N. van der 1901. *Kawi–Balineesch–Nederlandsch woordenboek*. deel 3. Batavia: Landsdrukkerij.

Vatter, E. 1932. *Ata Kiwan: unbekannte Bergvölker im tropischen Holland, ein Reisebericht*. Leipzig: Bibliographisches Institut.

Vavilov, N.I. 1926. Centry proishoždeniä kuljturnyh rastenii. *Trudy po prikladnoiĭ botanike, genetike i selekcii* 16(2), 139–248.

Vavilov, N.I. 1927. Geografičeskie zakonomernosti v raspredelenii genov kuljturnyh rastenii. *Trudy po prikladnoĭ botanike, genetike i selekcii* 17(3), 411–428.

Verheijen, J.A.J. 1984. *Plant Names in Austronesian Linguistics*. NUSA 20. Jakarta: Badan Penyelenggara Seri NUSA.

Vérin, P., C.P. Kottak and P. Gorlin, 1969. The glottochronology of Malagasy speech communities. *Oceanic Linguistics* 8, 26–83.

Vogel, J.Ph. 1918. The Yupa inscriptions of King Mūlavarman. *Bijdragen tot de taal-, land- en volkenkunde van Nederlandsch Indië* 74, 167–232.

Vogel, J.Ph. 1920. De giften van Mūlavarman. *Bijdragen tot de taal-, land- en volkenkunde van Nederlandsch-Indë* 76, 431–434.

Vogel, J.Ph. 1926. *Indian Serpent-Lore, or the Nagas in Hindu Legend and Art*. London: Arthur Probsthain (facsim. repr. 1972. Varanasi/Delhi: Indological Book House).

Vuuren, L. van 1908. De handel van Baroes, als oudste haven van Sumatra's Westkust, verklaard; en voor de toekomst beschouwd. *Tijdschrift van het Koninklijk Nederlandsch Aardrijkskundig Genootschap* 25, 1289–1402.

Wang Gungwu, 1958. The Nanhai Trade: a study of early history of Chinese trade in the South China Sea. *Journal of the Malayan Branch of the Royal Asiatic Society* 31(2), Kuala Lumpur: MBRAS.

Watson, A.M. 1983, *Agricultural Innovation in the Early Islamic World: the diffusion of crops and farming techniques, 700–1100*. Cambridge/London/New York/New Rochelle/Melbourne/Sydney: Cambridge University Press.

Wessels-Mevissen, C. 1991. Adichchanallur reconsidered – a typological study of south Indian 'megalithic' pottery, *Beiträge zur allgemeinen und vergleichenden Archäologie* 11, 13–56.

Wheatley, P. 1961, *The Golden Khersonese: studies in the historical geography of the Malay Peninsula before A.D. 1500*. Kuala Lumpur: University of Malaya Press.

Wheatley, P. 1983, *Nāgara and Commandery: origins of the Southeast Asian urban traditions*. Department of Geography Research Paper 207–208. Chicago: University of Chicago.

Wilder, G.P. 1931. *Flora of Rarotonga*. Bernice P. Bishop Museum Bulletin 86. Honolulu: Bishop Museum (facsim. repr. 1971. New York: Kraus).

Wolff, J.U. 1974. Proto-Austronesian *r and *d. *Oceanic Linguistics* 13, 77–121.

Wolff, J.U. 1982. Proto-Austronesian *c, *z, *, and *T. In *Papers from the Third International Conference on Austronesian Linguistics*. A. Halim, L. Carrington and S.A. Wurm (eds), 1–30. Pacific Linguistics C-75. Canberra: Australian National University.

Wolters, O.W. 1967. *Early Indonesian Commerce: A study of the origins of Śrīvijaya*. Ithaca: Cornell University Press.

Yen, D.E. 1977, Hoabinhian horticulture? The evidence and the questions from northwest Thailand. In *Sunda and Sahul: prehistoric studies in Southeast Asia, Melanesia and Australia*. J. Allen, J. Golson and R. Jones (eds), 567–599. London/New York/San Francisco: Academic Press.

Yule, H. and A.C. Burnell, 1903. *Hobson-Jobson: being a glossary of Anglo-Indian colloquial words and phrases and of kindred terms; etymological, historical, geographical, and discursive*, new edition, edited by William Crooke. London: Murray (facsim. repr. 1989. Calcutta: Rupa).

Yuncker, T.G. 1943. *The Flora of Niue Island*. Bernice P. Bishop Museum Bulletin 178. Honolulu: Bishop Museum.

Yuncker, T.G. 1959. *Plants of Tonga.* Bernice P. Bishop Museum Bulletin 220. Honolulu: Bishop Museum.

Zide, N.H. 1976. Munda and non-Munda Austroasiatic languages. In *Current Trends in Linguistics, 5.* T. Sebeok (ed.), 411–430. Linguistics in South Asia. The Hague/ Paris: Mouton.

Zorc, D. 1982. Where, O where, have the laryngeals gone? Austronesian laryngeals re-examined. In *Papers from the Third International Conference on Austronesian Linguistics, vol. 2.* A. Halim, L. Carrington and S.A. Wurm (eds), 111–144. Pacific Linguistics C-75. Canberra: Australian National University.

Index